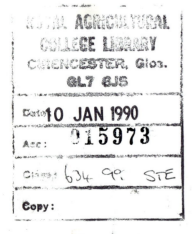

Agroforestry:
a decade of development

Agroforestry
a decade of development

Edited by

Howard A. Steppler

and

P.K. Ramachandran Nair

International Council for Research in Agroforestry
Nairobi

Published in 1987 by the
International Council for Research in Agroforestry
ICRAF House, off Limuru Road, Gigiri
P.O. Box 30677, Nairobi, Kenya

√ISBN 92 9059 036 X

Design and typography: Justice G. Mogaki, P.O. Box 74611, Nairobi

Copy editing: Caroline Agola, P.O. Box 21582, Nairobi

Typesetting: Arrow Stationers, P.O. Box 62070, Nairobi

Production co-ordination: P.K.R. Nair and Richard C. Ntiru, ICRAF

Printed by Printfast Kenya Limited, Lusaka Close, Off Lusaka Road,
    P.O. Box 48416, Nairobi

Dedicated to

John G. Bene (1910–1986)
Chairman of the Committee which recommended
the establishment of ICRAF, and
first Chairman of its Board of Trustees, 1977–1979

and

Walter Bosshard (1926–1986)
Chairman of the Board of Trustees, 1981–1985

ACKNOWLEDGEMENTS

The financial assistance rendered by the Canadian International Development Agency (CIDA) and the Dutch Ministry for Development Co-operation for the production of this book is gratefully acknowledged.

Contents

List of acronyms and abbreviations

AFRENA	Agroforestry Research Networks for Africa (of ICRAF)
BAIF	Bharatiya Agro Industries Foundation (India)
CARE	Cooperative for American Relief Everywhere
CATIE	Centro Agrónomico Tropical de Investigación y Enseñanza
CAZRI	Central Arid Zone Research Institute (Jodhpur, India)
CGIAR	Consultative Group on International Agricultural Research
CIDA	Canadian International Development Agency
CILSS	Comité Permanent Inter-états de Lutte Contre la Sècheresse dans le Sahel
CNRS	Centre National de la Recherche Scientifique (France)
COLLPRO	Collaborative Programmes (of ICRAF)
CSE	Centre for Science and Environment (New Delhi, India)
CSIRO	Commonwealth Scientific and Industrial Research Organization (Australia)
CTFT	Centre Technique Forestier Tropical (France)
D & D	Diagnosis and design
FAO	Food and Agriculture Organization (of the United Nations)
IARC	International Agricultural Research Centre
IBPGR	International Board for Plant Genetic Resources
ICAR	Indian Council of Agricultural Research
ICRAF	International Council for Research in Agroforestry
ICRISAT	International Crops Research Institute for the Semi-Arid Tropics
IDRC	International Development Research Centre
IITA	International Institute of Tropical Agriculture
ILCA	International Livestock Centre for Africa
IRRI	International Rice Research Institute
IPI	International Potash Institute
IUCN	International Union for the Conservation of Nature and Natural Resources
IUFRO	International Union of Forestry Research Organizations
MPT	Multipurpose tree
NAS	National Academy of Sciences (USA)
NEH	North-Eastern Hill (Region, of India)
NFTA	Nitrogen Fixing Tree Association
OAU	Organization of African Unity
OFI	Oxford Forestry Institute
ORSTOM	Office de la Recherche Scientifique et Technique Outre-Mer (France)
PCARR	Philippine Council of Agriculture and Resources Research
PICOP	Paper Industries Corporation of the Philippines
R & D	Research and development
SIDA	Swedish International Development Authority
T & V	Training and visit
TAC	Technical Advisory Committee (of the CGIAR)
UN	United Nations
UNDP	United Nations Development Programme
UNU	United Nations University
USAID	United States Agency for International Development
USDA	United States Department of Agriculture
WRI	World Resources Institute (Washington, D.C.)

Preface

This volume is part of the celebrations of the tenth anniversary of the establishment of the International Council for Research in Agroforestry (ICRAF).

Our authors are leaders in their fields and active in the promotion of agroforestry. Some are scientists actively engaged in research in a particular facet of agroforestry; some are active in the application of agroforestry as a land-use system; still others are concerned with the social and economic issues of the benefit/cost of agroforestry in development. We are deeply indebted to them for their dedication to agroforestry which is clearly shown by the thoughtfulness and insight in each paper.

The authors demonstrate—no doubt unintentionally—the newness of the discipline, for the reader will quickly discover differences in the definition of the term agroforestry as used by the different authors. We have not attempted to restrict the authors by forcing a single definition upon them. Nor, we hope, have we been overzealous in attempting to force the papers into a common mould. We believe that the shades of meaning in their use of the word agroforestry are both good and bad—good in that we have not closed our minds to the opportunities and benefits of dialogue with colleagues who can bring in new ideas and generate different approaches; bad in that it may hinder progress by dissipating our energies over too broad a field.

The authors raise several issues and concerns which, in our judgement, resolve into two basic problems. First, many of the concerns which have been identified would appear to be appropriate for an international organization such as ICRAF, but their implicit requirement for new technology would necessitate a major re-interpretation of the mandate of ICRAF. The other problem is that there are more issues raised than can be addressed effectively by one organization—and the list continues to grow. There is one ineluctable conclusion: the need for co-operation among the many institutions—national, regional and international—to ensure that maximum effort can be brought to bear on seeking solutions to the problems.

The book is divided into five sections. Chapters 1 and 2 are an introduction, with Chapter 2 presenting some projections into the future as well as a retrospective look at ICRAF. Chapters 3, 4 and 5 present some perspectives on agroforestry from the ecological, the institutional and the developmental viewpoints. Chapters 6, 7, 8, 9 and 10 describe the prominent agroforestry systems in some particular regions as seen by residents of each region or by persons with many years' experience there. These chapters clearly project the diversity as well as the importance of agroforestry in these different areas. Chapters 11, 12 and 13 cover problems associated with the measurement, impact and transfer of the technology of agroforestry interventions. These chapters should make clear the complexity and interdisciplinary nature of agroforestry, whether one is concerned with research, evaluation or transfer. Finally, Chapters 14, 15, 16 and 17 discuss some research findings and proposals for research activities in four areas of agroforestry, namely, systems, nutrient enrichment, germplasm evaluation and tree-component improvement, all of which ultimately come together as management approaches.

The opinions, ideas and agendas for research are those of the authors and do not reflect or imply the policy of ICRAF.

The editors accept responsibility for the selection of topics covered in this volume. We realize that there are many more subjects which might have been considered appropriate, but space and time constraints did not permit us the luxury of including them. In this context, we would like to draw the reader's attention to the publication of a special issue of *Agroforestry Systems* (Vol. 5, No. 3), which coincides with the publication of this book. This issue of the journal includes 12 articles written by ICRAF staff, and summarizes a decade of ICRAF's work.

We wish to thank the staff of ICRAF who have given many hours to the realization of this book in reviewing papers, typing manuscripts and in consultations over a myriad details. In the final analysis we, the editors, accept responsibility for any errors which have crept in, some of which might have been avoided had we not been working under such severe time pressure.

H.A. Steppler
P.K.R. Nair

Nairobi, July 1987

SECTION ONE

Introduction

The history of agroforestry

K.F.S. King

Director
Bureau of Programme Policy and Evaluation
United Nations Development Programme (UNDP)
1 UN Plaza, New York 10017, USA

Formerly: Director-General,
ICRAF, Nairobi, Kenya.

Throughout the world, at one period or another in its history, it has been the practice to cultivate tree species and agricultural crops in intimate combination. The examples are numerous. It was the general custom in Europe, at least until the Middle Ages, to clear-fell derelict forest, burn the slash, cultivate food crops for varying periods on the cleared areas, and plant or sow tree species before, along with, or after the sowing of the agricultural crop. This "farming system" is, of course, no longer popular in Europe. But it was still widely followed in Finland up to the end of the last century, and was being practised in a few areas in Germany as late as the 1920s (King, 1968).

In tropical America, many societies have traditionally simulated forest conditions in their farms in order to obtain the beneficial effects of forest structures. Farmers in Central America, for example, have long imitated the structure and species diversity of tropical forests by planting a variety of crops with different growth habits. Plots of no more than one-tenth of a hectare contained, on average, two dozen different species of plants each with a different form, together corresponding to the layered configuration of mixed tropical forests: coconut or papaya with a lower layer of bananas or citrus, a shrub layer of coffee or cacao, tall and low annuals such as maize, and finally a spreading ground cover of plants such as squash (Wilken, 1977).

In Asia, the Hanunoo of the Philippines practised a complex and somewhat sophisticated type of shifting cultivation. In clearing the forest for agricultural use, they deliberately left certain selected trees which, by the end of the rice-growing season, would "provide a partial canopy of new foliage" to prevent excessive exposure to the sun "at a time when moisture is more important than sunlight for the maturing grain". Nor was this all. Trees were an indispensable part of the Hanunoo farming system and were either planted or conserved from the original forests to provide food, medicines, construction wood and cosmetics, in addition to their protective services (Conklin, 1953).

The situation was little different in Africa. In southern Nigeria, yams, maize, pumpkins and beans were typically grown together under a cover of scattered trees (Forde, 1937). In

Zambia, in addition to the main crop in the homestead, there were traditionally numerous subsidiary crops that were grown in mixture with tree species (Anon., 1938). Indeed, the Yoruba of western Nigeria, who have long practised an intensive system of mixed herbaceous, shrub and tree cropping, explain that the system is a means of conserving human energy by making full use of the limited space laboriously won from the dense forest. They compare the method to a multistoreyed building in a congested area in which expansion must perforce be vertical rather than horizontal. They also claim that it is an inexpensive means of combating erosion and leaching, and of maintaining soil fertility (Ojo, 1966). As they picturesquely described it, "the plants eat and drink, as it were, not from one table, but from many tables under the same sky" (Henry, 1949).

These examples indicate the wide geographical coverage of the system and its early origins. What is more important perhaps, they clearly point to the fact that the earliest practitioners of what has now become known as agroforestry* perceived food production as the system's *raison d'être*. Trees were an integral part of a farming system. They were kept on established farmland to support agriculture. The ultimate objective was not tree production but food production.

By the end of the nineteenth century, however, the establishment of forest plantations had become the dominant objective wherever agroforestry was being utilized as a system of land management. This change of emphasis was not, at first, deliberate. It began fortuitously enough in a far-flung outpost of the British Empire. In 1806, U Pan Hle, a Karen in the Tonze forests of Thararrawaddy Division in Burma, established a plantation of teak through the use of what he called the "taungya" method† and presented it to Sir Dietrich Brandis (Blanford, 1958). Brandis is alleged to have prophesied that "this, if the people can ever be brought to do it, is likely to become the most efficient way of planting teak" (Blanford, 1958).

The taungya system spread to other parts of Burma, Schlich recording in 1867 that he had been shown a taungya teak plantation in its second year in the Kabaung forests of the Taungoo Division.

From these beginnings, the practice became more and more widespread. It was introduced into South Africa as early as 1887 (Hailey, 1957) and was taken from Burma to the Chittagong area in India in 1890 and to Bengal in 1896 (Raghavan, 1960).

It must not be imagined that once introduced, the system was practised continuously in India. It was abandoned both in Bengal and in the Chittagong, and was not resumed until 1908 and 1912, respectively. In the second decade of the twentieth century, however, the system became more and more popular with foresters as a relatively inexpensive method of establishing forests, and as Shebbeare (1932) puts it, it "became a full and rising flood". In 1920 it was adopted in Travancore (now Kerala), in 1923 in the United Province (now Uttar Pradesh), and in 1925 in the Central Provinces (now Madhya Pradesh) (Raghavan, 1960).

This period also saw its wider dispersal in Africa, and today it is practised in varying

---

* One of the first definitions of agroforestry reads as follows: "Agroforestry is a sustainable land management system which increases the yield of the land, combines the production of crops (including tree crops) and forest plants and/or animals simultaneously or sequentially on the same unit of land, and applies management practices that are compatible with the cultural practices of the local population" (Bene *et al.,* 1977; King and Chandler, 1978).
† Taungya is a Burmese word which literally means hill cultivation (*taung* — hill, *ya* — cultivation).

degrees in all the tropical regions of the world.* Teak is, of course, not the only forest species which is being established by the use of this agroforestry method. Indeed, the evidence suggests that if the system is utilized for the sole purpose of establishing forest plantations, that is only until the first closure of the forest canopy is attained, then it may be used in the establishment of forest plantations of most species.

It cannot be overemphasized, however, that for more than a hundred years, in the period 1856 to the mid-1970s, little or no thought appears to have been given, in the practice of the system, to the farm, to the farmer, and to his agricultural outputs. The system was designed and implemented solely for the forester. Indeed, some have asserted that in many parts of the world, local farmers were exploited in pursuit of the goal of establishing cheap forest plantations (King, 1968). Be that as it may, it was often stated that the socio-economic conditions that were necessary for the successful initiation of the system were land hunger and unemployment. It was sometimes said that another essential prerequisite was a standard of living which was low enough to border on poverty.

It is perhaps not surprising that nowhere in the relatively extensive literature which relates to this period are the positive soil-conservation aspects of the system mentioned, let alone emphasized. As the sole purpose of the exercise was to establish forests (which it was thought protected soils by their very existence), and as it was the undoubted policy of most forestry administrations to remove the farmer from the forest estate as soon as possible, the problems of man-induced soil erosion did not loom large in the thought processes of those tropical foresters who were involved with the system.

In order to fully appreciate the implications of this state of affairs, four factors must be clearly understood. First, it was considered that the forest estate should be inviolable. Secondly, it was perceived that the threat to the forest estate came mainly from peasants, particularly those who practised shifting cultivation. Thirdly, it was recognized that in many instances it would be advantageous to replace derelict or low-yielding natural forests with forest plantations. And fourthly, it had been demonstrated that the establishment of forest plantations was a costly business, especially because of their long gestation period, i.e., the long delays before returns were obtained from the initial investment.

So the ruling philosophy was to establish forest plantations whenever possible through the utilization of available unemployed or landless labourers. These labourers, in return for the forestry tasks which they were called upon to undertake, would be allowed to cultivate land between the rows of the forest-tree seedlings and would be permitted to retain their agricultural produce. This is, of course, a simplification of a system which varied from country to country, and from locality to locality. Nevertheless, it is a fair representation of its bare bones.

---

* The terms used to describe the system vary enormously. In German-speaking countries it is called *baumfeldwirtschaft, brandwirtschaft,* or *waldfeldbau.* In francophone countries it is referred to as *cultures sylvicole et agricole combineé, culture intercalaires, la méthode sylvo agricole, la système sylvo-bananier,* and *plantation sur culture.* The Dutch name is *Bosakkerbouw.* In Puerto Rico it is called the *parcelero* system, and in Brazil *consorciacao.* The name in Libya is *tahmil,* in the Philippines *kaingining,* in Malaya *ladang,* in Kenya the *shamba* system, in Jamaica *agricultural contractors' system,* in Sri Lanka *chena* and in Tanzania the *licensed cultivator system.* In India it is variously described as *dhya, jhooming, kumri, Punam, taila,* and *tuckle.* In the greatest number of countries in the world it is called taungya. In 1968, King (1968) suggested that the genetic term agrisilviculture be generally employed. From 1977, when the deliberations for establishing the International Council for Research in Agroforestry began, the term agroforestry began to become popular.

As a result of these preoccupations with the forests and the forest estate, the research which was undertaken was designed to ensure that little or no damage occurred to the forest-tree species; that the rates of growth of the forest-tree species were not unduly inhibited by competition from the agricultural crop; that the optimum time and sequence of planting of either the tree or agricultural crop be ascertained in order to ensure the survival and rapid growth of the tree crop; that forest species that were capable of withstanding competition from agricultural species be identified; and that the optimum planting-out espacements for the subsequent growth of the tree crop be ascertained.

In short, the research which was conducted was undertaken for forestry by foresters who, it appears, never envisaged the system as being capable of making a significant contribution to agricultural development, and indeed of becoming a land-management system (as opposed to a narrow forestry system) in its own right.

It would appear at first glance that a quite disparate set of factors has contributed to the now general acceptance of agroforestry as a system of land management that is applicable both in the farm and in the forest. Among these factors were re-assessment of the development policies of the World Bank by its President, Robert McNamara; a re-examination by the Food and Agricultural Organization of the United Nations of its policies pertaining to forestry; the establishment by the International Development Research Centre (IDRC) of a project for the identification of tropical forestry research priorities; a re-awakening of interest in both intercropping and farming systems; the deteriorating food situation in many areas of the developing world; the increasing spread of ecological degradation; and the energy crisis.

At the beginning of the 1970s, serious doubts were being expressed about the relevance of current development policies and approaches. In particular, there was concern that the basic needs of the poorest of the poor, especially perhaps the rural poor, were neither being considered nor adequately addressed. McNamara (1973) had stated the problem quite clearly:

> Of the two billion persons living in our developing member countries, nearly two-thirds, or some 1.3 billion, are members of farm families, and of these are some 900 million whose annual incomes average less than $100....for hundreds of millions of these subsistence farmers life is neither satisfying nor decent. Hunger and malnutrition menace their families. Illiteracy forecloses their futures. Disease and death visit their villages too often, stay too long and return too soon.
>
> The miracle of the Green Revolution may have arrived, but, for the most part, the poor farmer has not been able to participate in it. He cannot afford to pay for the irrigation, the pesticide, the fertiliser, or perhaps for the land itself, on which his title may be vulnerable and his tenancy uncertain.

It was against this backdrop of concern for the rural poor that the World Bank actively considered the possibility of supporting nationally oriented forestry programmes. As a result, it formulated a new Forestry Sector Policy paper which is still being used as the basis for much of its lending in the forestry sub-sector. Indeed, its social forestry programme, which has expanded considerably over the last decade or so, not only contains many elements of agroforestry but is designed to assist the peasant and the ordinary farmer to increase food production, and to conserve the environment as much as it helps the traditional forest services to produce and convert wood.

It is perhaps not unnatural that, on the appointment in 1974 of a new Assistant Director-General with responsibility for Forestry, FAO made a serious assessment of the forestry projects which it was helping to implement in the developing countries, and of the policies which it had advised the Third World to follow. It soon became clear that although there had been notable successes there also had been conspicuous areas of failure. As Westoby (1978) so aptly expressed it,

> Because nearly all the forest and forest industry development which has taken place in the underdeveloped world over the last decades has been externally oriented...the basic forest products needs of the peoples of the underdeveloped world are further from being satisfied than ever....
>
> Just because the principal preoccupation of the forest services in the underdeveloped world has been to help promote this miscalled forest and forest industry development, the much more important role which forestry could play in supporting agriculture and raising rural welfare has been either badly neglected or completely ignored.

FAO therefore redirected its thrust and assistance in the direction of the rural poor. Its new policies, while not abandoning the traditional areas of forestry development, emphasized the importance of forestry for rural development, the benefits which could accrue to both the farmer and the nation if greater attention was paid to the beneficial effects of trees and forests on food and agricultural production, and advised land managers in the tropics to "eschew the false dichotomy between agriculture and forestry" (King, 1979). They also stressed the necessity of devising systems which would provide food and fuel and yet conserve the environment.

As a result of this change in policy, FAO prepared a seminal paper "Forestry for Rural Development" (FAO, 1976) and, with funding from the Swedish International Development Authority (SIDA), organized a series of seminars and workshops on the subject in all the tropical regions of the world, and formulated and implemented a number of rural forestry projects throughout the developing world. In these projects, as with the World Bank's social forestry projects, agroforestry plays a pivotal role (see Spears, this volume). FAO also utilized the Eighth World Forestry Congress, which was held in Jakarta, Indonesia in 1978, to focus the attention of the world's leading foresters on the important topic of agroforestry. The central theme of the Congress was "Forests for People", and a special section was devoted to "Forestry for Rural Communities".

To these two strands of forest policy reforms, which evolved independently in an international funding agency and in one of the specialized agencies of the United Nations, was added a Canadian initiative which, some affirm, might transform tropical land use in the coming years.

In July 1975 the International Development Research Centre commissioned John Bene* to undertake a study to:

1.   Identify significant gaps in world forestry research and training;
2.   Assess the interdependence between forestry and agriculture in low-income tropical

---

* John Bene, who died in 1986, was an indefatigable Canadian to whose organizational and persuasive ability the early funding, establishment and success of the International Council for Research in Agroforestry is mainly due. (*Editors' note:* This book is dedicated to John Bene and Walter Bosshard.)

countries and propose research leading to the optimization of land use;
3.   Formulate forestry research programmes which promise to yield results of considerable economic and social impact on developing countries;
4.   Recommend institutional arrangements to carry out such research effectively and expeditiously; and
5.   Prepare a plan of action to obtain international donor support.

John Bene appointed an advisory committee* and regional consultants† to make recommendations on the forest research needs of the tropics. Professor L. Roche, one of the consultants, organized a workshop on tropical forestry research and related disciplines at the University of Reading. The proceedings of that workshop, along with the advice tendered by the other consultants, the advisory committee and a number of individuals and institutions who were consulted by Bene and his team, formed the basis for the report (Bene *et al.,* 1977) which was eventually submitted to the International Development Research Centre.

Although the initial assignment stressed the identification of research priorities in tropical forestry, Bene's team came to the conclusion that first priority should be given to combined production systems which would integrate forestry, agriculture and/or animal husbandry in order to optimize tropical land use. In short, there was a shift in emphasis from forestry to broader land-use concepts because the latter were perceived as being of both more immediate and long-term relevance.

Professor Roche was at that time Professor of Forestry at the University College of North Wales, Bangor. However, previously he had been Professor of Forestry at the University of Ibadan, Nigeria, where FAO and the United Nations Development Programme (UNDP) had assisted in the establishment of a Forestry Department in the early 1960s. One of the publications of that Department was a 1968 monograph on agrisilviculture (King, 1968), which undoubtedly influenced the thinking of Roche and of Bene and his team (Roche, 1976). Be that as it may, the Canadian International Development Agency (CIDA) had joined with IDRC to arrange a fact-finding meeting on agrisilviculture in Ibadan in 1973, and IDRC had followed this up with a research project in West Africa to discover how to make the forest-fallow phase of one type of agroforestry system more productive.

How was the agroforestry research that was proposed by Bene and his team to be undertaken? Bene and his colleagues stated in the report:

> It is clear that the tremendous possibilities of production systems involving some combination of trees with agricultural crops are widely recognized, and that research aimed at developing the potential of such systems is planned or exists in a number of scattered areas. Equally evident is the inadequacy of the present effort to improve the lot of the tropical forest dweller by such means.
>
> A new front can and should be opened in the war against hunger, inadequate shelter, and environmental degradation. This war can be fought with weapons that have been in the arsenal of rural people since time immemorial, and no radical change in their life style is required.

---

* A. Lafond, L.G. Lessard, J.C. Nautiyal, D.R. Redmond, R.W. Roberts, J. Spears and H.A. Steppler.
† J.D. Ovington, F.S. Pollisco, L. Roche and A. Samper.

This can best be accomplished by the creation of an internationally financed council for research in agroforestry, to administer a comprehensive programme leading to better land use in the tropics.

The report went on to suggest that the objectives of such a council should be the encouragement and support of research in agroforestry; the acquisition and dissemination of information on agroforestry systems; and the promotion of better land use in the developing countries of the tropics.

It recommended that the specific objectives of the proposed council might be:

1. To assemble and assess existing information concerning agroforestry systems in the tropics and to identify important gaps in knowledge;
2. To encourage, support, and co-ordinate research and extension projects in agroforestry in different ecological zones, aimed primarily at filling such gaps;
3. To support research that seeks to identify and/or improve tree species currently underused with respect to wood and/or non-wood products, to enhance the economic value and productivity of agroforestry systems;
4. To support research on agroforestry systems that will bring greater economic and social benefit to rural peoples without detriment to the environment; and
5. To encourage training in agroforestry and in the science of the tree species that form part of agroforestry systems.

The report advised that in order to attain these objectives, the activities of the council might include:

1. The collection, evaluation, cataloguing and dissemination of information relevant to agroforestry;
2. The organization and convening of seminars and working groups to collect, discuss, evaluate and disseminate information concerning agroforestry;
3. The promotion of teaching of the principles of agroforestry at all levels of the education system;
4. The encouragement of the orientation of forestry and agricultural teaching so that they make a stronger contribution to better land use; and
5. The demonstration, publication, and dissemination of research results and other relevant information.

It was apparent that, despite the growing awareness of the need for factual information on which agroforestry systems might be effectively based, very little research was being undertaken. The research that was being conducted was haphazard, unplanned and unco-ordinated. The IDRC Project Report therefore recommended the establishment of an internationally financed organization, now known as the *International Council for Research in Agroforestry* (ICRAF), which would support, plan and co-ordinate, on a world-wide basis, research in *combined* land-management systems of agriculture and forestry.

This proposal was generally well received by international and bilateral agencies and, at a meeting of potential donors and other interested agencies in November 1976, a steering committee was appointed to consider the establishment of the proposed Council in further detail.

The Steering Committee met in Amsterdam early in April and again in June 1977. It decided to proceed with the establishment of ICRAF along the lines proposed in the Bene/IDRC Report. It approved a draft charter for ICRAF and elected a Board of

Trustees.* It appointed IDRC as the Executing Agency for ICRAF until such time as the Council became a full juridical body. It decided that the permanent headquarters of ICRAF should be in a developing country, the selection of which would be left to the Board of Trustees, including the Director-General. And it accepted the kind offer of the Government of Netherlands to provide temporary headquarters facilities for ICRAF at the Royal Tropical Institute, Amsterdam, pending the completion of arrangements for the Council's location. ICRAF maintained an office at the Institute from August 1977 to July 1978 when it moved to its permanent headquarters in Nairobi, Kenya (King and Chandler, 1978).

At the same time as these hectic institution-building activities were being undertaken, there was renewed and heightened interest in the concepts of intercropping and integrated farming systems. It was being demonstrated, for example, that intercropping may have several advantages over sole cropping. Preliminary results from research that was being conducted in different parts of the world had indicated that in intercropping systems more effective use was made of the natural resources of sunlight, land and water; that intercropping systems might exercise beneficial effects on pest and disease problems; that there were advantages in growing legumes and non-legumes in mixture; and that, as a result of all this, higher yields were being obtained per area even when multi-cropping systems were compared to sole-cropping systems.

A significant workshop on intercropping was held in Morogoro in Tanzania in 1976. And it became obvious then that although a great deal of experimentation was being carried out in the general field of intercropping, there were many gaps in our knowledge. In particular, it was felt that there was need for a more scientific approach to intercropping research, and it was suggested that there should be greater concentration on crop physiology, agronomy, yield stability, nitrogen fixation by legumes, and plant protection.

Concurrently, IITA was extending its work on farming systems to include agroforestry, and many research organizations had begun serious work on, for example, the integration of animals with plantation tree crops such as rubber, and the intercropping of coconuts (Nair, 1979).

This congruence of men and of concepts and of institutional change provided the material and the basis for the development of agroforestry since then. Although many individuals and institutions have made valuable contributions to the understanding and expansion of the concept of agroforestry since the 1970s, it is perhaps true to assert that ICRAF has played the leading role in collecting information, conducting research, disseminating research results, pioneering new approaches and systems, and in general, by the presentation of hard facts, in attempting to reduce the doubts still held by a few sceptics.

Today, agroforestry is taught as a part of forestry and agriculture degree courses in many universities in both the developing and developed world; and specific degrees in agroforestry are already offered in a few. Today, instead of agroforestry being merely the handmaiden of forestry, the system is being more and more utilized as an *agricultural* system, particularly for small-scale farmers. Today, the potential of agroforestry for soil conservation is generally accepted. Indeed, agroforestry is fast becoming recognized as a system which is capable of yielding both wood and food and at the same time of conserving and rehabilitating ecosystems.

---

* John G. Bene, Chairman (Canada); M.S. Swaminathan, Vice-Chairman (India); Kenneth F.S. King, Director-General (Guyana); Jacques Diouf (Senegal); Robert F. Chandler (USA); Joseph C. Madamba (Philippines); Jan G. Ohler (Netherlands).

REFERENCES

Anon. 1938. *Report on the financial and economic position of Northern Rhodesia.* British
    Government, Colonial Office, No. 145.
Bene, J.G., H.W. Beall and A. Côte. 1977. *Trees, food and people.* Ottawa: IDRC.
Blanford, H.R. 1958. Highlights of one hundred years of forestry in Burma. *Empire Forestry Review*
    37(1): 33–42.
Conklin, H.C. 1957. *Hanunoo Agriculture.* Rome: FAO.
FAO. 1976. *Forests for research development.* Rome: FAO.
Forde, D.C. 1937. Land and labour in a Cross River village. *Geographical Journal.* Vol. XC, No. 1.
Hailey, Lord. 1957. *An African survey.* Oxford: O.U.P.
Henry, J. 1949. Agricultural practices in relation to soil conservation. *Emp. Cotton Growing Rev.*
    Vol. XXVI (1).
King, K.F.S. 1968. *Agri-Silviculture.* Bulletin No. 1, Department of Forestry, University of Ibadan,
    Nigeria.
————. 1979. Agroforestry. In *Agroforestry: Proceedings of the Fiftieth Symposium on Tropical
    Agriculture,* 1978. Amsterdam: Royal Tropical Institute.
King, K.F.S. and M.T. Chandler. 1978. *The wasted lands.* Nairobi: ICRAF.
McNamara, R.S. 1973. *One hundred countries, two billion people.* New York: Praeger.
Nair, P.K.R. 1979. *Intensive multiple cropping with coconuts in India.* Berlin: Verlag Paul Parey.
Ojo, G.J.A. 1966. *Yoruba culture.* University of Ife and London Press.
Raghavan, M.S. 1960. Genesis and history of the *Kumri* system of cultivation. *Proceedings of the
    Ninth Silviculture Conference,* Dehra Dun, India, 1956.
Roche, L. 1976. Priorities for forestry research and development in the tropics. Report to IDRC,
    Ottawa, Canada.
Shebbeare, E.O. 1932. Sal. Taungya in Bengal. *Empire Forestry Review* 12 (1).
Westoby, J. 1975. Forest industries for socio-economic development. *Y Coedwigwr,* No. 31.
Wilken, G.C. 1977. Integrating forest and small-scale farm systems in Middle America. *Agro-
    ecosystems* 3:291–302.

2

ICRAF and a decade of agroforestry development

Howard A. Steppler

Chairman, ICRAF's Board of Trustees

In the 1960s and early 1970s there was increasing concern for the forested lands of the tropics (Eckholm, 1976). It was clearly recognized that they were under severe pressure. Some thought that commercial exploitation was the problem; others that fuelwood needs were the culprit; while still others believed that shifting cultivation was the root cause. The president of the International Development Research Centre (IDRC), located in Ottawa, Canada, engaged Mr John Bene in 1975 to study the problem. Bene assembled a small team in Canada, an advisory committee and recruited experts in the various continents, to prepare studies pertinent to their area. The culmination of these various activities, including extensive travel by Bene, was the publication in 1977 of a report entitled *Trees, Food and People* (Bene *et al.*, 1977).

Bene and his co-authors recognized that the solution to the problems besetting tropical forests arose from population pressure exerted through the need to produce food and fuelwood. They were prophetic in their choice of sub-title for the report, "Land management in the tropics", for that was precisely the nature of their recommendation, although it was not immediately apparent. In the report, they identified some 23 tropical forestry problems. Of these, nine could be considered as dealing with the more traditional forestry problems. One clearly recognized the need to accommodate agriculture and the remainder encompassed problems related to land use, policy and environmental impact. Bene and his co-authors recognized that the key issue lay at the interface of forestry and agriculture. It is not evident whether they coined the word agroforestry to identify that interface. What is clear, however, is the prominence and widespread use accorded the term since their publication.* Their most significant recommendation was to establish an International Council for Research in Agroforestry (ICRAF).

Thus an old practice was institutionalized for the first time.

In the first years of its operation, ICRAF directed its attention to assembling the contemporary knowledge of agroforestry. Several international conferences and workshops

---

* It is interesting that the term "agroforestry" does not appear in the titles of the 54 works cited by Bene *et al.*; rather, "agrisilviculture" is used.

were held (Nair, 1987a), of which four are particularly worth mentioning here: one dealing with soils research in agroforestry (Mongi and Huxley, 1979); the second with international co-operation (Chandler and Spurgeon, 1979); the third treated plant research and agroforestry (Huxley, 1983); while the fourth addressed the problem of education in agroforestry. A fifth was held much later and was concerned with land tenure problems.

The Board of Trustees realized by 1980 that, while the collation of information on agroforestry was an important activity for ICRAF, it was not sufficient. ICRAF would need to develop a much sharper focus than envisaged in its charter and mandate if it was to meet expectations. Thus, in 1981, the Board adopted a strategy (Steppler, 1981; Steppler and Raintree, 1983) which set the Council on a path to develop a diagnostic methodology to determine relevance of agroforestry interventions in particular situations. Further, the diagnostic methodology was expected to identify the kind of intervention most appropriate for the situation at hand.

This strategy and focus have served the Council since its adoption. It was based on the Cycle of Technology Development (Steppler, 1981; Steppler and Raintree, 1983) and is basically concerned with phases I and II of that cycle (Figure 1). In 1984, an external review panel examined ICRAF's total operations. The panel confirmed the wisdom of the choice

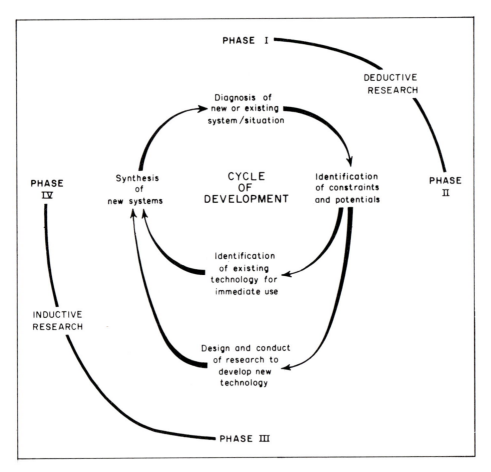

Figure 1   The cycle of technology development

of strategy and focus when it stated: "The Panel believes that this restricted interpretation (of the mandate) has been appropriate and necessary during these initial years" (Cummings et al., 1984).

The panel went on to recommend that the Council should move into a mode of extending and testing its methodology and assisting in the generation of new technology — essentially phases III and IV of Figure 1.

It should be pointed out that by the time of the external review, it had become clear to the Council, both the staff and the Board, that ICRAF's role was much more than that envisaged by Bene. The research that the Council had undertaken in developing its diagnostic methodology had shown that agroforestry as a land-use system was capable of many beneficial effects and with multi-product output; Bene had been right in his choice of sub-title.

The Council had also initiated activity through its Collaborative Programmes to reach out and to respond to the many requests that it was receiving, both from countries and from donor agencies (Torres, 1987). This was not, however, easy.

As previously mentioned, ICRAF, when established, was the first institution dedicated to agroforestry. Similar institutions did not exist at the national level. It is to the credit of the foresters that agroforestry was seen by them to be an essential development — albeit to secure the forest. The agriculturist did not recognize the situation, since loss of forest was not creating a problem for them; rather, the result was more land for agriculture. Thus, one of the first and critical functions that the Council undertakes when entering an area for a collaborative programme is to nurture at the national level the awareness of the contribution of agroforestry and the need to develop a national mechanism to be a focal point for agroforestry activities. The diagnosis and design methodology has a key role to play in this process.

Agroforestry was beset with much anecdotal material. It was clear that we were dealing with an old practice, but what was not clear, however, was the degree of diversity that might be in use. Thus, one of the projects launched in the early 1980s was systematically to inventory agroforestry systems (Nair, 1987b). The project was announced in *Agroforestry Systems* and several other international journals. Subsequent issues of *Agroforestry Systems* have carried articles describing specific systems. Nair (1985) published a first approximation of a classification of described agroforestry systems.

Four facts emerge from this preliminary compilation of systems: first, there is a bewildering array of agroforestry systems worldwide, and we have but scratched the surface; secondly, there are relatively few rigorous experimental data pertaining to performance of agroforestry systems; thirdly, the number of tree species — multipurpose trees — being used in various systems is in excess of 2,000; and fourthly, the systems vary from relatively simple with two or three components, to the complex homegardens which may contain upwards of 50 species plus animals and fish (Fernandes and Nair, 1986; Soemarwoto, this volume).

There is no question but that ICRAF should continue to catalogue agroforestry systems. Only through such activity can we build our body of knowledge relating to current practice. The objective in continuing the inventory is not so much to identify systems with which we wish to experiment. Rather, it is to record the kinds of systems used and where, the rationale for their use, the kinds of output in order that new or modified systems shall be designed to achieve the same goals. This is not to rule out the possibility of introducing systems which would have additional features, for example, halt soil erosion or improve soil fertility.

The problem of lack of experimental data or even of production data for the various systems is a serious gap in our knowledge. This is partly a reflection of the newness of agroforestry — it was until recently literally considered as a subsistence form of land management — and partly that there are essentially no experimental techniques applicable to agroforestry systems. Most statistical techniques and experimental designs have been developed for monocultures. There is even relatively little experimental work done with annual crop mixtures. The closest approximation to agroforestry systems would be found with perennials in forage mixtures, but even this is not as complex as agroforestry where one is dealing with at least one tree species and a crop species. ICRAF recognized this problem and has for some years been investigating different experimental designs at its Field Station. Further, it has published several working papers,* many of which deal specifically with this problem.

An additional dimension is the question of appropriate impact/output measurement. Monoculture and forage mixtures are relatively easy to measure — the output is clearly identified. With an agroforestry system we have multiple outputs. That of the agricultural component will probably be easy to measure — yield being the most visible output. The tree component may be much more elusive. There will be visible outputs, for example, fuelwood, building poles, fruits; others which affect crop production, such as leaf mulch and fixed nitrogen, and yet other nearly invisible ones which could have an effect on the entire system, such as recycling of nutrients from subsoil, control of erosion, increased infiltration rates.

The problem has several dimensions: first there is the need to determine which outputs/impacts shall be measured; secondly, the need to specify the baseline against which measurements will be made; thirdly, since many outputs and particularly impacts are liable to be qualitative, the need to quantify all measurements; and fourthly, the need to develop some common quantified measure of output which can be applied to any system in order that systems can be compared. For the latter, the most obvious such measure is the economic return. While this should be one of the measures, it is urgent to establish some other measure of the "value" of a system, such as the constancy or sustainability of the system.

The third factor identified was the plethora of multipurpose tree species that are candidates for experimentation. In many respects multipurpose tree research is in a very primitive state compared to agriculture. There is one species of the genus *Leucaena* (see Brewbaker, this volume) which is relatively well studied, although it pales when compared to wheat or maize — and this is but one out of literally hundreds of candidate species. The great majority of the species are represented by a single collection; there is virtually no information on the genetic variability which exists within a species. Thus, for example, there is a small stand — about 100 trees — of *Acacia albida* on the ICRAF Field Station at Machakos, Kenya. This stand was established from a seed lot and shows great variability in rate of growth, type of growth, retention of leaves, rate of leafing out — a mere indication of the wealth of variability which probably exists in the species. The same is no doubt true of many others, either within a provenance or between provenances.

There are two major tasks facing us with respect to the multipurpose tree dilemma. The

---

* As of 1 March 1987, some 48 working papers have been produced by ICRAF. These cover topics such as the diagnostic and design methodology, experimental techniques, economic and social studies, bibliographies and soils and agroforestry.

first task is rapidly to screen the hundreds of species and genera and to classify them on a limited number of criteria. These criteria must be agreed upon by all researchers undertaking the task. This screening should also have an objective to identify the most promising candidates within each class/group for more intensive study. The second task is to initiate the more detailed studies of the selected candidate species arising from the first task. Again, we must be pragmatic in our approach since we cannot indulge in the luxury of exhaustive physiological studies or cytogenetic research — that will come as the need arises. Work at this stage would be accelerated if one sought a specific ideotype of multipurpose tree to fit a particular niche. The challenge is great and the need for ingenuity and pragmatism in pursuing the research is most desirable — indeed essential. As an input to this whole process, two recent publications are most helpful.

Poore and Fries (1985), in reviewing the status of *Eucalyptus* as a candidate for agroforestry, indicate the kinds of questions which must be answered for *Eucalyptus* and, hence, for any candidate species. In so doing, they also demonstrate the difficulty in finding discrete unequivocal answers. Beer (1987), in discussing shade trees for three commercial tree crops (coffee, cacao and tea), lists 20 disadvantageous, 16 advantageous and 21 desirable characteristics for shade trees. Again, many of these are pertinent for any multipurpose tree in an agroforestry system.

The fourth factor is concerned with the very great number of systems, both large and small, which have been identified by the earlier-mentioned Agroforestry Systems Inventory. Some of the concerns and features relevant to this problem were discussed when discussing the second factor. I would like, however, to address a slightly different aspect.

Any one system undergoing experimentation would include, at a minimum, a tree species and a crop species. Each of these could have variation in genotype and management such as spatial arrangement, maturity type for the crop and harvesting methods (e.g., lopping and coppicing timing for the tree). It quickly becomes apparent that we are dealing with a multifactor design with many combinations. As we add species of trees and/or crops or introduce animals, the experiment grows in size logarithmically. Thus, a single experiment to elaborate the interrelations within an agroforestry system could be very large. Consider comparing systems and we have another almost quantum leap in size. There is a great challenge to develop experimental designs and test methodology applicable to agroforestry research.

The other dimension to the problem is the fact that we have combined long-lived woody perennials with annuals, short-lived perennials and/or animals. Ideally, experiments should continue for the life of the longest-lived component; this could be upwards of 40 years and we cannot wait that long. Thus, we must also devise tests and methods of prediction which will have acceptable levels of confidence in predicting long-term effects. In my judgement, the volume of agroforestry research will increase with time and in the near future there will be the need for these statistical tools.

To date, most of the agroforestry species are known on the basis of one or very few collections (von Carlowitz, 1986). Evidence from economic crops suggests that higher productivity will be obtained from non-indigenous species rather than indigenous ones (Harlan, 1959).

Thus, two consequences are suggested: first to increase the number of collections evaluated for each major candidate species, sampling over as wide a range of environments as possible. The other conclusion that I would draw is that we should, if at all possible, test species from other continents. There should be an active exchange programme of agroforestry germplasm initiated as soon as feasible. This latter means also that suitable quarantine provision should be in place to facilitate the safe international movement of plant material.

It is evident from this brief analysis of ICRAF over the past decade that there have been three milestones for it and for agroforestry. First and foremost, ICRAF was formed in 1977 — agroforestry was institutionalized. Secondly, ICRAF adopted a rigorously defined strategy in 1981; this set it on a very clear path of development within the discipline of agroforestry. And, thirdly, ICRAF, in 1984, was given a clean bill of health by the external review panel and commended for the work accomplished up to that date. It was urged to extend its activities to the many countries seeking its assistance, and this is what it is pursuing vigorously as it enters its second decade. I have discussed many research issues which have arisen during the past ten years. I should now like to address some other issues which have not yet been mentioned and, finally, to suggest some courses of action.

As has been mentioned, agroforestry was initiated by foresters. For that reason, and in addition because it deals with trees, it is frequently inextricably enmeshed with forest policy. This is particularly true if agroforestry is deemed by our diagnostic and design analysis to be the most appropriate land-use system for an area within a forest mandated area. This may even be true when trees are planted on non-mandated land. Forest policy almost invariably runs counter to an envisaged agroforestry use with cropping and/or livestock. The major context of forest policy is conservation and protection with industrial wood as the product. Hence, continuous outputs as expected from agroforestry are contrary to such policy. There is a great need to study forest policy with a view to suggesting amendments which would meet the objectives for forestry and not hinder agroforestry. ICRAF could play a central catalytic role here.

Policy issues also extend to such matters as marketing of the expected products, with pricing policies which will encourage the use of desirable practices. Land tenure must also favour the use of long-term land-use systems which tend to be sustainable and, hence, encourage the farmer to invest labour in trees. ICRAF undoubtedly has a role to play here in this general debate by objectively assessing various policy and tenure alternatives.

Educational and training programmes in agroforestry are key requirements as we move into this second decade of agroforestry development. There are an increasing number of universities, both in the tropical areas and in the temperate, that are beginning to offer programmes in agroforestry; some are at the undergraduate and some at the post-graduate level. Further, and most encouragingly, students from both agricultural and forestry backgrounds are beginning to prepare for careers in agroforestry. As was stated earlier in this paper, ICRAF held a workshop on education in agroforestry in December 1982, the proceedings of which will soon be published. Sufficient to say that there is no clear picture of the optimum curriculum in agroforestry and such may, in fact, never emerge. Rather, we shall probably have three avenues of development for agroforesters: one would be via an undergraduate programme in agroforestry, which might be in a faculty of forestry or of agriculture; a second could be via an undergraduate degree in agriculture with post-graduate training in agroforestry; and a third could be the converse, namely undergraduate in forestry and post-graduate in agroforestry. Of course, any person with a first degree in agroforestry might also pursue post-graduate studies. There is no basis upon which to judge one avenue as being better than another, although certain situations might favour a particular sequence of study.

ICRAF has as part of its mandate a responsibility to provide training in agroforestry. It has already provided much training in the use of its diagnosis and design methodology (Zulberti, 1987). As it moves into a collaborative research mode, it will expand its training offerings to include research methodologies and experimental techniques. The majority of these training offerings are relatively short courses and not designed to provide advanced degrees.

The discussion has been concerned with the preparation of people for a career in agroforestry research. Of equal or possibly greater importance is the need to prepare people for the transfer/extension of agroforestry technologies as they become available. The problem of education of these people has not been addressed. They will face complex situations where solutions will involve appropriate MPTs, crops, their combined management and spatial arrangement and possibly even animals. It is expecting a lot to anticipate that one person could be prepared and have at his/her fingertips the bank of data needed to respond to the challenges. Are we looking at a generalist in the field backed up by a group of specialists responding to his requests, or can we even anticipate a generalist with a computer terminal linked to a data bank which will provide the answers to the complex questions — the data bank being constantly updated by researchers? Is it possible that each extension worker is trained in micro diagnosis and design methodology? These questions require consideration now. Although extension is outside of ICRAF's mandate, some agency should at least initiate the process of examination — and ICRAF would seem most appropriate.

Trees often have deep cultural significance and, hence, their retention and management may not be based entirely on pragmatic decisions. Farmers who live on the knife edge of success or failure at the whim of climate and pests often have different priorities from those of the scientist. It is essential that in dealing with these complex agroforestry systems the socio-economic and cultural factors are recognized. Evaluation of these is a part of the diagnosis and design approach developed by ICRAF (Raintree, 1987). One must ensure, in addition, that these same criteria are considered at various stages of development and testing of new technology.

There is a massive research agenda in agroforestry — not only what has been set down in the preceding paragraphs of this paper, but also to be found in the succeeding chapters of this volume, where the various authors have frequently directly or indirectly indicated problem areas pertinent to their topic. There is much more than can be done by any one institution, and certainly much beyond the capacity and mandate of ICRAF as envisaged over the next decade. To move forward and to meet the expectations means that there must be a sharing of responsibilities for the agenda and a pooling of resources and of information for the sake of expediency and efficacy.

An examination of the research requirements, both within and without the agenda, clearly indicates that the activities can be classified according to whether they are:

(a) Appropriate for an international or regional organization that the information resulting from the research transcends national boundaries, tends not to be location-specific; or

(b) Appropriate for a national organization since the results are location-specific and immediately transferable to the ultimate user, the farmer.

The national research systems do not at this time have specific agroforestry research units, although as a consequence of the ICRAF activity in collaborative programmes many are developing agroforestry co-ordinating committees. However, at the national level any or all of the sectors of the agricultural research system — crops or animals, the forestry research sector (particularly tree nursery activities) and the university — may take part in the agroforestry research programme. All activities from location-specific to basic non-location-specific can be carried out at the national level. However, every national system should undertake the testing at farm level of new systems and the validation of components of the system proposed for its use. Any additional work "up-stream" from this adaptive/applied research would depend on resources available for its use.

At present there are two groups of institutions on the international scene: ICRAF, which constitutes the first group of those dedicated to agroforestry, and the international agricultural research centres, IARCs, which constitute the second group.

This latter group is composed of some 13 centres, nine of which deal with primary production of food commodities with either a regional or world responsibility. This group of 13 is funded by the Consultative Group for International Agricultural Research (CGIAR) (Baum, 1986). There are many other international and regional organizations (for example, ICRAF) outside the CGIAR but funded by the international community and engaged, or with a potential for engagement, in agroforestry. One such organization is the International Union of Forestry Research Organizations (IUFRO) which has held several sessions on agroforestry at its international conferences.

The discussion of needs in agroforestry has centred on the tree component and on the system *per se* — the much neglected and virtually unknown subjects. As we begin to refine our information, we shall be seeking genotypes of the other components, particularly of the crops to better fit the system. These crops are in most cases the commodities which are the mandate of the previously mentioned IARCs. Thus, there will be added demand to the IARCs to develop these appropriate genotypes. Further, there are some specific systems such as alley cropping (see Kang and Wilson, this volume) or the interaction between animals and browse, which would best be done by the appropriate IARCs.

Turning now to the research on multipurpose trees *per se,* it would appear that ICRAF is the most appropriate institution to co-ordinate the activities if not actually to undertake them. As with any plant species, there is much basic work to be done upon which to build the more applied research. It is the former, along with the development of the relevant methodologies, which seems most appropriate for ICRAF.

There are three other functions which are also critical and which would most naturally fall within the ambit of ICRAF.

The first of these is to act as the focal point for information emanating from the various research activities. This is not new to ICRAF, but its central role in this must be reinforced. As activities multiply at the national and international level, this role will increase in importance. Naturally, the concomitant activity is the dissemination of that information to the users.

A second role is to ensure that new and essential areas/problems in agroforestry research are addressed. This means maintaining a constant watch on developments in both the research field and in the application field to identify these new challenges. One would envisage that the continuous and logical use of the "Cycle of Development" (Figure 1) with a constantly improved diagnosis and design methodology would be a major source of such information. The problems uncovered could conceivably run the gamut from the biological to the socio-economic problems, from policy to sustainability to statistical techniques.

Thus, the third role would be to seek out the partners to undertake the investigations identified in role two. Many of these will be highly specialized areas of investigation, with some of the problems having a global connotation, while others are more regional. It would, therefore, seem expedient to begin the process of anticipating the kinds of problems likely to be encountered and to seek the collaborative partners for the undertakings — one might even initiate some preliminary studies.

Finally, before euphoria completely clouds rational thinking and attainable expectations for agroforestry, let us return to reality. Agroforestry will not save the world — it is not the panacea for the ills of land misuse. There are undoubtedly many benefits to be gained from an agroforestry intervention but there may — in fact probably will — be costs. Labour requirements may be higher, production of some selected component may drop, new problems, such as bird damage, might even emerge. The title of ICRAF's first publication, *The Wasted Lands* (King and Chandler, 1978) may have both raised expectations that agroforestry would correct the problems of these areas and at the same time denied to ICRAF opportunities to work in other areas. In either case, ICRAF has moved beyond these boundaries, but now let us give it the opportunity in this second decade to prove its capability to address and correct problems, remembering that it cannot be all things to all people.

REFERENCES

Baum, W. 1986. *Partners against hunger.* Washington, D.C.: International Bank for Reconstruction and Development.

Beer, J. 1987. Advantages, disadvantages and desirable characteristics of shade trees for coffee, cacao and tea. *Agroforestry Systems* 5: 3–13.

Bene, J.G., H.W. Beall and A. Cote. 1977. *Trees, food and people: Land management in the tropics.* Ottawa: IDRC.

Chandler, T. and D. Spurgeon (eds.). 1979. International cooperation in agroforestry. Proceedings of an Expert Consultation. Nairobi: ICRAF.

Cummings, R.W., J. Burley, G.T. Castillo and L.A. Navaro. 1984. Report of the External Review Panel of the International Council for Research in Agroforestry, September-December, 1984. Nairobi: ICRAF.

Eckholm, E.P. 1976. *Losing ground: Environmental stress and world food prospects.* New York: Norton and Co.

Fernandes, E.C.M. and P.K.R. Nair. 1986. An evaluation of the structure and function of some tropical homegardens. *Agricultural Systems* 21: 179–210.

Harlan, J.R. 1959. Plant exploration and the reach for superior germ plasm for grasslands. In H.P. Sprogue (ed.), *Grassland* Publication 53, American Society for the Advancement of Science, Washington, D.C.

Huxley, P.A. (ed.). 1983. *Plant research and agroforestry.* Nairobi: ICRAF.

King, K.F.S. and M.T. Chandler. 1978. *The wasted lands.* Nairobi: ICRAF.

Mongi, H.O. and P.A. Huxley (eds.). 1979. *Soils research in agroforestry.* Nairobi: ICRAF.

Nair, P.K.R. 1985. Classification of agroforestry systems. *Agroforestry Systems* 5: 97–128.

―――――. 1987a. International seminars, workshops and conferences organized by ICRAF. *Agroforestry Systems* 5: 375–382.

―――――. 1987b. Agroforestry systems inventory. *Agroforestry Systems* 5: 301–318.

Poore, M.E.D. and C. Fries. 1985. The ecological effects of eucalyptus. FAO Forestry Paper 59. Rome: FAO.

Raintree, J.B. 1987. The state of the art of agroforestry diagnosis and design. *Agroforestry Systems* 5: 219–250.

Steppler, H.A. 1981. A strategy for the International Council for Research in Agroforestry. Nairobi: ICRAF.

Steppler, H.A. and J.B. Raintree. 1983. The ICRAF research strategy in relation to plant science research in agroforestry. In P.A. Huxley (ed.), *Plant Research and Agroforestry.* Nairobi: ICRAF.

Torres, F. 1987. The ICRAF approach to international co-operation. *Agroforestry Systems* 5: 395–410.

von Carlowitz, P.G. 1986. *Multipurpose tree and shrub seed directory.* Nairobi: ICRAF.

Zulberti, E. 1987. Agroforestry training and education at ICRAF: Accomplishments and challenges. *Agroforestry Systems* 5: 353–374.

SECTION TWO

Perspectives on agroforestry

3

The promise of agroforestry for ecological and nutritional security

M. S. Swaminathan

Director-General
International Rice Research Institute (IRRI)
P.O. Box 933, Manila, Philippines

President
International Union for Conservation of Nature and Natural Resources (IUCN)
Gland, Switzerland

Contents

Introduction

From the dawn of civilization, sustainable food security has been a major human goal. FAO defines food security as "physical and economic access to food for all people at all times". I have repeatedly stressed the need for enlarging this concept to cover all aspects of balanced nutrition as well as clean drinking water so that all human beings have an opportunity for the full expression of their innate genetic potential for physical and mental development (Swaminathan, 1986). Also, I have pointed out that enduring food and nutrition security can be built only on the foundation of ecological security, i.e. the security of the basic life-support systems of land, water, flora, fauna, and the atmosphere (Swaminathan, 1981). It is in this context that I wish to assess the role of agroforestry systems in helping us to achieve sustainable nutritional and ecological security.

Thanks to new technologies that emphasize the cultivation of genetic strains of crops that respond to irrigation and good soil-fertility management, many tropical and subtropical (developing) countries in Asia and Latin America have made good progress in food production since the mid 1960s. Many traditionally food-deficit or food-importing countries have become self-sufficient and even food-surplus countries. What is even more significant is that increases in food production have come largely from increases in productivity rather than increases in cultivated area. Because many developing countries, particularly those of south and south-east Asia are population rich but land poor, this is an important gain. Today world grain stocks have increased to more than 450 million tonnes.

Despite such a satisfactory global situation, scientists and planners are worried. For them, increasing the pace of food production to keep pace with unabated population growth in the tropics and subtropics is still an unfinished task. Although most countries of the world are in the process of demographic transition, the progress toward the final stage of this transition is lagging behind dangerously in Africa, the Indian subcontinent, Latin America, the Middle East, and south-east Asia (Brown and Jacobson, 1986). It is predicted that between 1980 and 2000, world population will increase by 1.7 billion. Ninety percent of this growth will occur in the developing countries. This tremendous increase will require at least 50–60 percent greater agricultural output than in 1980. What then should be the appropriate strategy for increasing food production?

Now it is sufficiently clear that any increase in food production has to come primarily from raising the productivity of currently tilled soils rather than from bringing new land resources into farming. In fact, a large portion of currently tilled marginal areas will have to be phased out of agriculture for economic and ecological reasons. Land for agriculture is a shrinking resource. Because some land is being taken out of production all the time and diverted to uses such as roads, housing, and industry, health care of the soil is a priority task.

The carrying capacity of land in many developing countries is already overstretched. According to a recent FAO study, 54 of 117 developing countries did not have sufficient land resources to meet the food needs of their 1975 populations at low levels of input use (Higgins *et al.,* 1983). These critical countries, covering an area of 2.2 billion ha, in 1975 had 278 million people in excess of the population supporting capacity of the land. By AD 2000, at the same level of inputs, the number of critical countries will increase to 64 and the population in excess of the land's potential carrying capacity may be over 500 million. Even if input use is raised to the intermediate level, which may not be easy considering the external indebtedness of many developing countries, 36 countries will still be in a critical situation with 141 million people above the carrying capacity of the land.

Modern agricultural production technology has raised the hope that hunger can be eliminated and the carrying capacity of the land increased through better use of cubic volumes of soil, water, and air. Nevertheless, the ecological sustainability and economic viability of new technologies are increasingly at stake. The rising populations of humans and animals, with their ever expanding food, fodder, and feed needs, exerts great pressure on the stabilizing elements of agro-ecosystems. As productive land becomes scarce, marginal farmers are pushed into fragile crop lands and forest areas unsuitable for modern agriculture. If the present trend of population growth persists, forest and pasture lands will be further reduced. Figure 1 projects these relationships for the Himalayas, a very delicate agro-ecosystem (Shah, 1982).

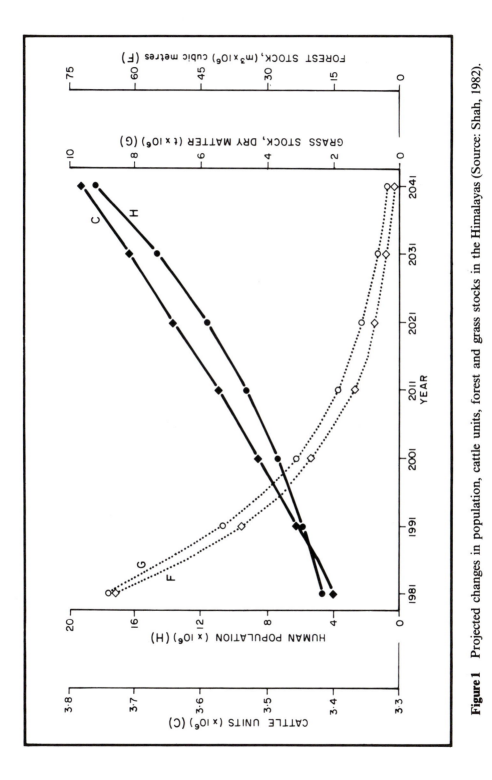

Figure 1 Projected changes in population, cattle units, forest and grass stocks in the Himalayas (Source: Shah, 1982).

Unscientific land-use practices on such marginal soils lead to many problems, notably soil erosion. Higgins *et al.* (1983) estimated that if soil erosion continued at its 1983 rate, loss in rain-fed cropland in the developing world would range from 9.7 percent to 35.6 percent, leading to an overall 28.9 percent decrease in crop production (Table 1) by the year 2000.

Table 1   Projected effects of unchecked soil erosion on productivity (1983-2000)

	Africa	South-west Asia	South-east Asia	South America	Central America	Global av (%)
Decrease in area of rain-fed cropland (%)	16.5	20.0	35.6	9.7	29.7	17.7
Decrease in rain-fed crop productivity (%)	29.4	35.1	38.6	22.6	44.5	28.9

Source: Higgins *et al.*, 1983.

A major cause of soil erosion is deforestation. Table 2 indicates the huge gap between deforestation and tree plantation in the tropics where the problem is most acute. The World Resources Institute has estimated that 160 million hectares of upland watershed in the Himalayas and Andean range, and in the Central American, Ethiopian and Chinese highlands, have been seriously degraded due to human interference (WRI, 1985). Cherrapunjee, once the wettest area in the world and covered by dense tropical forest, is now practically devoid of vegetation. Overcutting for fuelwood and overgrazing in arid and semi-arid areas, combined with non-sustainable resource-use patterns triggered by commercial greed or careless technology, have accelerated desertification. Such activities directly affect agriculture. Extensive deforestation results in raised river beds, which reduces their water-carrying capacity, and consequently their irrigation potential. In India, for example, the National Commission on Floods has projected that an irrigation potential of almost 60,000 ha may be lost every year because of siltation.

Table 2   Annual deforestation and plantation projections for the tropics (1981-1985)

Region	Annual rates of deforestation (ha x 10³)			Annual rates of plantation	Plantation: deforestation ratio
	Tree formations				
	Closed	Open	All		
Tropical America (23 countries)	4,339	1,272	5,611	535	1:10.5
Tropical Africa (37 countries)	1,331	2,345	3,676	126	1:29
Tropical Asia (16 countries)	1,826	190	2,016	438	1:4.5
Total (76 countries)	7,496	3,807	11,303	1,099	1:10

Source: FAO, 1982.

Shifting cultivation, long practised all over the tropical highlands, has also contributed to deforestation. At the beginning of this century, shifting cultivation cycled in 30–40 years, but now it cycles in as few as 3–5 years due to increased population pressure. An important offshoot is the reduced availability of fuelwood — a major source of energy in the rural areas of developing countries. If the gap between harvesting and tree planting remains as it is today, fuelwood shortage may become an even more serious problem than food availability. A Study Group of the Planning Commission of the Government of India estimated in 1982 that to meet the fuelwood demand in AD 2000, at least 3 million ha need to be planted every year with fast-growing fuelwood trees (Swaminathan, 1982). The increasing distance between villages and forests has increased the time needed for fuelwood collection, thus depriving farm women and children of time which could have been utilized in other productive activities.

Although the evidence is still inconclusive, extensive cutting of the tree cover may contribute to the increased level of carbon dioxide in the atmosphere. The accompanying increase in global temperature could directly affect agricultural production. That the global mean surface temperature actually increased during the last 100 years has recently been proved by comprehensive estimates of temperature based on calibrated ocean data and land measurements (Jones *et al.,* 1986). A series of papers contained in the publication *State of the World — 1987,* published by the World Watch Institute, provides a grim picture of the emerging global ecological scenario. Climate change carries a global price tag of $200 billion for irrigation adjustments alone in the coming decades (Brown, 1987).

It is obvious that the maintenance of tree cover is of utmost importance for ecological and economic sustainability of food-production systems. Agroforestry involving the integrated cultivation of woody perennials, crops, and animals provides one answer to our quandary. A typical agroforestry system allows symbiotic economic and ecological interactions between the woody and non-woody components to increase, sustain, and diversify the total land output. Some of the dominant agroforestry systems are: (a) shifting cultivation, (b) taungya afforestation, (c) homegarden, (d) silvopastoral, (e) agrisilvicultural, and (f) windbreaks and live fences (Nair, 1985). Farming systems that incorporate perennial trees and shrubs have the advantage of producing fuelwood, fruit, fodder, and other products along with annual crops. In addition, they decrease the farmer's exposure to seasonal environmental variations and, over the long-term, maintain and improve soil health.

The following sections give a brief account of agroforestry systems, some recent successes, and the potential of these systems for increasing food and environmental security.

Traditional systems of agroforestry

Different patterns of agroforestry were common in the early days. For many upland farmers, agroforestry was a way of life. Shifting cultivation, for example, is believed to have originated in the Neolithic period around 7000 BC (Sharma, 1976). In this system, still common in many hilly areas of tropical Asia, Africa, and Latin America, trees and agricultural crops are arranged sequentially in time and space. Its sustainability in the past was due to low population pressure and availability of large tracts of undisturbed forests. Today, shifting cultivation promotes soil erosion and land degradation. Inasmuch as we have alternative methods of soil fertility restoration, shifting cultivation is no longer necessary.

Homegarden, or homestead, is another common agroforestry system (Soemarwoto, this volume). In this system, tall trees are intercropped with medium shrubs and short annual crops to produce a variety of foods and green manure besides reducing soil erosion. Intercropping in coconut and oil palm plantations is also common. Farmers generally plant smaller trees such as coffee and cacao, and banana underneath the palms.

To arrest land degradation due to shifting cultivation, a fairly successful system called taungya was developed in the mid-1800s in Burma. In this system, the government gave land to shifting cultivators and allowed them to grow trees and agricultural crops together. When the tree canopy closed and precluded further agricultural cropping, farmers were shifted to another site. Meanwhile, the abandoned site developed into a fully-fledged forest. Taungya was later adopted by many countries of Asia, Africa, and Central America. (see King, this volume.)

Many of these systems have now given way to subsistence agricultural systems in several developing countries. Because subsistence farming practices are not ecologically sustainable and often not economical, interest in agroforestry is increasing.

Recent trends in agroforestry

With the growing realization that agroforestry is a practical, low-cost alternative for food production as well as environmental protection, forest departments of many countries are integrating agroforestry programmes with conventional silviculture. Forest research institutes and agricultural research centres are increasingly developing programmes for agroforestry research, training, and education. The UN Conference on Desertification held in Nairobi in 1977 stressed the significance of agroforestry systems for meeting the food, fuel, fodder and fertilizer needs of rural communities without causing ecological harm. The establishment of ICRAF in 1977 was a significant milestone in the history of agroforestry research. ICRAF for the first time provided a global professional organization for stimulating and supporting scientific and developmental interest in silvopastoral, "silvo-horticultural", agrisilvicultural and other systems of land management.

Agrosilvopastoral systems

Among recent developments, the most important has been the realization of the importance of multipurpose, woody, leguminous trees and shrubs in low-input farming systems. These legumes, such as various species of *Leucaena, Sesbania, Gliricidia, Acacia,* and *Prosopis,* are capable of providing the food, fodder, fertilizer and fuel needs of rural populations. The trees also diversify income, dominate over weeds, reduce soil erosion, and improve soil structure and fertility. Many of these species are widely adapted. For example, *Sesbania* can tolerate a wide range of soil environments — saline, alkaline, and waterlogged.

Many agrosilvopastoral systems have been proposed in recent years. Among these, alley farming is one of the most important. In this system, food crops are grown in alleys formed by hedgerows of trees or shrubs (see Kang and Wilson, this volume). The hedgerows are cut back at the time of planting crops and are kept pruned to prevent shading the crops (Figure 2). Pruned foliage is allowed to decompose in the alleys and the nutrients released increase grain yields of interplanted crops (Table 3). The foliage is also used to feed livestock. Simultaneously, the trees provide many other by-products such as fuelwood and stems for staking viney crops.

Figure 2 Alley cropping is a promising agroforestry technology for not only the humid lowlands but also the subhumid to semi-arid zones, as this photo from Machakos, Kenya (about 700 mm of rainfall, bimodal, five rainy months) shows (Photo: P.K.R. Nair, ICRAF).

Table 3 Main season grain yield of maize alley cropped with *Leucaena leucocephala* as affected by application of leucaena prunings and nitrogen

N rate (kg/ha)	Leucaena prunings	Yield by year (t/ha)				
		1979	1980	1981*	1982	1983
0	Removed	-	1.0	0.5	0.6	0.3
0	Retained	2.1	1.9	1.2	2.1	0.9
80	Retained	3.5	3.3	1.9	2.9	3.2
LSD 0.05		0.4	0.3	0.3	0.4	0.8

* Maize crop seriously affected by drought during early growth.

Source: Kang *et al.*, 1984.

Alley farming thus appears to be a low-cost, sustainable agricultural technology and an attractive alternative to the prevailing shifting cultivation and bush-fallow system.

Multi-level plantations and homegardens

Multi-level plantations and homegarden systems, common in smaller landholdings and designed to increase food production, are fairly analogous to a rain forest with a multilayered canopy. The systems and their component crops vary with the location. In the humid tropics, plantations of coconut, oil palm, and rubber have increased during the last few years due to greater demand for vegetable oils and rubber products. New opportunities have become available with the development of high-yielding hybrids between tall and dwarf coconuts, which have a potential similar to that of oil palm. During the initial and later years, palms in the plantation do not make use of all the available sunlight, space, and water (Nair, 1979). These plantations, therefore, offer opportunities for intercropping (Figure 3). Cacao in Malaysia, cassava in India, banana in Jamaica, and pineapple in the Philippines are now commonly intercropped with coconut. A significant amount of cacao in Sri Lanka is now grown under rubber.

Figure 3 Intercropping and multistorey cropping with plantation crops is a very common practice in south-east Asia. It is also being tried as a strategy for crop diversification in other places. The photo shows cacao, black pepper (*Piper nigrum*) and leucaena under Pejibaye palms (*Bactris gasipaes*) in Bahia State, Brazil (Photo: P.K.R. Nair, ICRAF).

Another form of multi-level associations is typified by the homesteads of Indonesia (Michon *et al.,* 1986) and Kandy gardens of Sri Lanka (Jacob and Alles, 1987). The species used are selected on the basis of their economic value. Tree crops such as coconut, acacia, and mango form the upper stratum; medium-sized trees such as guava, coffee, avocado, *Sesbania,* papaya, and banana compose the middle stratum; and annuals such as pineapple, pigeon pea, chilli, onion, ginger, beans, and tomato occupy the lowest stratum. Generally, in villages away from the market, all plant residues are recycled. In villages close to the market, often the emphasis is on producing cash crops. In several villages of Java, the houses are completely hidden by the surrounding homestead gardens. The soil erosion in this system is minimal as evidenced by the sharp contrast between heavily eroded land outside the village and well-preserved soils in the village homesteads in the Solo River basin of Central Java (Soemarwoto and Soemarwoto, 1984; Soemarwoto, this volume).

Farm forestry

In several countries, some very successful farm forestry projects have begun to increase the rate of reforestation and to augment the supply of timber, fodder, fruit, and fuelwood. In the Gujarat State of India, the Forestry Department started a project in the early 1970s. The system was gradually accepted by farmers because it was less labour-intensive and labour requirements were spread over the year. Tree farming with eucalyptus has now become so popular that irrigated, fertilized fields are also being used for this purpose (CSE, 1985). Today, at least 10 percent of Gujarat's farming families are involved in farm forestry. Similar success has been achieved in Haiti, Kenya, Senegal, and Nepal (WRI, 1985). Such tree monocultures on farmland may not, however, always represent a major ecological advantage. Trees such as eucalyptus do not provide fodder or mulch and may consume large quantities of water. There is also the risk that such programmes may increase rural unemployment. Hence, proposals for monoculture with tree species have to be carefully examined for their potential impact on soil, water, and employment.

In the Philippines, an industry-related agroforestry system has become popular. In the project launched by the Paper Industries Corporation of the Philippines (PICOP) in 1967, 20 percent of the land is used to raise agricultural crops and 80 percent for tree farming with *Albizia falcataria,* a fast-growing tree for paper pulp in an eight-year rotation (Veracion, 1983). The scheme provides farmers with a continuous source of food and income. PICOP guarantees the purchase of wood, provides help in acquiring land, assists in obtaining loans and *Albizia* seedlings, and furnishes technical help (see also Arnold, and Spears, this volume). The scheme has been able to meet all its objectives — to meet pulpwood requirements, to curb deforestation, and to increase the small farmer's income.

Agroforestry in arid and semi-arid areas

In arid and semi-arid environments, agroforestry systems help to provide greater insurance against weather abnormalities. Many multipurpose trees such as *Prosopis cineraria, Zizyphus rotundifolia, Casuarina* spp., *Tecomella undulata, Acacia tortilis,* and *Dalbergia sisoo* thrive in arid areas. Crops accompanying these trees may not show any significant reduction in grain yield (Government of India, 1986). Perennial shrubs such as *Sesbania* and *Cajanus cajan* also show promise for producing food, fodder, and fuelwood. Alley cropping can be successfully practised in many wetter areas.

Windbreaks and live fences are other options available in agroforestry for dry areas. *Leucaena leucocephala,* when planted as a windbreak, increases the grain yield of agricultural crops and moisture availability in soil by reducing surface run-off and evaporation. In Niger, millet yields increased by 23 percent when neem trees were planted as windbreaks. Vast tracts of sand dunes have been stabilized in Senegal by planting trees in and around the farms. In Maroua, Cameroon, *Cassia siamea* trees were planted across lowland plains as a shelterbelt to reduce soil erosion and increase agricultural output. Although the shelterbelt reduced soil erosion, yields of sorghum and cotton, the principal crops in the region, decreased, particularly in a 30m-wide strip on either side of the tree rows. This was not due to competition between the trees and crops but to the combined effect of reduced air turbulence and undisturbed heating of the ground raising the temperature. Therefore care needs to be taken to leave sufficient gaps in the tree fences to allow optimal air movement.

Agroforestry in problem soils and wastelands

Large areas in the tropics are affected by salinity, alkalinity, acidity and waterlogging. Unscientific land-use practices have led to a further increase in the area affected by toxicities and deficiencies. Such degraded lands can often be reclaimed by agroforestry while providing poor farmers with some income. Many species of trees can grow well in these problem areas where most agricultural crops cannot. The various species of *Sesbania,* for example, can grow successfully in saline, alkaline, and even waterlogged soils. In the coastal areas of Gujarat, India, extensive areas have been planted to *Prosopis juliflora.* In West Bengal, India, the government has leased out marginal degraded forest wastelands to landless farmers. These farmers are provided sufficient incentives and inputs to practise agroforestry, leading to increased tree plantations in the area. Large tracts of eroded wastelands in the Loess Plateau of China have been reclaimed by planting trees and using legumes as ground cover. Cheaper techniques such as planting tree seedlings in pits to which gypsum has been added can also be very useful in expanding agroforestry.

There is a need for formulating land-use policies based on sound principles of ecology and economics in such areas. The Indian example, where the government has formed a National Land-Use and Wasteland Development Council headed by the Prime Minister, can be followed in countries with similar problems.

Potential: the African opportunity

We are currently witnessing a good deal of optimism about what agroforestry can accomplish for food production and environmental protection. Generally, most countries in Asia and Latin America are able to meet their food requirements. In contrast, most sub-Saharan African nations face complex technological problems arising from the fragility of soils, scarcity of water, diversity of crops and pests, and climate variability. More than 40 percent of Africa's people live in countries where grain yields are lower than they were a generation ago. The loss of tree cover in closed forests and in savannas is extensive. In many countries wood collection for fuel and other uses exceeds the sustainable yield of remaining accessible forests. A recent World Bank study of seven West African countries covering five rainfall zones showed that in ecozones having the lowest rainfall, agricultural and fuelwood demands equal or exceed sustainable yields (World Bank, 1985). Another finding was that in all countries and in all zones, the sustainable carrying capacity of the forests was much less than that of croplands and grazing lands (Table 4).

Africa today is witnessing gradual shifts in its ecological zones. The recent drought and consequent famine in Ethiopia and other countries made Africa the focus of world attention and concern. In tourist literature, Ethiopia is often described as a country with 13 months of sunshine. It is ironic that agriculture, which is essentially a solar-energy-harvesting enterprise, is so poor in these countries. Restoring the African tree cover is essential to the restoration of the hydrological cycle and to the recovery of agriculture (Brown and Wolf, 1985). Widespread introduction and promotion of agroforestry can go a long way towards sustainable resource management and ecological and economic rehabilitation of Africa.

Table 4   Measures of sustainability in seven African countries* by ecological zones, 1960

	Food			Fuelwood		
Zone	Agriculturally sustainable population	Actual rural population	†Food disparity	Fuelwood-sustainable population	Actual total population	†Fuel disparity
			(millions)			
Sahelo-Saharan	1.0	1.8	-0.8	0.1	1.8	-1.7
Sahelian	3.9	3.9	0.0	0.3	4.0	-3.7
Sahelo-Sudanian	8.7	11.1	-2.4	6.0	13.1	-7.1
Sudanian	8.9	6.6	2.3	7.4	8.1	-0.7
Sahelo-Guinean	13.8	3.6	10.2	7.1	4.0	3.1
Total	36.3	27.0	9.3	20.9	31.0	-10.1

* Burkina Faso, Chad, Gambia, Mali, Mauritania, Niger, and Senegal. The given ecological zones are delineated by amounts of rainfall.

† Number of people in excess of or less than the agriculturally sustainable (in terms of food/fuelwood) population, expressed in millions.

Source: World Bank, 1985.

However, given the diversity of climates, farming systems, and economic conditions in various regions of Asia, Africa, and Latin America, one might wonder if changed land-use practices, such as those needed in agroforestry, will be an economically and ecologically sustainable alternative in increasing food production and protecting the environment in all ecozones. The potential of agroforestry in the ecological rehabilitation of upland, deforested, and already eroded watersheds is beyond doubt. It should also be a viable alternative to slash-and-burn and to bush-fallow systems of agriculture. However, greater acceptance of agroforestry in these areas would need efficient agrosilvopastoral systems capable of meeting the subsistence farmer's requirement for food, fodder, fuel, and some cash income.

Agroforestry can also play a greater role in reclaiming wastelands and wasted lands and in increasing food production in problem soil areas. It should also be appropriate in maintaining the long-term soil health of poor or average quality lowland soils. Agrisilvicultural or agrosilvopastoral systems such as alley farming can become very successful. Windbreaks and silvopastoral systems can help in mitigating drought-associated risks in arid and semi-arid regions.

The crucial question, however, is whether land-use practices should be changed to accommodate agroforestry in presently good quality, fertile, highly productive, resource-rich farms. In this age of a highly dynamic market and consequent changes in farming systems, introduction of trees into these areas may lead to inflexibility and many management problems. However, it will be desirable to encourage tree plantation on farm boundaries, canal bunds, and poor patches of the farmland.

In view of the enormous potential of agroforestry for promoting sustainable production, there is a need to identify and remove the technological and socio-economic constraints limiting the spread of agroforestry. Some of the important challenges that require immediate attention are now discussed.

Challenges

Biological constraints

To sustain agroforestry, it is important to strengthen our research efforts. Such low-cost and ecologically sound technologies should not receive low inputs of scientific and financial resources. There is an immediate need to extensively survey existing agroforestry systems to determine the interaction between component species, to classify the trees used, and then to refine the systems in view of soil, climate, and socio-economic limitations. Clearly, an interdisciplinary approach is warranted. Earlier, agroforestry systems were predominantly based on economic principles. Future systems, however, will have to overcome physio-logical (canopy structure), biological (pests and diseases), and ecological (sustainability and environment protection) constraints besides being economically sound. Detailed studies on the competition and complementarity between trees and understorey agricultural crops for solar radiation, space, and soil factors are needed. The enormous experience gained in intercropping annual crops can be very useful. The tallest component of agroforestry systems, the tree, should have foliage tolerant of strong light and high evaporative demand; the shorter components should have foliage adapted to shade and relatively high humidity. It is very important to consider the microclimatic changes that agricultural crops have to face under the trees. The entire process of selection and breeding of crops and crop varieties should take this into consideration. Similarly, agroforestry systems should avoid below-ground competition for water and nutrients by ensuring that component species have non-overlapping root systems.

The incorporation of deciduous trees such as *Dalbergia sissoo* into agroforestry systems can often be very useful. The natural abscission of leaves during autumn enriches the soil while the availability of solar radiation under the tree increases. Growing short-duration, high-yielding crops during this period of abundant sunshine and nutrients will be very productive. Alternatively, we should consider the use of growth regulators to induce partial defoliation of the trees when the radiation requirement of the understorey agricultural crop is at its peak. This may also help reduce the labour required for pruning in systems such as alley cropping. However, studies on the feasibility and practicality of such methods are needed.

Diversity in agroforestry systems is very important for their ecological sustainability. Extensive plantations with a single strain of *Leucaena leucocephala* in the Philippines and elsewhere has led to severe psyllid pest epidemics, damaging more than 50 percent of the trees (Lapis, 1986). Brewbaker (1985, this volume), drew attention to the genetic vulnerability to pest attack of single variety plantations of *Leucaena leucocephala*. Similarly, overdependence on a single genotype of the stem-nodulating *Sesbania rostrata* may lead to pest and disease outbreaks. Therefore, it is necessary to identify and describe more nitrogen-fixing tree species as well as genotypes of *S. rostrata*. Many such trees have been catalogued by the National Academy of Sciences (1979) and ICRAF (1986). Eucalyptus plantations have significantly increased in recent times due to their importance in pulp and paper manufacture. In many areas these can be replaced by fast-growing kenaf, *Hibiscus cannabinus,* another excellent source of raw material for paper manufacture.

Lack of suitable germplasm can delay future research and development efforts in agroforestry. National, regional, and global germplasm banks for preserving seeds of tree species are needed. Ecological sustainability of agricultural practices can be promoted only by spreading awareness that conservation is development.

Pest and disease control through agroforestry has been rarely studied. Today, integrated pest management involving non-overlapping pest crops and conservation of natural enemies is very important. Trees can, for example, provide a physical barrier to flying insects. In Samoa, there is a conspicuous reduction in cacao-leaf damage caused by the root beetle, *Andoretus versutus,* when it is intercropped with trees (Newton and Thomas, 1983). The role that multipurpose tree species such as neem (*Azadirachta indica*), known to be an effective pest-control agent, can play in agroforestry should be determined.

There is also a need to resolve silvicultural problems. It is important to raise the ecological adaptation of tree crops. A major problem with many tree species is the difficulty of establishing them and their slow initial growth. Some species need scarification of seeds for germination. We must examine alternative methods of establishment and propagation. Foresters have considerably improved the techniques of vegetative propagation for hardwood trees. These can be applied to nitrogen-fixing trees as well. Success in vegetative propagation and clonal selection will allow production of a large and continuous supply of plantation stock.

The stem- and root-nodulating shrubby legume *Sesbania rostrata* (Dreyfus and Dommergues, 1980; Dommergues, this volume) has the capacity to grow and fix nitrogen in waterlogged soils. The possibility of transferring this stem-nodulating habit to other legume species by genetic engineering should be explored to increase their adaptability.

Many tree crops, such as eucalyptus, could be unsuitable for agroforestry simply because their foliage and roots produce allelopathic toxins. Physiological and biochemical studies to control the production of these toxins should be initiated.

Last, methods should be developed to reduce the time taken to develop agroforestry systems. Research in agroforestry is long-term and does not promise major returns in the short run. Mistakes in agroforestry can, therefore, be costlier than mistakes made in agriculture.

Socio-economic constraints

The adoption of the agroforestry system of land use requires fundamental changes in approaches to farming. For a subsistence farmer this may involve, besides a change in farming practices, a change in diet or a change in marketing and labour-input requirements. Recent experience with Green Revolution technology has demonstrated the roles human ecology and sociology play in the acceptance and spread of technologies. We need to study the various socio-economic constraints and design appropriate strategies to convince the farmer that the short- and long-term payoff in adopting agroforestry will be considerable.

To promote agroforestry as a sustainable method of increased food production and environmental protection, we should develop and introduce the three mutually supportive and harmonious packages:

1. Economically viable, ecologically sound, and socially compatible technology;
2. Services and inputs to help farmers; and
3. Public policies that can stimulate and sustain the farmer's interest in agroforestry.

Success in cereal production in Asia and Latin America during the last two decades was due to the availability of mutually reinforcing agricultural packages (Swaminathan, 1986). The three major components of a symphonic agricultural system are briefly described below.

Package of technology

The proposed technology should aim to achieve the highest output possible per unit of land, water, time, and labour while disallowing any depreciation in the basic agricultural assets of land, water, flora, and fauna. The "cafeteria" approach, in which farmers can choose based on their capabilities and requirements, should be proposed. For the subsistence farmer, the proposed agroforestry technology should not only produce food, fodder, fertilizer and fuelwood, but some cash income. For the market-oriented farmer, the technology package should operate at still higher efficiency, both at the production and post-harvest level. A package of information should be built in to suggest the kinds of trees and agricultural crops, best combinations, management practices, costs and benefits, markets, and sources of financial and technical assistance.

In Africa, 75 percent of the food grown and eaten is produced predominantly by women. The proposed technology should also take note of the sex-related roles in food production.

Package of services

Equality of opportunity to appropriate technology should be the foundation of all agricultural extension and development planning. Designing and developing packages of essential services so farmers can take advantage of the new agroforestry technology is extremely important. Both government and private agencies should be active in providing seeds, seedlings, fertilizers, and, very important, credit. Regional seed and seedling banks should be established to ensure the timely availability of seeds and seedlings for farmers. Modern propagation methods can be used to produce quality stocks. The last service, credit, is essential because the payoff in agroforestry starts several years after the introduction of the scheme. Governments need to evolve innovative policies for an effective and timely input supply scheme.

Package of government policies

No agricultural or agroforestry technology can remain productive and sustainable without government support. One major area that requires government action is land reform. Agroforestry is a long-term practice so it will not be surprising if tenant farmers fail to adopt it. In an agroforestry project in the Philippines, it was noted that part-owners of land used their own area on the site for contour-hedges, indicating their acceptance of agroforestry but a reluctance to establish trees on land they did not own (Kent, 1985).

The concept of land reform should be enlarged to include not only ownership but regulations to prevent abuse of land. In parts of Africa, livestock reform to enable controlled grazing is equally important. Unscientific land-use practices have led to degradation of large tracts. In each country, there is a pressing need to set up a national task force for designing and promoting agroforestry systems, which should design appropriate components of agroforestry for all ecozones. This should be supported by suitable training programmes at various levels. The technique of training should be learning by doing. Mass media such as local language papers, radio, and television should be actively involved. Software should be developed for agroforestry education and communication. Extension agencies should have skilled and motivated workers to successfully protect and promote the interests of individuals and society in agroforestry.

The available food-grain surpluses in the world give us increasing opportunities to diversify agroforestry systems based on long-term sustainability criteria. For example, to reclaim eroded, marginal soils, one should avoid growing annual food crops in the new agroforestry systems. Subsistence farming families in remote isolated areas, hard pressed to earn their daily bread, can be persuaded to adopt ecologically sound land-use practices only if they are assured of the staple grain they need. Food security for the poor must first be ensured before the promotion of ecological security. Governments will have to build visible grain stocks in habitats characterized by fragile ecosystems. In countries where governments do not have their own stocks, special programmes such as "Food for scientific land use" or "Food for agroforestry" development could be initiated.

In the case of better-off farmers, opportunities for producer-oriented and remunerative marketing becomes essential to stimulate and sustain their interest in agroforestry. Here, input-output pricing policies become crucial. Enough incentives, such as support price for wood and other tree products, should be provided.

Conclusion

Eternal vigilance is the price of stable agriculture. The greatly increased population and its ever expanding needs for food and fuelwood in this century threaten agricultural stability. Political and commercial greed, the genuine needs of the poor for fuel and fodder, inappropriate technologies, and the absence of a systems approach in the design and implementation of agricultural and industrial projects in ecologically fragile areas, have all contributed to increased environmental deterioration. B.F. Skinner (personal communication) has rightly emphasized:

> Every new source from which man has increased his power on earth has
> been used to diminish the prospects of his successor. All his progress is
> being made at the expense of damage to the environment, which he
> cannot repair and cannot foresee.

It is time that we devote greater attention to economically and ecologically sustainable agricultural production systems where present economic progress and prospects for survival will not be in conflict. Fortunately, agroforestry systems are characterized by this happy blend and help us to exploit in a sustainable manner cubic volumes of soil and air and thereby give farmers the maximum return from the available soil, water, nutrient, and sunlight.

There is now an opportunity to design more efficient and ecologically sustainable agroforestry systems by putting the large food grain stocks of today to intelligent use. Agroforestry systems designed to overcome physiological, biological, ecological, and economic constraints can help to enhance production efficiency. We therefore need both greater support for agroforestry research and greater integration of agroforestry research into the mainstream of farming-systems research. Stimulating and helping to sustain a symphonic agroforestry system based on appropriate blends of political will, professional skill, and peoples' action will be a major challenge for ICRAF in its second decade.

Acknowledgements

The author is deeply indebted to Dr P.K. Aggarwal of the Multiple Cropping Department of IRRI for assistance in compiling material for this paper.

REFERENCES

Brewbaker, J.L. 1985. The genetic vulnerability of single variety plantations of *Leucaena*. *Leucaena Research Reports*. (Nitrogen Fixing Tree Association. Hawaii.) 6: 81–82.
Brown, L.R. (ed.). 1987. *State of the world*. Washington, D.C.: World Watch Institute.
Brown, L.R. and E. Wolf. 1985. Reversing Africa's decline. World Watch Paper 65, Washington, D.C.
Brown, L.R. and J.L. Jacobson 1986. Our demographically divided world. World Watch Paper 74, Washington, D.C.
Centre for Science and Environment (CSE). 1985. The state of India's environment. *The second citizen's report*. New Delhi: CSE.
Dreyfus, B. and Y.R. Dommergues. 1980. Non-inhibition de la fixation d'azote atmosphérique chez une légumineuse à nodules caulinaires *Sesbania rostrata*. *C.R. Acad. Sci.* (Paris) 291:767–770.
FAO. 1982. *Tropical Forest Resources*. FAO Forestry Paper No. 30. Rome: FAO.
Government of India, Ministry of Agriculture. 1986. Annual Report. New Delhi: Department of Agricultural Research and Education.
Higgins, G.M., A.H. Kassam, L. Naiken, G. Fischer and M.M. Shah. 1983. *Potential population-supporting capacities of lands in the developing world*. Rome: FAO.
ICRAF. 1983. An account of the activities of the International Council for Research in Agroforestry. Nairobi: ICRAF.
————. 1986 Multipurpose tree and shrub seed directory. Nairobi: ICRAF.
Jacob, V.J. and W.S. Alles. 1987. Kandyan gardens of Sri Lanka. *Agroforestry Systems* 5: 123–137.
Jones, P.D., T.M.L. Wigley and P.B. Wright. 1986. Global temperature variations between 1861 and 1984. *Nature* 322:430–434.
Kang, B.T., G.F. Wilson and T.L. Lawson. 1984. *Alley cropping: A stable alternative to shifting cultivation*. Ibadan, Nigeria: IITA.
Kent, T. 1985. Development and extension of the agroforestry/hillside farming programme in Zamboanga del Sur Development Project, Republic of the Philippines.
Lapis, E.G. 1986. Psyllids invade the Philippines. *Canopy Int.* 12(2):1, 10.
Michon, G., F. Mary and J. Bompard. 1986. Multistoried agroforestry garden system in West Sumatra, Indonesia. *Agroforestry Systems* 4: 315–338.
Nair, P.K.R. 1979. *Multiple cropping with coconuts in India*. Berlin/Hamburg: Verlag Paul Parey.
————. 1985. Classification of agroforestry systems. *Agroforestry Systems* 3: 97–128.
National Academy of Sciences (NAS). 1979. *Tropical legumes: resources for the future*. Washington, D.C.: NAS.
Newton, K. and P. Thomas. 1983. Role of the NFTs in cocoa development in Samoa. *Nitrogen-Fixing Tree Research Reports* 1:15–17.
Shah, S.L. 1982. Ecological degradation and the future of agriculture in the Himalayas. *Indian J. Agric. Econ.* 37(1):1–22.
Sharma, T.C. 1976. The pre-historic background of shifting cultivation. Proceedings of a seminar on shifting cultivation in North-East India. New Delhi: Indian Council for Social Science Research.
Soemarwoto, O. and I. Soemarwoto. 1984. The Javanese rural ecosystem. In T. Rambo and Percy E. Sajise (eds.), *An introduction to human ecology research in agricultural systems in Southeast Asia*. University of the Philippines at Los Banos, Philippines.
Swaminathan, M.S. 1981. *Building a national food security system*. New Delhi: Indian Environmental Society.

————. 1982. *Report of the fuelwood study committee.* Planning Commission, Government of
        India.
————. 1986. *Sustainable nutrition security for Africa: Lessons from India.* The Hunger Project
        Papers No. 5, October 1986. San Francisco: USA.
Veracion, A.G. 1983. Agroforestry: The Paper Industries Corporation of the Philippines' experience.
        In *Agroforestry in perspective.* Los Banos, Laguna, Philippines: PCARRD.
World Bank. 1985. *Desertification in the Sahelian and Indanian Zones of West Africa.* Washington,
        D.C.: World Bank.
World Resources Institute (WRI). 1985. *Tropical forests: a call for action.* Part I. Washington, D.C.:
        WRI.

4

Institutional aspects of agroforestry research and development

Bjorn O. Lundgren

Director
International Council for Research in Agroforestry (ICRAF)
P.O. Box 30677, Nairobi, Kenya

Contents

Introduction

The history of agroforestry as a science and as a focus for systematic development efforts is very short — fifteen years at the most (see King, this volume). In 1982 — the "middle ages" of this short history — the present writer was asked by the Office of Technology Assessment of the United States Congress to make an evaluation of the role of agroforestry in improving tropical lands. The forecasts contained in that report (Lundgren, 1982a) regarding the likely developments in and constraints to agroforestry over the coming five to ten years generally seem to have been correct. Interest in agroforestry is increasing rapidly among scientists, land-use experts and development professionals; resources for research and development are being made available from donors and national institutions at an unprecedented level (although they are still modest in absolute terms); concrete results from R & D programmes are just starting to emerge on a significant scale; and the next three to five years will see an information explosion in agroforestry. These developments in general, and the progress made in specific fields and regions, are highlighted in other contributions to this volume.

Another assessment contained in the report mentioned above was that the main constraints to a full realization of the potential of agroforestry were of an institutional nature and related to the rigid disciplinary compartmentalization which characterizes institutions working in the field of land use. There have been very few signs in the last five years of this situation changing for the better. On the contrary, it is more urgent than ever that these institutional questions are addressed at the highest possible levels, both in individual countries and internationally. If effective and relevant institutional arrangements are not developed for implementing agroforestry R & D programmes on a large scale within the next five to ten years, the risk is very real that the potential of agroforestry will never be fully realized.

This article deals with the institutional aspects of agroforestry. It presents the writer's personal thoughts and should certainly not be seen as an ICRAF policy statement.

Land-use institutions today

The basic institutional structures established to deal with the use of land in virtually all the countries of the world today originate from temperate Europe and North America. There, in the late nineteenth and early twentieth centuries, the modernization of agriculture and forestry, which was necessitated by and dependent upon the rapid industrialization, led to the gradual emergence of government and private institutions to support the land users. Crop production and industrial wood production, which were carried out on separate types of land, required different professional skills, had different aims, and very often were managed by different owners (farmers versus governments or private companies). It was entirely rational, therefore, that agricultural and forestry institutions developed independently of each other. In the few cases where it was deliberately planned that trees, crops and/or animals should interact in specific technologies or land-use practices, for example in windbreaks, shelterbelts, hedges, grazing in fruit orchards, or game management for meat production in forests, there was never any difficulty in establishing which institutional sector was "responsible" for the technology or practice. With very few exceptions, anything done on designated farmland, even if it involved tree growing, was (and is) the responsibility of the agricultural/horticultural sector, and any use of forest land, including game management, rational utilization of wild berries, mushrooms, etc., falls under the forestry sector.

As a result of these separate institutional developments, there are today different laws and policies governing agricultural and forest land use; there are separate training, education and research institutions; advice to land users is provided through separate extension services; agriculture and forestry normally fall under different ministries or, if they are under the same ministry, under separate departments.

Another important aspect of the land-use legacy from the industrialized countries is that all policies and disciplinary R & D efforts are aimed at *maximizing,* in a sustainable way, the output of products per unit of land — this applies as much to wheat, maize, milk and meat as it does to timber and pulpwood. Commercially oriented monocropping dominates the use of land and has been seen as very successful as markets for agricultural and forest products have continuously increased in volume over the last century. Subsistence use of land, in the sense of people being dependent on their own land for food, has virtually disappeared in industrialized countries.

When the European colonial powers established their administrations in Africa and elsewhere in tropical and subtropical regions, the institutional structures, policies and aims related to land use and development used in the home countries were simply copied in the colonies. This applied also to those countries which were not colonies, e.g., in Latin

America, where governments chose to adopt the industrialized countries' institutional structure in land use. The model has been continued after independence in all tropical countries and the post-war international organizations that were set up to assist the emerging nations in improving and rationalizing the use of their land resources are all oriented on conventional-discipline lines. There is little doubt that the concentration of R & D efforts on the particular commodities, technologies and practices which the mono-disciplinary approach leads to, has resulted in some remarkable success stories in tropical countries during this century. Cultivation of export crops such as coffee, tea, oil palm, fruits and spices forms the mainstay of many developing-country economies. Some countries have achieved self-sufficiency in industrial wood production by systematically building up plantations of exotic and indigenous trees, and, most remarkable of all, the "green revolution" of the last two to three decades has turned previously food-deficit countries into major grain exporters. Although all these developments have had their share of criticism (some of it justified, but most based on ignorance) from economists, social scientists and environmentalists, their technical and economic success is an undisputed credit to all the horticultural, silvicultural and agronomic scientists and R & D institutions behind their development.

In spite of the relative successes achieved in agriculture, forestry and other disciplines in the tropics and subtropics (developing countries), there are, quite obviously, many aspects of land use which are not successful. Food production per capita has been decreasing in most of Africa over the last 25 years; man-made desert conditions are spreading at an alarming rate; erosion and flooding, largely as a consequence of defective land use following deforestation, cause unprecedented loss of farmland; rural populations in more and more areas are affected by existing or imminent energy crises due to lack of fuelwood; and the effects of naturally occurring droughts, in terms of loss of human life and livestock, become more and more devastating with time. It is open to discussion whether the main causes of these conditions are demographic, political, technological, economic or environmental. It is generally agreed, however, that there is no single cause but rather a complex interaction among several factors — interactions which are different from place to place and country to country.

I have become increasingly convinced that a significant contributor to the failure to solve many important land-use problems is the inappropriateness of conventional-discipline-oriented institutions for identifying and addressing *real* problems in land-use systems in most tropical and subtropical (developing) countries. This particularly applies to the multitude of subsistence or mixed subsistence/cash farming and pastoral systems in which the vast majority of rural land users live and from which they eke out a living. They differ, of course, with ecological and socio-cultural conditions, from the purely nomadic pastoral systems of the arid to semi-arid zones to the sedentary mixed-farming systems on upland soils in subhumid areas and the shifting cultivation in the humid zones. Some features are shared by most of them, e.g., low cash incomes and, hence, little ability to invest in improvements requiring money, and marginal ecologies, such as infertile or erosive soils, or marginal climates. Minimizing the risk of crop failure or animal loss in these situations is a much more important concern than increasing yields. Often, land tenure is insecure or non-existent. Common lands, which often serve as important sources for fuelwood, building material, grazing, etc., are diminishing in area or are being degraded through overuse as a result of increasing populations.

With few exceptions, subsistence or near-subsistence farming systems are mixed in the sense that the farmer produces not only the bulk food crop from the land but also specialized food crops (vegetables, fruits, spices, etc.), animals for meat, milk or draught power, and, very often, trees and shrubs for fuel, fodder and building material. Cash income sometimes comes from specialized crops intended for sale but normally from surpluses of the "subsistence" crops and animals. Obviously, the relative importance of each of these components of the system varies, but they all serve to satisfy *basic* needs of the land user (food, shelter, energy, cash, etc.), and they all interact economically and/or ecologically in that they are managed by the same limited labour resource and they share the same farm environment (soil, water, topography). The subsistence land user's strategy and aims are to use his labour and land resources to *optimize,* with minimum risk, the production of various products and services required to satisfy all his basic needs.

Why are today's institutions inadequate?

The fundamental inadequacy of conventional-discipline-oriented institutions lies in the failure to acknowledge and understand these basic facts, strategies and aims, and in the inability to adapt to them. The aims, infrastructure, rationale and philosophy of these institutions, as well as the training of their experts, are geared to the *maximization* of individual components, be they food crops, cash crops, animals or trees. There is little understanding that the land user needs to share out his resources for the production of other commodities or services. Even if such an understanding appears to exist — the forester may generously agree that food production is essential and the agronomist may not disagree that fuelwood is needed — there is very little technical comprehension of the requirements for producing commodities outside one's own discipline.

The inability of technical institutions and experts to understand how social, religious, cultural and traditional beliefs and preferences can nullify a convincing cost/benefit analysis on, say, increased fertilizer use, upgrading of cattle or establishment of fuelwood plantations, has been so well documented by social scientists that there is no need to elaborate the point.

It is, quite obviously, this lack of understanding of the complexities of many land-use systems, which is built in to the very foundation of conventional institutions, that is the basic cause of the many failures and frustrations in trying to solve the land-use problems of developing countries today. Land users are generally not conservative or "primitive"; they are not adamantly unwilling to improve their lot; they are not opposed to increasing their yields or cash incomes; nor are they hostile to planting and caring for trees on their land — *provided* we understand how technology changes and improvements fit into *their* problems, priorities, beliefs and aims. The situation today can be summarized as in Figure 1, which is derived from an idea by R. Chambers (personal communication).

There are a large number of more or less narrowly specialized sectors, disciplines, institutions, scientists and experts, laws, policies, etc., all dedicated to maximizing (within limits known to and identified by themselves) "their" product, i.e., which look at *one* segment of the land-use system only.

There is *no* institution today which has both the mandate and the competence to identify solutions to land-use problems based on an interdisciplinary analysis of interactive constraints and potentials *within* land-use systems, and the power to assign resources in a way that will cut across institutional boundaries in order to implement such solutions.

There are, nevertheless, a few positive signs that awareness of the need to address the institutional constraints to real problem-solving is increasing. For example, the recently

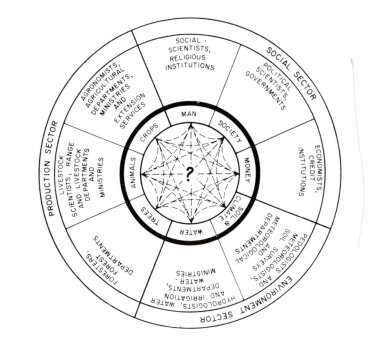

Figure 1 Institutional dilemma of land development? Disciplinary institutions look at a specific compartment of a land-use system only and normally do not consider the complex interactions within the system.

published report of the *World Commission on Environment and Development* (the "Brundtland Commission") (1987) concluded:

> The integrated and interdependent nature of the new challenges and issues contrasts sharply with the nature of the institutions that exist today. These institutions tend to be independent, fragmented, and working to relatively narrow mandates with closed decision processes.

Less encouraging, however, is the fact that virtually all recent policy and planning documents from international institutions, even if they pay lip service to the need for multidisciplinarity, integrated approaches and holistic views, end up making conventional recommendations such as increased use of fertilizers, irrigation and genetically improved crop varieties, increased tree planting, etc. None contain any critical analysis of the adequacy of existing institutions for addressing the totality of the problems and for contributing to their solution. This applies as much to agricultural (TAC, 1987) and forestry (Adams and Dixon, 1986; Carlson and Shea, 1986) research, as to development statements, again both agricultural (FAO, 1986a, 1986b, 1987; OAU, 1985) and forestry (FAO, 1985; WRI, 1987).

One reason for the slowness with which the institutional problems are being addressed has probably been the lure of the statistics on "global problems" which have emerged over the past 15 years, e.g., on environmental and developmental problems such as desertification, deforestation, the fuelwood crisis and declining food production. As a result of improved techniques of survey, monitoring and other statistical methods, it has been possible to break down an almost infinite number of complex, local problems into their component effects and to add up these effects into global perspectives. Since it is easy to express the magnitude of these problems in conspicuous and alarming global figures, such as the numbers of hectares of forests that are lost, or land that is turned into desert, or the distance people will have to walk every day to collect their fuelwood, it becomes imperative and even very attractive to politicians, decision-makers and institutions to look for simple solutions to these problems and to extrapolate the likely benefits to a global scale. Disciplinary-sector institutions have been very successful in exploiting this situation to increase resource allocation to themselves by promising simple solutions to what appear to be straightforward problems — more tree plantations will solve the fuelwood and deforestation problems; more fertilizers and irrigation will increase food production, and so on.

The fact, however unappealing and complex it may be, is that just as the "global problems" are the sum of the effects of a large number of local problems, the solutions can only be achieved by adopting an equal number of sound land-use practices and political and economic measures (Lundgren, 1985). Problems *must* be identified, diagnosed and solved where they occur. It does not help small-scale mixed farmers in district *x* of country *y*, to know that 300 million people in the developing world do not have adequate supplies to meet their fuelwood or protein requirements. There is an urgent need to re-think and re-evaluate the situation. The conclusions arrived at must direct the relevant institutions from the discipline-oriented maximization thinking that is fuelled by global statistics to the multidisciplinary optimization thinking that is geared to solving local problems.

Agroforestry as a catalyst for change

When agroforestry was institutionalized through the creation of ICRAF in 1977, there were very few people who thought of the subject as anything but an off-shoot from the forestry sector. Indeed, the early ideas and concepts originated with tropical foresters who were concerned about the poor contribution that the forestry sector made to the well-being of rural populations other than those directly involved in forestry operations. The long and basically positive experience of taungya-type agrisilvicultural systems on forest land had demonstrated to foresters that timber and food-crop production from the same land was possible (King, this volume). In the early stages, agroforestry was seen as the forestry sector's contribution to agriculture and many foresters still think of it in that way. There were no serious efforts to integrate forestry, or rather tree growing, into agricultural practices, let alone any critical analysis of whether the existing forestry institutions were competent to take trees outside the forests.

It was only in the early 1980s that agroforestry developed into a truly integrated and interdisciplinary approach to land improvement, mainly through ICRAF's conceptual and methodological work (Lundgren, 1987b; Steppler, this volume). A more objective definition of agroforestry than previous ones, and one that ICRAF has used since the early 1980s (Lundgren, 1982b), is as follows:

> Agroforestry is a collective name for all land-use systems and practices in which woody perennials are deliberately grown on the same land management unit as crops and/or animals. This can be either in some form of spatial arrangement or in a time sequence. To qualify as agroforestry, a given land-use system or practice must permit significant economic and ecological interactions between the woody and non-woody components.

This definition clearly underlines the integrated nature of the approach.

The aim of agroforestry is (or should be) to optimize the positive interactions between components in order to achieve a more productive, sustainable and/or diversified (in relation to the land users' need) output from the land than is possible with other forms of land use. It is obvious that, with this definition and aim, agroforestry as a science and practice *must* cut across conventional institutional areas and draw upon several disciplines in the social, production and environmental sectors (see Figure 1) if its full potential for improving land use is to be realized.

ICRAF has built up a truly multidisciplinary team of experts and scientists, representing all the relevant disciplines deemed necessary to study all aspects of a land-use system. Through this team, objective analytical methods of identifying problems in land-use (farming) systems, and potentials for their solution that are not restricted to agroforestry solutions, have been developed (Raintree, 1987). Using an analogy from Figure 1, this diagnostic and design (D&D) methodology means that instead of the individual experts standing outside the land-use system and observing their own disciplinary components of the system, the whole team, without any preconceived ideas about the nature of the problems and their potential solutions, "parachutes" into the middle of the complex system and tries to diagnose causal mechanisms behind the problems and interactions between the components of the system. From this diagnosis, technologies with a potential for solving the problems are designed. It is only at this stage that the role of different disciplinary sectors and institutions in developing these technologies can be defined.

When putting this methodology into practice in collaborative research programmes with national and international institutions, ICRAF has been faced with a host of problems related to the compartmentalization of conventional disciplinary institutions and professions, ranging from direct mistrust and lack of appreciation of each other's expertise to the enormous difficulty of arriving at decisions about resource allocation in programmes involving different institutions from different ministries.

In the Agroforestry Research Networks for Africa (AFRENA) programme, ICRAF has developed a model for inter-institutional collaboration at national and regional levels which, at least in its early stages, has been very successful (Torres, 1986). The key elements of the approach are to stimulate and assist (technically and, if necessary, financially) national forestry, agriculture and other institutions to work together in analysing land-use problems and designing research programmes to solve them, and to define their exact role in implementing such programmes. It is essential that participation in these activities is voluntary for the different institutions, that it is seen as professionally stimulating and enriching and that it does not initially "threaten" existing institutional power structures. The technically sound results across disciplinary boundaries that are starting to be produced from such semi-formal programmes show that this approach certainly has potential. Once their usefulness has been confirmed we will be ready to take the next steps required to achieve more appropriate institutional functions and structures for addressing the problems of the small-scale subsistence farmers and land users of the tropics and subtropics.

Agroforestry as a discipline has the potential for taking a leading and catalytic role in this process of change, because of its inherent integrative and multidisciplinary nature, its optimization rather than component-maximization aims, and because of the great interest shown in it today.

Land-use institutions tomorrow

It would be presumptuous to end this article by proposing an "ideal" structure and set of objectives for tomorrow's land-use-related institutions. No such ideal institution will ever exist because of the enormous variety in conditions, policies and institutions among regions, countries, economies, etc. Some general thoughts on directions may, however, produce some food for thought.

1.   In the short and medium term, and within existing institutional structures, collaborative programmes which cut across disciplinary boundaries and address concrete land-use problems must be encouraged and given more support, both at the national level and at the level of international agencies such as donors, UN bodies, and International Agricultural Research Centres (IARCs). It would be appropriate if the international institutions took the initiative in this. Unfortunately, such is not the case today: most international bodies and institutions are firmly entrenched in disciplinary thinking and actions. Although there are a few very encouraging exceptions, integrated approaches still seem far away. With the risk of over-generalizing facts and of stepping on some toes, I feel that it would be encouraging to see the following positive developments on the international land-uses scene:

     (a)   That the institutions within the Consultative Group of International Agricultural Research (CGIAR) accept the fact that trees and shrubs form integral parts of most small-scale food-producing farming systems in the tropical world and that their rational development can enhance both productivity, sustainability and diversity of food production;

     (b)   That FAO's Agriculture and Forestry Departments develop joint projects to show that the UN's main land-use-related agency has the interest, mandate and competence to look at crops *and* trees, and their interactions, in the same place and in the same land-use systems;

     (c)   That the forestry departments and advisors of bilateral and multi-lateral donor agencies, development banks, research and policy institutes, etc., accept the fact that tree growing on farmland (agroforestry) is *not* an exclusive forest-sector activity, and that the same donors and institutions ask themselves why they channel all their support to agroforestry through forestry, energy or environmental programmes rather than through agricultural programmes.

2.   It must be strongly emphasized that in the future we will still need specialized institutions and experts — the plea for multidisciplinary, inter-institutional approaches and integrated thinking must not lead to the formation of super institutions manned exclusively by generalists. Undoubtedly there will be a need for crop physiologists, maize breeders, tree geneticists, fertilizer experts, insect ecologists, and so on, and for specialized institutions that provide a working environment for such experts. Even the most "perfect" interdisciplinarily-derived research or development programme must, when being implemented, be broken up into its component parts, and then resynthesized. If there is inadequate expertise on the structure and function of the component parts, then the whole entity (the improved land-use system, or whatever) will never be fully operational and optimized.

3.   What will be needed are new institutional functions for problem identification, priority setting and resource allocation, without necessarily making fundamental changes of

structure. Such new functions can be created within or between existing land-use ministries and departments, or as independent bodies with existing institutions subordinate to them. It would probably be best to start by creating inter-institutional committees for planning but which will become more and more executive as experience is gained. Depending on the problems to be addressed, these committees can then create appropriate task forces of existing disciplinary institutions and resources.

These may not sound very fundamental changes but they will require fundamental rethinking among disciplinary scientists, institutions and decision-makers. The sooner this process starts, the faster can some of the key land-use problems of the world be solved.

REFERENCES

Adams, N. and R.K. Dixon (eds.). 1986. *Forestry networks*. Proceedings of the first network workshop of the forestry/fuelwood research and development project (F/FRED), 24–27 September 1986, Bangkok. Washington D.C.: Winrock International.
Carlson, L.W. and K.R. Shea (eds.). 1986. *Increasing productivity of multipurpose lands*. Proceedings of IUFRO research planning workshop for Sahelian and North Sudanian zones, Nairobi, January 1986. Vienna: IUFRO.
FAO. 1985. *Tropical forestry action plan*. Committee on Forest Development in the Tropics. Rome: FAO.
————. 1986a. A programme of action for African agriculture proposed by the Director-General. Rome: FAO.
————. 1986b. *African agriculture: the next 25 years*. Main report (ARC/86/3). Rome: FAO.
————. 1987. Summary programme of work and budget 1988-1989. Document COAG/87/5 presented to 9th Session of the Committee on Agriculture, 23 March-1 April 1987. Rome: FAO.
Lundgren, B. 1982a. The use of agroforestry to improve the productivity of converted tropical land. Report prepared for the Office of Technology Assessment of the United States Congress (mimeo).
————. 1982b. Introduction. *Agroforestry Systems* 1:3–6.
————. 1985. Global deforestation, its causes and suggested remedies. *Agroforestry Systems* 3:91–95.
————. 1987a. Agroforestry in third world countries. Paper presented to the IUFRO workshop on agroforestry for rural needs. New Delhi, India, 22–26 February 1987.
————. 1987b. ICRAF's first ten years. *Agroforestry Systems* 5: 197–218.
Organization of African Unity. 1985. *Africa's priority programme for economic recovery 1986–1990*. Addis Ababa: OAU.
Raintree, J.B. 1987. The state of the art of agroforestry diagnosis and design. *Agroforestry Systems* 5: 219–250.
Technical Advisory Committee (TAC). 1987. *CGIAR priorities and future strategies*. Rome: TAC/FAO.
Torres, F. 1986. Agroforestry research networks in tropical Africa: an ecozone approach. Paper presented to the First International Conference of Agricultural Research Systems, IFARD, Brasilia, Brazil, 6–11 September 1986.
World Commission on Environment and Development. 1987. *Our common future*. London: Oxford University Press.
World Resources Institute. 1985. *Tropical forests: a call for action*. Parts I-III. Washington, D.C.: WRI.

5

Agroforestry:
a development-bank perspective

John Spears

Forestry Advisor, World Bank
Washington, D.C., USA

Contents

Introduction

The period 1977–1986 has seen a sharp escalation in development banking support for forestry. The approximately US$ 2 billion invested in forestry by four of the multilateral development banks represents a 13-fold increase in the volume of lending compared with the previous decade. More important than the volume of lending, however, is the fact that the banks have revised their earlier forest policies to give much greater support to agroforestry, fuelwood, watershed protection, forest conservation, education, training, research and extension, in addition to the more traditional industrial plantation-oriented forestry projects that accounted for 80 percent of bank forestry investments in the 1960s and

---

Editors' note
The author has used the World Bank perspective and classification for projects, many of which were initiated before the term agroforestry was widely used. He has also included projects which are peripheral to many readers' perception of agroforestry.

early 1970s. Agroforestry lending, for example, during this last decade totalled US$ 750 million, representing a rise from 6 percent to 37 percent of total bank forestry investments.

Economic rates of return of agroforestry projects

In most cases the economic rates of return of the agroforestry projects financed by development banks have been significantly higher than those yielded by the industrial plantation forestry projects that characterized earlier bank lending programmes, as illustrated in Table 1.

Table 1   Comparative rates of return for forestry projects financed by development banks during the decade 1977–1986

Category of project	No. of projects reviewed	Range of economic rates of return (%)
Industrial plantation forestry	15	10–15
Agroforestry/fuelwood	27	15–30
Watershed rehabilitation	8	15–21

Source: World Bank Annual Reports.

As has been the Bank's experience in agriculture, the message is that investing in poverty-oriented agroforestry projects is a bankable activity that requires no special justification on humanitarian or environmental grounds. Agroforestry projects can ensure increased farm productivity and income for rural people on the one hand, as well as protection of the farming environment (particularly soil and water resources) on the other.

There are several clearly identifiable reasons why agroforestry projects have demonstrated fairly high financial rates of return:

(a) On-farm tree planting or local woodland protection and conservation by village communities usually requires a lower level of investment in supporting infrastructure (e.g., access roads, firebreaks and administrative staff) than would normally be required for government plantation or forest protection programmes. Particularly in small-farm situations much of this necessary infrastructure is already in place;

(b) In general, farmers have been quick to adopt fast-growing short-rotation genera such as *Leucaena, Sesbania, Grevillea, Calliandra, Eucalyptus* and *Albizia* with prospects of early financial return (in more favourable ecological situations 3–7 years) or which they recognize can benefit the farming system (e.g., by nitrogen fixation, provision of leaf mulch, fodder, fruit and other products);

(c) The use of low-cost tree-establishment techniques such as direct seeding, or use of the "basket" system for distributing seedlings as is widely practised in India, together with decentralization of planting among many thousands of smaller farmers has helped keep establishment costs down;

(d) Prices of poles, fuelwood and timber have, in general, been rising in real terms at a faster rate than those of other commodities. A combination of increasing scarcity and rising prices for building poles, fuelwood, fodder and timber make tree farming in favourable ecological conditions a profitable venture.

The underlying ecological framework for agroforestry is important. In general, opportunities for bank support for agroforestry schemes based on farm tree planting have been more obvious in tropical countries with reasonable rainfall or surplus irrigation water and reasonably fertile soil that favours fast tree growth and early returns. Developing investment programmes for agroforestry in semi-arid or arid-zone environments is proving a much slower business because of the length of time it takes to produce results and because of land-tenure and socio-economic problems associated with ensuring sustainable common-property resource management. This is especially difficult in situations where the same savanna woodlands are under pressure for production of charcoal as well as providing a critical dry weather grazing reserve for nomadic livestock (as is the case, for example, in many of the African countries situated in the Sudano-Sahelian zone).

Table 2 summarizes the cash flow and financial rate of return for a typical cash-crop-tree-farm operation in the Philippines where several thousand small farmers have taken up cash-crop-tree farming for production of pulpwood, poles, timber, charcoal or fuelwood.

Financial rates of return to the farmer from such investments have typically been in the range of 20–30 percent. Small farm woodlots of *Acacia mearnsii* and *Eucalyptus* spp. in Kenya covering about 60,000 ha are currently yielding surplus income (i.e., over and above the value of domestically consumed poles or fuelwood) of about US$ 0.5 million a year. For many rural families this revenue represents their only significant source of surplus farm income. Most of the food produced on the farm is needed for subsistence.

The Philippines and Kenya experiences cited above relate to situations where farmers are practising cash-crop-tree farming in discrete woodlots which are maintained under tree cover and operated on a rotational basis. The financial rate of return to the farmers' investment can be readily measured by quantifying the cash value of pulpwood, poles, fuelwood or other forest output.

High economic rates of return are also being demonstrated for agroforestry projects where trees are being interplanted with crops or included as an integral component of livestock-farming systems. In such situations the benefits from investment are based on measurement of the impact of the agroforestry tree/crop combination on both agricultural and forestry output. Evaluation of benefits must take into account both positive and negative effects of trees on crop yield or livestock output, as well as the cash value of forest products. An encouraging feature of development banks' experience of project appraisal in the last decade has been the improved perception of techniques of economic analysis that can help to take these sometimes less obvious benefits into account.

For example, Table 3 summarizes the results of a recent development bank appraisal of the economic viability of a shelterbelt/agroforestry programme in Northern Nigeria. If the analysis in this particular case had been confined only to consideration of wood benefits, the internal rate of return would have been of the order of 4.7 percent (which is below the opportunity cost of capital and would probably have led to the project being rejected for bank financing).

However, when the benefits of improved fodder availability, the positive impact of shelterbelts on crop yields, prevention of soil erosion and enhancement of soil fertility were taken into account, rates of return jumped to between 16 and 21 percent.

Table 2 Smallholder tree farming project in the Philippines: Summarized cash flow and financial rate of return for a 10 ha *Albizia falcataria* tree farmer (US$, 1980)*

Activity	Year															
	1	2	3	4	5	6	7	8	9	10	11	12	13	14	15	16
Accumulative planted area (ha)	–	4	8	10	10	10	10	10	10	10	10	10	10	10	10	10
Annual area harvested (ha)					1.25	1.25	1.25	1.25	1.25	1.25	1.25	1.25	1.25	1.25	1.25	1.25
Net cash flow																
Loan	640	640	320													
Net revenue							828	864	958	958	1034	1111	1111	1181	958	
Net cash flow																
Investment **	665	675	343													
Debt service							–	709	709	709	709	709	709	709	–	–
Net cash flow	(25)	(35)	(23)				828	155	249	249	325	402	402	472	958	958

* Financial rate of return computed over 25 years = 22 percent.
** Compounded at 12 percent.

Table 3   Appraisal of economic benefits of an agroforestry/shelterbelt/soil conservation
project in Northern Nigeria*

	Internal rate of return (%)
Agroforestry component	
Wood/fruit benefit alone	7.4
Wood/fruit benefits plus positive impact of trees on conservation of soil and crop yield	16.9
Shelterbelt component	
Wood/benefits alone (poles/fuelwood)	4.7
Wood benefits plus positive impact of shelterbelt on soil conservation and crop yield	21.8

* The original analysis includes a broad range of rates of return related to different assumptions about
the phasing of benefits level of crop yields and other variables. This table summarizes the high and
low ends of the analysis results.

Criteria for bank appraisal of agroforestry projects

Criteria for banks' appraisal of agroforestry projects must clearly go far beyond
consideration of a project's economic rate of return. Some of the other key factors taken
into account include:

Sociological aspects

Putting people first is a recent World Bank publication (Cernea, 1985) that focuses on the
increasing attention now being given by the banks to sociological aspects of project design
in agriculture, forestry and other types of development projects. It has become a standard
practice for the banks to incorporate, as a part of agroforestry project preparation, a
sampling of farmer or community attitudes to tree planting and forest conservation prior to
implementation. For example, farmers interviewed prior to bank involvement in a rural
development project in Yemen Arab Republic had a very clear picture of the reasons why
they felt trees were important for incorporation on their farmlands (see Table 4). The same
survey also clearly identified farmers' preferences over choice of species in addition to
potential end-use of different trees.

Clearly such preliminary surveys cannot substitute for the more in-depth diagnostic
methodology such as that developed by ICRAF as a basis for designing agroforestry
research programmes (Raintree, 1987). Nevertheless, sufficient information can be gained
in a relatively short period to identify people's preferences in tree species, the end
management objectives of tree planting and likely availability of land for it. In other words,
a start can be made by investing in expansion of existing and socially acceptable
agroforestry technologies pending the results of longer term agroforestry research which, in
time, will help to ensure further gains in productivity. A key point is the need for a flexible
approach to project targets that permits funds to be rapidly shifted from one category of
lending to another according to people's receptivity.

Table 4   Farmers' perception of the usefulness of trees in two regions of the Yemen Arab Republic

Purpose for tree planting	IBB Region % of households	TAIZZ Region % of households
Environmental (soil protection, leaf mulch, shade, etc.)	30	46
Fuelwood	23	17
Timber and poles	24	12
Fodder	5	6
Fruit production	4	6
Medicinal	4	4
Other	10	9

Land-tenure issues

Experience from a wide range of agroforestry projects financed by development banks in India, Kenya, Haiti, the Philippines, Thailand and Niger has clearly demonstrated that security of land tenure is a major incentive to investment in agroforestry and to protection of trees or woodlands. World Bank support for a recent project in Thailand, for example, is intended to assist the government in speeding up land titling programmes in which planting of trees will frequently be carried out by farmers as a first step in land consolidation and demarcation of their new farm boundaries.

Whilst the incentive of private land ownership can act as a powerful stimulus to farmers' interest in tree planting, it is not in any way a mandatory precondition for bank involvement. In situations of communal ownership of woodlands it is not inconceivable that various usufruct and harvesting rights could be subdivided among various claimants and satisfactory arrangements made for management of common property resources. Encouraging small-scale experiences along these lines have been reported for development bank projects financed in Nepal, Niger and Sudan. Nevertheless, as noted earlier, problems of ensuring sustained and effective management of common-property resources management are still a constraint to larger scale bank support of agroforestry in many arid-zone countries. How to overcome this problem is an issue that is receiving high priority in the Bank's research agenda.

Ecological sustainability of the proposed agroforestry farming system

The success of agroforestry projects will ultimately be judged by their long-term sustainability well beyond the life of any development-bank support. A basic principle of bank support for any agriculture or forestry investment is that it should build on an existing and well-proven, socially acceptable, technically and economically viable farming system. As has been stated elsewhere, agroforestry is not a new science invented by the development establishment. Agroforestry farming systems have been practised in Asia, Africa and Latin America for hundreds of years.

Long established, traditional agroforestry farming systems, such as preservation of *Acacia albida* trees in millet, sorghum or groundnut fields (a farming system practised in many West African countries; von Maydell, this volume), have provided a solid basis for expansion of development-bank support for agroforestry in several Sahelian-zone countries. In such situations, provided bank appraisal makes realistic judgements about, for example, the take-up rate of agroforestry practices, the area of woodland that can be brought under effective protection within a given time-frame, farmers' likely receptivity to credit, extension advice and other inputs, etc., the prospect that the farming system will be ecologically sustainable beyond the life of the development bank's support will be reasonably assured.

Of a more controversial nature has been the development banks' widespread support for agroforestry systems incorporating introduced species such as *Eucalyptus* and *Albizia* which have raised environmental concerns. The issue has been clouded by emotive criticism and the debate is at this time unresolved. The main arguments advanced by the banks for their support of *Eucalyptus* planting, for example, include, firstly, the fact that the farmers themselves frequently specify the use of *Eucalyptus* species because they are well aware of their fast growth and potential high yield of building poles, fuelwood and income.* The farmers interviewed about their perception of the usefulness of *Eucalyptus* species usually also cited their unpalatability to grazing animals and the fact that the species coppices very vigorously.

Eucalyptus woodlots established by farmers have already proven themselves sustainable over long periods of time. For example, as noted earlier, spontaneously planted *Eucalyptus* woodlots are a common feature of the landscape throughout Western Kenya, even in the most densely populated regions. Many of these woodlots have been in existence for 50 years or more and are now into their fourth or fifth coppice rotations.

Problems have arisen with *Eucalyptus* in situations where larger block plantations have been established, usually by government-sponsored agencies, on the slopes of catchment areas situated in arid environments. In such situations the transpiration by the trees can cause a reduction in water yield, and if planted too close together they can accelerate soil erosion. A conscious attempt is being made to avoid bank support for such programmes. *Eucalyptus* is not a good tree for intercropping because of its potential allelopathic effects on crops, but those effects can be largely avoided by widely spacing trees along farm boundaries. For boundaries and small woodlot planting on poor land it will probably continue to be a preferred species for farm planting in many developing countries. Clearly, great care needs to be taken to select appropriate situations for on-farm planting of *Eucalyptus* and to take into account the sensitivity of agricultural crops to the possible toxic effects of some *Eucalyptus* species. The scientific evidence relating to the positive and negative effects of *Eucalyptus* planting have been spelled out in a recent FAO publication (Poore and Fries, 1985).

What are development bank funds used for?

Of the approximately US$ 750 million invested by development banks in agroforestry during the last decade, about 70 percent has been for farm forestry planting or protection and upgrading of existing woodlands, and 30 percent for a range of education, training, extension, research and institution-building activities.

---

* In a sample of recently completed surveys of people's attitudes to tree planting covering a range of countries with different ecological and socio-economic conditions (India, Yemen, Zimbabwe, Haiti and Thailand), notwithstanding the listing of a wide range of indigenous species which farmers wished to plant, *Eucalyptus* species were given a high priority by most of the farmers interviewed.

In areas with favourable ecological conditions, the bank-supported agroforestry programmes have helped accelerate on-going, spontaneous cash-crop-tree farming and agroforestry by financing, for example:

(a) Sociological surveys of farmers' attitudes to tree planting and of potential cash markets carried out prior to or as an integral part of the project-implementation process;

(b) Decentralization of seedling production to involve more small farmers and schools in raising seedlings, and provision of seeds, pesticides, water pumps, etc.;

(c) Financial support for mass-media promotion of agroforestry programmes in rural areas, e.g., by ensuring a supply of audio-visual material for encouraging agroforestry in schools and villages and by financing film strips and radio programmes;

(d) Strengthening of government extension support for agroforestry planting. Special emphasis has been given to accelerated training of extension staff, with particular reference to encouraging closer co-operation between forestry and agricultural extension staff;

(e) Financing of monitoring and evaluation programmes for assessing farmer or community response to agroforestry investments, measuring their positive and negative impacts on food and forest production and monitoring the distribution of project-financed inputs (such as seedlings) to try and ensure that smaller farmers or poor communities are the main beneficiaries (FAO, 1966; World Bank, 1984);

(f) Strengthening support for both agroforestry education and training and forestry research with special reference to improved agroforestry research. The World Bank, for example, provides part of ICRAF's core budget and the Bank is deliberately incorporating into its project design various technologies identified by ICRAF as having potential for increasing the productivity of farming systems (e.g., introducing leguminous species in place of non-leguminous trees, selection of spacing systems that offer possibilities for improving both forestry and crop yield, and replacing slow-growing species with faster-growing multipurpose trees that can produce a wider range of products suitable for indigenous consumption).

Bank support has also been provided for strengthening of basic forestry research in national research institutions focused on high-priority topics that can lead to likely productivity gains and improved rural incomes. Jointly with the United Nations Development Programme (UNDP) and other agencies, Bank support has been made available to the International Union of Forestry Research Organizations (IUFRO)'s "Special Program for Developing Countries" which is playing a key role in strengthening regional research networks in Africa, Asia and Latin America. An example of the type of research being supported is multipurpose tree breeding and improvement programmes that can help to ensure access to improved planting stock for farmers practising agroforestry.

In arid and semi-arid countries where the conditions are much less favourable for new tree planting, the current emphasis of development bank support is shifting away from forest plantation and community woodlot approaches towards support for schemes that would encourage local and village community involvement in more effective protection and management of existing savanna woodlands. Such investments usually aim to improve technical packages for increasing the productivity of arid-zone savanna woodlands (e.g., via more effective fire protection, adjustment of timing of lopping and harvesting of trees to encourage more vigorous regrowth and encouragement of rotational grazing schemes which would permit adequate regeneration of trees in areas being rested). Bank support in countries such as Niger, Mali and Senegal has also been made available for planting trees in family compounds and around homesteads, with species for which local people have a high preference such as *neem* (*Azadirachta indica*) and *gao* (*Acacia albida*). Support has also been given to irrigated tree planting around agricultural perimeters, particularly in situations where effluent water is available and irrigated forests will not seriously compete with food crops (e.g., in the Sudan).

Examples of successful agroforestry projects

The word "success" should be viewed in this context with caution. As noted earlier, criteria for success in agroforestry go well beyond the disbursement period of a bank loan (typically five years). Nevertheless, it has been possible in several instances to record positive changes in a relatively short period of time. The three examples given below from the Philippines, India and the Sudan illustrate a range of alternative approaches to development-bank support of agroforestry projects and highlight their more successful aspects to date.

The project in the Philippines is a traditional cash-crop-tree farming operation where trees are being grown in discrete woodlots for a clearly identifiable pulpwood market. In India the National Social Forestry Project works mainly through the existing government Forest Service network in the different states. In the case of the Sudan Western Savanna Development Project the agroforestry package was financed through a combination of Ministry of Agriculture Forestry Department and local tribal institutions as a part of a wider integrated rural development approach incorporating agriculture, livestock and forestry components.

The Philippines PICOP Pulpwood Project

This programme gained significant momentum in 1972 when the Paper Industries Corporation of the Philippines (PICOP), a pulp and paper company, entered into an arrangement with the Development Bank of the Philippines (DBP) to develop a loan programme for smallholder tree farming. Loans were offered to smallholders to finance 75 percent of the costs of plantation development and maintenance. Farmers with titled property (typically ranging from 2 to 10 hectares) could receive loans at a 12 percent rate of interest, and farmers with unsecured property could receive loans at a 14 percent rate of interest. Provision was made in the tree-farm plan for part of the farm area to be maintained under food crops. PICOP continued to guarantee a minimum purchase price for smallholder production, but allowed farmers to sell wood to other outlets if they could get a better price. The scheme proved quite popular with farmers. By 1981, the programme supported 3,800 participants and covered 22,000 ha. About 30 percent of these farmers had taken advantage of the credit programme.

There were several weak areas in the project design (particularly lack of adequate support for harvesting of the wood). Nevertheless, the overall lesson that emerged was that, given adequate market incentives and security of land tenure, development-bank assistance aimed at improving availability of seedlings, extension advice and access to credit could help significantly to accelerate this type of agroforestry scheme. The key factor which triggered farmers' response was the prospect of making handsome profits from tree growing. As Table 2 clearly illustrates, financial rates of return for participating farmers in that project typically exceeded 20 percent.

India National Social Forestry Project

This project is an expansion of the previous Bank-supported social forestry programmes covering several different states. In terms of tree planting area agroforestry (mainly farm forestry tree planting) is by far the largest component (nearly half a million equivalent hectares of planting spread over some 3 million farm holdings). The farmers themselves control the choice of species and use of the product.

Most plantings are taking place around individually owned farm boundaries, bunds, around the homestead and along water channels. The main direct cost to the local forest departments is for seedling production. A substantial part of the investment is providing support for extension and monitoring and evaluation of project progress.

The project also included a component for tree and fodder-grass planting on eroded agricultural wasteland covering a total of some 40,000 hectares. The state forest departments assist in the establishment of wasteland plantations, and provide incentive payments during initial months to the farmers or communities owning the land in order to compensate them for income foregone while they work on the agroforestry plantations. Improved (grafted) orchards consisting of fruit-bearing bushes (e.g., *Zizyphus mauritania*) are being planted on farmers' land. Farmers sell the fruit and use loppings for fuel. As a requirement for support under the wasteland reclamation category, the Forest Department would have to ascertain that the land was seriously eroded or in imminent danger of erosion, and hence of concern for conservation, which would justify higher government investment than under farm forestry. To give planting and other technical advice to farmers, the state forestry departments make extensive use of the existing agricultural extension service.

One of the more obviously successful aspects of these experiences in India has been the very positive response to schemes aimed at decentralization of seedling production and attempts to involve farmers, schools and villagers in raising nursery stock under guaranteed buy-back contracts. Table 5, extracted from a recent Bank supervision report on an agroforestry project in Gujarat State in India, illustrates the point. As can be noted from

Table 5 Evolution of school and farmer involvement in seedling production programmes in Gujarat state, India

	1980/81	1981/82	1982/83	1983/4
Department nurseries				
No. of nurseries	349	526	597	659
No. of plants (millions)	48.4	79.7	121.8	168.2
School nurseries				
No. of nurseries	109	217	677	801
No. of plants (millions)	0.8	2.2	7.3	14.7
Farmer nurseries				
No. of nurseries	—	—	563	884
No. of plants (millions)	—	—	21.1	27.9

Source: Supervision Mission Report on Bank-Financed India Gujarat Social Forestry Project (1985).

this table the number of non-Forest Department nurseries increased 15 times during the project period and the share of seedlings produced by farmers and schools from 2 percent to 25 percent.

Sudan's Western Savanna Development Project

This project was based on a more integrated approach, including involvement of local communities in the management of natural forest and savanna woodland resources, resettlement of people in the Sudanian-Guinean zone, continued and expanded research into more drought-resistant, high-yielding millet and sorghum varieties, and fast-growing tree species suitable for drought-prone environments.

This project focused on measures to improve the way of life of some 25,000 families in the area. A settlement programme was initiated with the dual purpose of opening up presently blocked stock-routes and increasing crop production. The opening up of stock-routes was vital for the annual migration of the transhumant families and their livestock. The project included settlement and crop-production programmes with major emphasis on extension services, programme-specific adaptive research, and pilot crop-marketing schemes. Range management encouraged local populations to set aside specific areas of woodland as dry-weather grazing reserves. Livestock-development components included veterinary services, livestock-related adaptive research, and non-formal education of livestock producers. The project also included improved maintenance and further development of water facilities, project monitoring and evaluation.

The original appraisal report identifies one of the main causes of the decline in per-unit animal production and the deterioration of the range in the project area as being an increase in the cattle population beyond the sustainable capacity of the range. The report considered that deferred grazing, the introduction of improved species and establishment of browse reserves, plus improved cattle management, would be ineffective without control of grazing pressure.

The range-demarcation schemes aimed at rectifying the overgrazing situation by granting exclusive tenure of dry-season grazing areas to small groups of pastoralists while maintaining the transhumant movements of cattle during the remainder of the year. This strategy was based on the assumption that, under such a scheme, there would be an incentive to equate cattle numbers with the long-term sustainable carrying capacity of the range. The range demarcation schemes were tested with three groups of livestock owners:

(a) A small group of transhumant pastoralists who made up the total population around a small well-field;
(b) A similar group of transhumant pastoralists that graze an area within a larger community area's well-field; and
(c) A group of sedentary livestock owners.

A recent evaluation of the project noted that with the introduction of the Range Improvement Programme increased livestock production was attained through grazing land enclosures and rotational grazing at Umm Belut. Herd performance was monitored over the project period. Results indicated that the density of plants on the grazing land increased. Some improvement in the calving rate was achieved together with a reduction in calf and adult mortality.

Currently the United States Agency for International Development (USAID), and a number of other donors are examining the scope for replicability of these experiences on a larger scale.

Problem areas in agroforestry projects financed by development banks

Cost recovery and incentive policies

Currently there is a wide variation in the incentive and cost-recovery policies being applied by the development banks in different regions and countries, and a lack of clear policy guidelines. In the State of Gujarat in India, for example, the World Bank has endorsed a free seedling distribution programme for the past seven years — a programme which has been widely acclaimed as having had a very positive influence on farmers' willingness to plant trees. By contrast, under a Bank-financed project in the State of Uttar Pradesh, India, a charge is levied on all seedlings. Paradoxically, there has also been a significant increase in tree planting in that State although it is not clear whether this is taking place on the scale recorded in Gujarat or whether smaller farmers are benefiting to the same extent.

In the east African region, there is a similar variation in seedling sale, distribution and cost-recovery policies. In Kenya, for example, the third phase of a Bank-financed project will raise the price of seedlings to all farmers from US$ 1 per hundred to US$ 6 per hundred over the next five years. By contrast, in Burundi, a range of seedling charges is in force based on the premise that poorer farmers located in areas remote from cash-tree-crop markets may need to be induced to plant trees by issue of free seedlings, whereas those farmers sited close to attractive cash markets for poles or fuelwood can afford to pay a higher price.

This issue of the cost-effectiveness of alternative subsidy/incentive policies for encouraging reforestation is poorly understood. The range of different approaches being tried by the various development banks includes credit, guaranteed access to markets, and guaranteed prices (e.g., the Philippines PICOP project), free or subsidized seedlings (India, Kenya, Bangladesh, Rwanda, Burundi, Zimbabwe, Haiti), and improved security of tenure (Philippines and India's Gujarat and West Bengal States). Other types of incentive such as revolving funds, tax rebates or similar fiscal incentives have not yet been tried in World Bank-financed projects although they are widely used elsewhere. World Food Programme aid has been a major factor in stimulating reforestation initiatives in many countries and the value of food-aid support for such programmes currently exceeds US$ 150 million a year. The scope for more systematic integration of food aid into World Bank-supported agroforestry planting is under review. Evaluation of various incentive approaches and of their relative cost-effectivness is the subject of a planned research project being proposed for financing by Rockefeller Foundation and the World Bank.

Forestry extension

Another major area of institutional concern has been the weakness of bank-financed forestry extension programmes. As noted earlier, a considerable shift in emphasis of forestry has taken place over the last decade towards farm forestry and other types of social forestry programme.

Dealing directly with farmers and local communities has necessitated a considerable change in the nature, organization, and attitude of traditional forest departments which, in many developing tropical and subtropical countries, have been primarily concerned in the past with protecting and "policing" government-owned forest reserves and keeping people and livestock out of the forests.

Some of the earlier attempts by the banks to address this issue failed to make any significant impact, partly because in project design too little attention was paid to the importance of ensuring a clear link between forestry and on-going agricultural extension programmes. In response to that issue, the World Bank in 1982 assisted the Government of India to carry out an in-depth review of forestry education, research and extension with special reference to alternative options for strengthening of forestry extension. The main extension approach recommended for wider adoption was the well-proven training and visits (T & V) system that has made a positive impact on agricultural crop yields in several South and South-East Asia region countries. Individual state forestry department staff are being retrained to provide specialist technical advice to existing agricultural extension staff, who, through more systematic T & V farm visits, concentrate on a few critical impact points such as choice of species, planting, espacement, depth of planting, and improved systems of farm-tree management (e.g., pollarding or coppicing).

Attempts to transplant this experience to African countries (e.g., in Malawi, Uganda and Kenya) have so far produced mixed results and suggest the need for a more flexible approach to extension which would take into account the wide range of traditional agroforestry farming systems. Extension needs to build on farmer and local-community experiences and their perceptions of the usefulness of trees in the farming system rather than exclusive dependence on a single formula such as the T & V system. Where the latter would appear to be of particular relevance is in cash-crop-tree farming situations where simple technological changes such as introduction of improved planting material, or changes in planting espacement and harvesting practices offer a reasonable prospect of early gains in productivity and income.

Ensuring that the benefits of bank support reach the smaller farmer and the landless

A significant proportion of the banks' agroforestry investment has benefited cash-crop-tree farmers who were well placed to take advantage of the availability of seedlings, extension advice, improved marketing opportunities, credit and other inputs. Through monitoring and evaluation and adjustments to government policies on seedling distribution, attempts have been made, with some success, to ensure that smaller farmers are assured a high proportion of the benefits from subsidized seedling programmes, but the track record is far from statisfactory.

A high priority issue for the future is how to ensure a more equitable sharing of the benefits from the banks' agroforestry investments. Experiments with "group" tree farming on government land, such as those being tried by the governments of Gujarat and West Bengal States of India, are one promising approach to involving the rural poor and landless in agroforestry.

In looking ahead to the 1990s, perhaps the greatest challenges that the development banks face are:

1. How to find more cost-effective ways of channelling funding directly to farmers, schools and local communities (i.e., with more emphasis on investment in non-governmental local institutions and less emphasis on support for government forestry agencies); and

2. How to widen the support for agroforestry systems to give greater emphasis to intercropping situations aimed at increasing farm productivity and sustaining crop yields (i.e., in addition to providing support for cash-crop-tree farming).

A recently appraised project for Uganda, for example, will build on the success of a past non-government organization (NGO)-supported agroforestry programme with part of the project funds being disbursed directly to NGOs and local communities with the technical advice and support of the government forest service. The project will cover a balanced range of agroforestry planting and forest protection activities focused both on improvement of productivity of the existing farming system as well as generation of cash income.

Conditions in bank agroforestry lending

Perhaps the most widely misunderstood aspect of development bank involvement in agroforestry projects has been in the use of policy-based lending as a way of tackling fundamental issues such as inappropriate government fuelwood/charcoal pricing policies, the need for assurance that project benefits will flow to poor segments of society, and for assurance that government forestry or agricultural agencies will receive adequate and timely local budget support for implementation of project activities. The incorporation of loan covenants dealing with such issues is a standard practice in all project agreements. Their formulation takes into account local sensitivities on such issues and the terms of the loan agreements are a matter for negotiation between the bank and government prior to loan effectiveness. A review of the range of policy-based interventions that have been applied in bank-funded projects over the last decade is beyond the scope of this chapter but it is important to re-emphasize that the underlying purpose behind these policy interventions is to assist in creating a more favourable socio-economic and investment climate for agroforestry. In the case of the Philippines PICOP project, for example, key loan covenants required, as a precondition for bank involvement, that satisfactory arrangements be concluded between PICOP and the farmers over price guarantees for the pulpwood produced and assurance of land title for farmers participating in that scheme.

REFERENCES

Cernea, M. (ed.). 1985. *Putting people first.* Washington, D.C.: World Bank.
FAO. 1966. Monitoring and evaluation of participatory forestry projects. FAO Paper 60. Rome: FAO
Poore, M. E. D. and C. Fries. 1985. The ecological effects of eucalyptus. FAO Forestry Paper 59. Rome: FAO.
Raintree, J. B. 1987. The state of the art of agroforestry diagnosis and design. *Agroforestry Systems* 5 (Special Issue) (in press).
World Bank. 1984. Operational guide to monitoring and evaluation of social forestry in India. Washington, D.C.: World Bank.

SECTION THREE

Prominence and importance of agroforestry in selected regions

6

The development of agroforestry in Central America

Gerardo Budowski

Head, Program of Natural Resources and Quality of Life
University of Peace
San José, Costa Rica

Formerly: Head, Renewable Natural Resources Department
Centro Agronómico Tropical de Investigación y Enseñanza (CATIE)
Turrialba, Costa Rica

Contents

Introduction

This chapter deals with the development of agroforestry in Central America, including all the countries of the isthmus from Panama to Guatemala and Belize, with some inroads into tropical Mexico. Nowhere in Central America is there yet a government department dealing specifically with agroforestry — although many individual projects within different departments include this topic. CATIE, a regional organization covering most of the Central American countries, has the strongest contingent of agroforesters in the region. This chapter, therefore, is focused on CATIE's agroforestry activities. Although there will be some discussion on historical background, the main aim is to concentrate on the latest research; earlier research can be found in the literature.

Agroforestry in Central America, as in other parts of the world, is a very old practice, but it was not known by that name in the literature until the mid-1970s. When the science of **agroforestry** began taking shape, it was of course realized that a number of studies had already been made on the subject in the region without their ever having been labelled as agroforestry. Moreover, when the new term agroforestry was coined, there was some confusion about the exact boundaries of agroforestry, i.e., what could be considered as agroforestry and what not. This type of discussion still lingers, and is complicated by the Spanish terminology because there is — or was — no precise translation of the term forestry; in fact the words "foresteria" and "agroforesteria" were not used until the late 1970s and here too opinions were far from uniform.*

Agroforestry as used here implies a number of techniques that all include the combination, either simultaneously or sequentially, of trees and food crops, trees and livestock (trees on pastures or for fodder), or all three elements (Combe and Budowski, 1979).

This chapter will review the following agroforestry practices in Central America with emphasis on CATIE's research:

> Taungya
> "Shade" trees in fields of coffee, cacao or other crops
> Trees and annual crops
> Alley cropping
> Trees and livestock
> Live fences and windbreaks
> Shifting cultivation (or shifting grazing) with managed fallow
> Mixed homegardens.

There will be some consideration of the methodological and socio-economic aspects as well as of extension, documentation and training.

The history of agroforestry in Central America

No systematic research has been carried out on the history of agroforestry in Central America. However, various techniques of mixing trees with food crops were well known to the pre-Colombian Indians, particularly the managed fallow in shifting cultivation, the tropical mixed homegardens, the mixing of trees and crops along ditches (chinampas) and elsewhere that are well described from Mexico (Gliessman, 1981) and Guatemala (Wilken, 1977). Combinations such as heavily pruned pines associated with food crops, and shade trees in cacao fields, have also been reported from that period. But many of these combinations have still not been properly documented.

The earliest documented mention of agroforestry in Central America may possibly be that by Cook (1901) who had recognized the various beneficial effects (for instance the nitrogen-fixing properties and the input of organic matter) of shade trees, particularly legumes, in coffee plantations. Holdridge (1951) described the decades-old practice of planting alders (*Alnus acuminata*) in pastures in the highlands of Costa Rica. This striking feature of the land-use system was also described by Budowski (1957), who reported on the successful windbreaks made up of cypress, *Cupressus lusitanica,* also in the highland dairy region, as well as of laurel (*Cordia alliodora*) in pastures in the wet lowlands, both in Costa

---

* In 1984, one group within the agroforestry programme of CATIE sent a cable to ICRAF asking how ICRAF translated agroforestry into Spanish. The answer came swiftly: "agroforesteria". That ended that argument.

Rica. Perez (1954) used growth rings on the trunks of laurel to calculate the best management cycle for laurel in pastures and found it to be approximately 18 years. Budowski (1959) also suggested the use of shade trees, including *Cordia alliodora,* as a good management practice for coffee.

For taungya, a literature review for "Tropical America" was prepared by Budowski (1956), which included the previous work in the then British Honduras, now Belize. Gonzalez de Moya (1955) and Aguirre (1963) reported promising results in managing a secondary forest derived from an abandoned coffee stand. Live fences in Costa Rica were probably described first by Lozano (1962), a student at CATIE, and formal research ideas on this topic were later suggested by Budowski (1977) and Sauer (1979).

The Spanish equivalent of the term "agro-silvo-pastoral systems" appeared first in 1976 when it was introduced by this author at CATIE as part of one of the three programmes within the Renewable Natural Resources Department. In early 1977, a request from the International Development Research Centre (IDRC) led to a contract for CATIE to present a "programme of work for agroforestry in the humid tropics" (Budowski, 1977), which, jointly with a similar programme for the dryer areas, was considered by the Committee that recommended the creation of ICRAF. These programmes suggested a series of promising lines of research, such as association of trees and food plants, taungya, live fence-posts, shade trees in coffee and cacao fields, nitrogen-fixing timber trees in pastures, use of trees for fodder for cattle as well as tree belts alternating with crops or pastures.

The creation of ICRAF in 1977 was an enormous boost to the programme at CATIE which has since benefited greatly from the scientific and educational output of ICRAF, including successful co-operative efforts in surveys and training programmes.

Research on agroforestry systems and components at CATIE

At CATIE, thesis research by graduate (M.S.) students on agroforestry-related topics had been a regular feature even in the early days. Examples dealt with the local species of alder (*Alnus* sp.) (Alvarez, 1956), live fences (Lozano, 1962), and *Cordia alliodora* in pastures (Perez, 1954; Marinero, 1962). Subsequently, when agroforestry was institutionalized, student theses on agroforestry subjects increased considerably: 24 such theses have been produced to date. Three CATIE publications, all issued in 1979, describe the research efforts: these are, (a) the proceedings of a workshop on agroforestry systems in Latin America (De las Salas, 1979), (b) the "Field Guide" describing all forestry research plots of CATIE (in Spanish) edited by Combe and Gewald (1979); and, (c) an international symposium entitled "Forestry and its Contribution to Development in Tropical America" in San José (Chavarria, 1979), where 10 out of a total of 36 papers dealt with agroforestry subjects, almost all of them from CATIE personnel.

Efforts in agroforestry at CATIE increased considerably from 1979. These included organizing workshops, training courses and seminars in agroforestry at the headquarters as well as in Mexico, Colombia, Guatemala and Honduras, assistance to training activities in the Dominican Republic, Thailand, Nigeria, Mexico, the United States of America, Indonesia, Peru, Honduras and Ecuador, and in-service training for students from several countries with United Nations University (UNU) scholarships. A significant effort has also been made in documentation, and close relations have been established with ICRAF in this area too. CATIE staff are also involved in consultation jobs in agroforestry in many tropical countries throughout the world. Recently a Spanish book *Sistemas Agroforestales* (the English translation of which is "Agroforestry Systems"), was produced by OTS (Organization of Tropical Studies, a consortium of mostly US universities and scientific institutions) and CATIE (OTS/CATIE, 1986), which is an excellent training material for agroforestry in Latin American countries.

The resource limitations of M.S. theses were reflected in the initial results of agroforestry research at CATIE and in some of the obvious differences between desired research and actual execution. For example, several M.S. theses were prepared on the taungya system, measuring initial growth of trees, labour costs and yields, etc., in replicated blocks of various combinations with food crops, in comparison with plots of food crops alone (Combe and Gewald, 1979; Budowski, 1983a). Although these results proved valuable and worthy of M.S. degrees, the ideas were not being taken up by the farming community because of the lack of adequate promotional and extension activities. Later, however, a pulp and paper company (Celulosa de Turrialba) used the research for their pine plantations.

A more in-depth approach was taken by the CATIE staff researchers. Not constrained by the 12-month time limit of M.S. theses, the staff researchers studied in detail the existing local agroforestry practices — some of them decades old. The quantification of these practices, design for improvement, and the transfer of well-studied and validated practices to other areas where ecological as well as socio-economic factors indicated high chances of success, became the focus of such research, and these too had student involvement.

Agroforestry initiatives elsewhere in Central America were slow. This was mainly because agroforestry was generally ignored by local government or university programmes until CATIE graduates and trainees returned to their countries and began building up local programmes. Moreover, and quite understandably, many of the first papers of the newly-trained agroforesters from CATIE concentrated on describing — if not "discovering" — the various agroforestry practices found in the various countries or regions. Examples of such descriptions are papers by Castillo and Beer (1983) for the Kuna Indians in Panama, Budowski (1981) and Fournier (1981) for Costa Rica, Martinez (1982) and Leiva and Lopez (1985) for Guatemala, Campos Arce (1982) for hillsides, and Rodriguez *et al.* (1983) for Guatemala, Honduras and Panama. This compilation of existing practices was later published in the earlier-mentioned book *Sistemas Agroforestales* (OTS/CATIE, 1986).

Research on specific agroforestry combinations

Taungya
The results and recommendations of a review on taungya by Budowski (1983c) have been widely used for the many short agroforestry courses at CATIE. These and other results (Combe, 1981b; Budowski, 1983a; Fernandez, 1978) have been used in reforestation projects in Costa Rica involving species such as *Eucalyptus deglupta, Cordia alliodora, Gmelina arborea, Pinus caribaea* and *Terminalia ivorensis*. Several fuelwood species have also been established successfully using maize as an associated crop in Guatemala (Detlefsen *et al.*, 1984).

Shade trees in coffee and cacao fields
Coffee and cacao are very valuable cash crops throughout Central America. The presence of shade trees in these fields (Figures 1 and 2) has therefore attracted considerable attention by agroforesters. This is partly due to the fact that shade trees had until then received very

little research attention, particularly in those aspects not related to shade *per se*. More important is that two major diseases have recently affected these crops: the coffee rust (*Hemileia vastatrix*) and the cacao pod-rot or moniliasis (*Monilia rohrerii*), and in both cases shade management (reduction) has been advocated as an important aspect of disease control. In the case of coffee, complete removal of shade and use of large amounts of fertilizers, pesticides and herbicides, and closer planting have been strongly advocated and promoted through several incentives. This strategy has proven to be relatively successful for large enterprises but disastrous to small farmers who have difficulty getting credit and access to modern technologies. Hence smallholder-oriented agroforestry research has concentrated on quantifying the multiple benefits of management practices which involve shade trees, especially legumes (Budowski *et al.,* 1984). The genus *Erythrina* is particularly prominent in this context. Espinoza (1984, 1985, 1986a,b) studied the coffee stands in the Acosta-Puriscal region of Costa Rica, an area of small farmers, where 117 coffee stands were surveyed in detail. The study clearly demonstrated the many benefits these trees provide to the farmers such as fuelwood, fruits and construction wood.

At CATIE a long-term replicated experiment known as "La Montaña" was established by the Plant Production Department (Enriquez, 1983). It includes the use of different combinations of shade trees as practised by farmers in Costa Rica. Carefully monitored since 1977, it has been a good experimental site for researchers, students and trainees. A significant amount of data on organic matter, nutrient cycles (of N, P, K, Ca and Mg), litter fall, water infiltration, etc., are now available from this experiment, using two shade trees: *Erythrina poeppigiana* (periodically cut back) and *Cordia alliodora* (a valuable timber tree, periodically thinned), (Alpizar, 1985; Alpizar *et al.,* 1985; Alpizar *et al.,* 1986; Enriquez, 1983; Fassbender *et al.,* 1985; Heuveldop *et al.,* 1987; Jomenez, 1986). The coffee/shade tree combinations were also studied in other conditions, notably small farmers' plots close to CATIE. Emphasis was on leaf fall, nutrient balance, soil acidity, etc., comparing the two trees (*Erythrina poeppigiana* and *Cordia alliodora*) (Figures 1 and 2) versus only one (*Erythrina poeppigiana*) over coffee (Beer, 1983, 1987; Glover and Beer, 1984, 1986). *Cordia alliodora* is also much used in cacao fields and was investigated for its growth, volume and yields, notably by Combe *et al.* (1981a,b) and Somarriba and Beer (1986). Other species that have showed similar promise include *Cedrela odorata,* a very valuable timber tree (Sabogal, 1982, 1983), *Alnus acuminata* (Combe *et al.,* 1981a, Combe, 1982; Fournier, 1981) also a nitrogen-fixing tree, and *Grevillea robusta* in Guatemala (Villatoro, 1986). Environmental influences on the coffee/shade tree combinations were also studied by Barradas and Fanjul (1984), who showed various protective actions when shaded coffee was compared with non-shaded coffee.

It is now well recognized that the use of shade trees implies a multitude of biological, environmental and socio-economic benefits (Budowski *et al.,* 1984). Although the amounts of nitrogen fixed are still a matter of discussion and investigation, 40 kg N ha^{-1} yr^{-1} for *Inga* trees in coffee fields in Mexico (Roskoski, 1982) is considered an acceptable figure. However, wide variations have been reported depending on the methodology used for quantification. Nitrogen fixation, the effects of different management practices and the relation with endomycorrhiza for *Erythrina* were also investigated at CATIE, notably by Lindblad and Russo (1986), and Russo (1983a, 1984). Studies by Russo and Budowski (1986) have shown that periodic (once or twice a year) pollarding of *Erythrina* trees results in the addition of substantial amounts of organic matter and nitrogen to the soil. The genus *Erythrina* is presently the subject of a special project (financed by IDRC) at CATIE. Thus, CATIE investigations have made significant contributions to the understanding of the role of shade trees.

Figure 1   Coffee with two additional strata of trees at CATIE, Turrialba, Costa Rica: 600 m elevation; about 2,600 mm rainfall. Immediately above coffee is the fast-growing and nitrogen-fixing legume, *Erythrina poeppigiana* (large leaves), established by large cuttings for "shade", and pollarded twice a year. The tall trees, "laurel", *Cordia alliodora,* are highly priced for timber. During the dryer months of January to April, they lose their leaves and compete little for water from the soil. This is an extremely common practice up to 800 m elevation. (Photo: G. Budowski.)

Figure 2 Cacao under a fairly dense stand of laurel (*Cordia alliodora*) trees at Madre de Dios, Siquirres, Costa Rica: 200 m elevation; around 4,000 mm annual rainfall. The fast growth, small and light crown, good form and self-pruning ability of laurel make this an ideal tree for agroforestry combinations. Yield of laurel standing timber in this plot has been estimated at about 15 m³ ha⁻¹ yr⁻¹. The laurel trees were established by natural regeneration. Harvesting is done at about 20–25 years when the trees reach 40–50 cm dbh. (Photo: G. Budowski.)

Trees and annual crops

This area has been little explored except for *Cordia alliodora* in sugar-cane fields (Fournier, 1981; Somarriba and Beer, 1986). The opening of various types of forest (primary, secondary and old coffee groves with shade trees) over cardamom (*Elettaria cardamomum*), a valuable spice cultivated in Guatemala (Leiva and Lopez, 1985) and Honduras, may also be mentioned here although cardamom is not an annual.

The use of *Cordia alliodora* in sugar-cane fields deserves some comment since sugar cane is a sun-loving plant. *Cordia* trees are common in sugar-cane fields of smallholders who sell their sugar cane to local factories. The presence of trees does indeed depress sugar production but this is adequately compensated by the value of the timber, an important consideration for the small farmers who prefer to harvest timber for local needs on their own land rather than to buy it from the sawmill where it is very expensive. Moreover the *Cordia* trees grow very fast and have a preferred straight form, since they are self-pruning and always possess a relatively small and open crown that becomes leafless during the dry season. Hence they intercept only a small proportion of the incoming radiation and cast only little shade on the understorey species; their moisture uptake in drier months is also relatively less, unless of course the density of trees is too high.

The desirable characteristics of "agroforestry trees" when associated with annual or perennial crops were recently analysed by Budowski (1987a). He analysed the uses of such trees for goods and services, growth habits, morphological and physiological features, methods of propagation, production of biomass, response to pruning, coppicing ability, litter decomposition, tolerance to fire, relationship to soil fertility and texture, capacity to fix nitrogen, efficiency of their mycorrhizae and tolerance to pests and relationship with local wildlife.

Alley cropping

This agroforestry technique, well publicized by IITA in Ibadan, Nigeria (see Kang and Wilson, this volume), was only tested in recent years at CATIE, notably by Donald Kass of the Plant Production Department. The trees used for alley cropping were the leguminous *Erythrina poeppigiana* and *Gliricidia sepium,* both propagated by large cuttings and well known for their ability to fix nitrogen and the relatively large amounts of crude protein (20–30 percent) in their leaves. The associated crops in these experiments were maize, millets, cassava and common beans.

Kass *et al.* (1983) summarized the beneficial effect of these trees, which were periodically lopped, as supply of organic matter and nutrients to the associated crops, suppression of weeds and action as a mulch. The lopped branches and leaves could also be carried from the nurse trees to nearby crops: in one study, *Erythrina* provided 8,000 kg dry matter (with 3–4 percent nitrogen) per year (Kass *et al.,* 1983). In another experiment with *Erythrina* and maize, better yields were obtained at various spacings of *Erythrina* in comparison with control plots of maize only. However there was a slight increase in acidity and a decrease in the C/N ratio in the plots where *Erythrina* nurse trees were used (Alavez, 1987).

Trees and livestock

As mentioned earlier, the associations of local alder trees (*Alnus acuminata*) in the highlands (Figure 3) and laurel trees in pasture in the humid lowlands of Costa Rica had been reported decades ago (Holdridge, 1951; Perez, 1954; Budowski, 1955 and Alvarez, 1956). Recent studies give detailed analyses of the advantages and limitations of alder in pasture (Combe, 1981a; Combe *et al.,* 1981a; Garriguez, 1983), using various examples from Costa Rica.

Figure 3   Native alder trees (*Alnus acuminata*) planted by natural seedlings collected along creeks at Las Nubes de Coronado, a dairy region, Costa Rica: elevation 1,800 m; rainfall about 2,500 mm per year; good volcanic soil. The trees fix nitrogen and are planted at 10–15 m spacing. Growth is very fast and the cutting cycle is 16–22 years when trees reach about 40 cm dbh. The grass is Kikuyu grass, *Pennisetum clandestinum,* from Africa, but now widely naturalized. (Photo: G. Budowski.)

Figure 4 Forest or pasture? A silvopastoral combination derived from a secondary forest which itself was derived from pasture that had been abandoned at Siquirres, Costa Rica: elevation 300 m; rainfall about 4,000 mm. About 12 tree species are found, each selectively retained and with relatively valuable timber. Many trees have been harvested in recent years and grass has become established allowing moderate grazing (Photo: G. Budowski.).

The practice of associating trees with pastures is widespread in various climatic zones throughout Central America (see Figure 4), notably the wet lowlands, lowlands with a marked dry season, and highlands. Scientific investigations on these practices have mostly dealt with feeding trials with loppings of trees — mostly legumes — and some associations of leguminous N-fixing trees such as *Erythrina* spp. with grass (Russo, 1984) as well as other species (Ruiz, 1983). Non-leguminous trees such as *Guazuma ulmifolia* that are common in areas with a marked dry season have also been found to be promising since they can be managed on a coppice system every 2–3 years, combining forage production with fuelwood from the branches (Russo, 1984).

An interesting experiment at CATIE using *Erythrina poeppigiana,* established by large cuttings in a stand of king grass (a hybrid of *Pennisetum purpureum* x *P. typhoides*), showed promising initial results: the presence of periodically lopped *Erythrina* trees did not influence the yield of pasture in comparison to control (no tree) plots. Moreover, the total biomass production was 35 percent higher than in the control and the production of protein increased by 193 percent. The presence of *Erythrina* also increased the amount of protein in the pasture (Rodriguez, 1985).

A series of laboratory analyses showed high nutrient content in the branches of *Gliricidia sepium* and *Erythrina poeppigiana* which are usually fed to cattle (Espinosa, 1984). Leguminous live fences that need to be lopped periodically also showed a high content of protein. However, possible toxicity problems in cattle need to be monitored. On the other hand, there was no such toxicity-problem with goats and early results were very promising (Borel, 1986). For goats, *E. poeppigiana* feed compared favourably with *Dolichos lablab* (Benavides, 1983; Samur, 1984). A 67 percent substitution of soya meal by *Erythrina poeppigiana* leaves in the diet of calves showed the highest economic benefit (a gain of US$ 0.63/kg) (Pineda, 1986).

The "La Montana" experiment at CATIE (Enriquez, 1983) compared pasture production of *Cynodon plectostachyus* in unfertilized replicated plots with and without *Erythrina poeppigiana* and *Cordia alliodora* trees. The results (Table 1) showed that the presence of *Erythrina* resulted in the production of a greater amount of biomass that is also richer in protein; the weeds were also suppressed.

Table 1 Comparison of yields of *Cynodon plectostachyus* pasture alone and associated with *Cordia alliodora* and N-fixing *Erythrina poeppigiana*

	Grass alone	Grass + Cordia	Grass + Erythrina
Biomass produced (kg dry matter ha^{-1} yr^{-1})	5,931	5,090	10,420
% of Graminae (in relation to weeds)	44	80	90
% protein in Graminae	7.7	7.9	10.1
Digestibility of Graminae (%)	45.1	.47.3	46.9
Production of protein (kg ha^{-1} yr^{-1})	656	468	1,113

Source: Bronstein, 1984.

Research on grazing under *Pinus caribaea* in the large plantations of Celulosa de Turrialba (near CATIE) showed that grazing was possible when the pines were 3–8 years old, the carrying-capacity curve of animal units/ha showing a maximum of 1.4 animal units for a 5-year-old plantation (Somarriba *et al*, 1986). *Cordia alliodora* trees in pastures in the wet lowlands of Costa Rica showed a growth of 4.2–10.8 m yr^{-1} at a density of 150–208 trees ha^{-1}. The growth of *Cordia* was superior in non-grazed associations, for instance in cacao groves (Combe *et al.*, 1981b).

Guava (*Psidium guajava*) and pasture is a very popular association at middle elevations in Costa Rica (Somarriba, 1982, 1985, 1986; Somarriba and Beer, 1985; Sequeira, 1984). There was considerable variation among the guava trees, pointing to the need for eliminating low-yielding trees. The fruits are eaten by cattle and many seedlings germinate from the excreta.

Live fence-posts and windbreaks

CATIE has been a pioneer in the study of live fence-posts, raised by planting large cuttings of trees to which barbed wire is attached (Budowski, 1987b) (Figure 5). The practice is widespread in Central America, Mexico, the Caribbean and northern parts of South America. Its advantages and disadvantages were discussed by Budowski (1982b). Over 90 species are used in Costa Rica alone, of which a dozen make up about 0.5 percent of all types of fences. The most common species are *Gliricidia sepium, Erythrina berteroana, E. costaricensis, Bursera simaruba, Spondias purpurea, Diphysa robinioides, Jathropha curcas, Yucca elephantipes* and *Croton glabellus* and there is now considerable empirical knowledge about the establishment and management techniques and uses. These live fence-posts not only serve as support for barbed wire but they also produce fruits, fodder,

Figure 5 *Gliricidia sepium,* one of the most popular live fence-post species at La Francia, near Siquirres, Costa Rica: elevation 200 m; rainfall about 3,500 mm. The leaves contain about 25% protein and the wood is used for fuel, but the most important products are the large branches used to establish new fence-posts on which barbed wire will be attached. (Photo: G. Budowski.)

fuelwood and, above all, new fence posts. There is at present a world-wide interest in live fences and Costa Rica, mostly through CATIE, has been a centre of research and a source of seeds and information on the subject.

Research on this aspect was initiated in the 1960s (Lozano, 1962) and has progressed considerably in the past seven years. Most of the research was aimed at describing species for live fences, methods for species establishment and for estimation of yields of both fuel and fodder from the posts. Only a few species in a few ecological zones have been examined so far, and the field is wide open for an array of promising research areas and extension work.

Sauer (1979) identified 57 species of trees producing live fence-posts in Costa Rica, of which 26 are common. He has drawn attention to the decreasing diversity of the species used as live fence-posts, and the tendency to substitute live fences with other materials. It is much cheaper to collect local stakes and plant them as live fence-posts than to use other fencing materials; moreover, live fences last much longer and are economically attractive because of their production of more fences as well as fodder and fuelwood. As the realization of these advantages is becoming more widespread, the tendency to substitute live fence-posts with other materials is gradually becoming reversed. Live fences have also been described from the Mexican humid tropics (Alavez, 1983). Gross (1983) described tree species for live fence-posts in the Nicoya Peninsula in Costa Rica.

A methodology to evaluate yields of fuelwood from *Gliricidia* live fence-posts was designed by Rose and Salazar (1983). The empirical knowledge of farmers concerning choice of large cuttings and planting and management methods for *Gliricidia sepium* has also been used in extension (Baggio, 1982). Costa Rica has become a centre for collection of germplasm of *Gliricidia* (Sumberg, 1985) (also see Burley, this volume). In order to co-ordinate the research efforts on *Gliricidia* world-wide, an international conference on *Gliricidia* was held at CATIE in June 1987.

Beliard (1984a) experimented with three pruning intervals for *Gliricidia sepium,* and measured the volume and quality of products (fodder *vs.* forage). As expected, the shortest rotation produced the largest amount of forage while the longer intervals yielded high fuelwood returns. These results are of considerable interest to farmers who want to manage their live fences according to local needs or market values. Volume tables for *Gliricidia,* based on length of branches and diameter at the base of branches, have also been produced (Beliard, 1984b; Beliard and Mora, 1984) for both forage and fuelwood, with conversion factors to transform each into dry matter. Similarly, for eight-month-old fence-posts of *Erythrina berteroana,* a 100 m fence line yielded 1,107 branches from 169 live posts, with 319 kg of dry matter. The protein content was 4.2 percent for fresh leaves and 1.3 percent for the stems (Budowski *et al.,* 1985). The initial performance of *Calliandra calothyrsus,* widely planted in Indonesia for fuel and fodder, was also evaluated at CATIE (Baggio and Heuveldop, 1982). However, this species has a limitation in that it must be established by seed and its large stem cuttings do not sprout.

In the "La Esmeralda" farm in the dairy highland region of Costa Rica, cypress (*Cupressus lusitanica*) windbreaks have been planted since the early 1930s and are now yielding excellent returns. A modern sawmill has been successfully installed on the farm. The whole farm, with its sawmill and its *Cupressus* stands, has been heralded as a success story, and the farm is visited regularly by hundreds of students and researchers. Evaluation of fuelwood yield of cypress windbreak was carried out at CATIE (Salazar, 1984 as reported in OTS/CATIE, 1986).

Shifting cultivation (or shifting grazing) with managed fallow

This aspect of agroforestry deserves some comment. The fallow is rarely labelled as "managed", although this may prove to be a very superficial appraisal when we learn more about the farmers' practices. A good example is some of the fallows from the Central American Indian communities. They may look disorderly and untended but the local inhabitants have a good knowledge of fallow management and they occasionally harvest and tend certain trees in the fallows (Castillo and Beer, 1983; Nations and Nigh, 1980; Nations, 1981). Budowski (1983) summarized the evolution of a fallow, derived from an abandoned pasture and later successfully converted into a highly productive secondary forest.

Most research on fallow regeneration carried out at CATIE compares either different forms of fallow resulting from slash and burn to secondary forest with an evaluation of changes above and below the soil surface (Berish, 1983; Ewel, 1969, 1971, 1976; Ewel *et al.,* 1981), or biological processes and biomass yields in manipulated successional stages that are compared with control plots without intervention (Ewel and Babbar, 1981; Ewel *et al.,* 1981, 1982; Babbar, 1983; Berish, 1983). Altogether this is a relatively unexplored field which deserves much more consideration, including enrichment techniques (with valuable timber species) for the fallow (Leiva, 1982), and even fertilization of the fallow as carried out by Harcombe (1977) on CATIE plots.

Tropical mixed homegardens

Knowledge of homegardens is still in its infancy in Central America, although the practice has been strongly advocated, notably by Lagemann (1981). There may possibly be a new surge of interest after the second International Workshop on Tropical Mixed Homegardens to be held in San José, Costa Rica in early September 1987.

A detailed world-wide literature review on homegardens by Brownrigg (1985) cites only a few references referring to Central America. A later review for tropical America by Budowski (1985) contains more entries. Detailed case studies on mixed homegardens have been described from Santa Lucia, Guatemala (Anderson, 1950), Orotina, Costa Rica (Maffioli and Holle, 1981) and from various other areas of Costa Rica (Price, 1983). The latter also suggests promising lines of research. A questionnaire for surveying tropical homegardens has also been produced by CATIE (Huerto Casero, 1982).

Other aspects of agroforestry research at CATIE

Methodological aspects

CATIE's contributions to research methodologies on various aspects of agroforestry are summarized in Table 2.

Table 2   CATIE's contributions to research methodologies in agroforestry

Subject matter	Reference
Characterization and evaluation	Maydell, 1981; Lageman and Heuveldop, 1983
Survey techniques, general	Heuveldop and Espinoza, 1984
Site-specific survey for fuel	Proyecto agroforestal CATIE/GTZ, 1983; Rose and Salazar, 1983; Jones and Campos, 1983
Use of trees on farms	Martinez, 1982
Permanent plot establishment	Beer, 1984a; Beer, 1984b
Fuel production from guava in pastures	Somarriba, 1982
Biomass measurements	Russo, 1983b
Soil measurements in agroforestry	Fassbender, 1984; Bornemisza, 1983; De las Salas and Fassbender, 1981
Modelling soil changes in agroforestry	Fassbender et al., 1985; Fassbender et al., 1987
Diagramming an agroforestry farm	Bertisch, 1983
Establishment of an agroforestry production model	Cedeno and Chavelas, 1980
Agro-ecosystem methodology	Jones, 1983

Socio-economic considerations and extension

Many socio-economic aspects are incorporated in the earlier-mentioned studies on systems and components. For instance, many aspects of plantations of trees established by taungya are compared, in economic terms, with the establishment of the same species of trees through normal planting and tending operations (Combe and Gewald, 1979; and Spiegeler, 1981). Reiche (1983) described various ways of characterizing the economic aspects for small agroforestry farmers, while Espinoza (1986a, b) described socio-economic factors of the Acosta-Puriscal area in a dry region of Costa Rica, and Maldonado (1986) made a similar analysis for the Taque-Taque area in a wet environment in Costa Rica.

Extension aspects of agroforestry systems were discussed by Beer (1984b) and Beer and Somarriba (1984) who described the step-by-step approach to improving agroforestry systems for small farmers; this aspect was also discussed in a more general approach by Dulin (1982) in Honduras. Since very often, agroforestry amounts to incorporating trees in the existing farms, special emphasis has been given to stimulating farmers to plant trees (Beer, 1985; Clarkin, 1982). A notable effort towards this end was also achieved by a Peace Corps volunteer in Honduras (Rodbell, 1986).

Documentation

The large number of publications issued by CATIE was made possible by building up a special section on forestry documentation and information in tropical America (INFORAT). One of the first tasks of INFORAT was the publication of a bibliography on tropical agroforestry with 680 entries (Combe *et al.,* 1981). This computerized information bank is periodically up-dated and freely disseminated to students, trainees and others.

Training

Professional education in agroforestry in Central America was reviewed by Budowski (1982a) and a special manual on teaching methods for use in intensive short courses was issued by CATIE (Major *et al.,* 1985). A landmark for training is, of course, the earlier mentioned textbook (OTS/CATIE, 1986). A textbook on soil aspects in agroforestry systems was also produced at CATIE (Fassbender, 1984), while Beer and Somarriba (1984) compiled a series of papers to serve as a support for the organization of short courses in agroforestry.

The increasing interest in agroforestry training is evidenced by the attendance at the short agroforestry courses, now offered at even more frequent intervals in the Central American region by various bilateral aid organizations and agencies as well as CATIE. Agroforestry will be a special study area at the Graduate School of the University of Costa Rica, beginning in 1988. It will also be taught and investigated at the University of Peace, located close to San José, Costa Rica. Agroforestry is also becoming a regular course at the Universities in Guatemala and Honduras.

The transition from theory to practice may still have a long way to go, but at least a good number of trained local personnel are now available and there is the large diversity of existing agroforestry practices throughout the region waiting to be analysed, quantified and, in most cases, improved. The basic materials are there and one can reasonably expect the next few years to witness a considerable increase in agroforestry development in Central America.

Acknowledgement

The author is particularly indebted to Alexander Imbach, a graduate student at CATIE, who prepared an annotated bibliography on agroforestry. Special credit is also due to the many staff members at CATIE, past and present.

REFERENCES

Aguirre, A. 1963. Estudio silvicultural y económico del sistema taungya en condiciones de Turrialba. Tesis Mag. Agr. Turrialba, Costa Rica, IICA.

Alavez, L. S. 1983. Estudio preliminar de los cercos vivos en la ganadería de Teapa, Tabasco. Tesis Ing. Agr. Chapingo, M.xico, Universidad Autónoma de Chapingo.

————. 1987. Efecto del poró (*Erythrina poeppigiana* (Walpers) O.F.) Cook plantado a cuatro espaciamientos, sobre la producción de maíz (*Zea mays* L.) en un sistema de cultivo en franjas (alley cropping). Tesis Mag. Sc. Turrialba, C.R., Programma UCR/CATIE.

Alpizar, O.L.A. 1985. Untersuchungen über den Stoffhaushalt einiger agroforstlicher Systeme in Costa Rica. Dokt. Diss. Göttingen, Georg-August-Universität. Federal Rep. Germany.

Alpizar, O.L.A., H.W. Fassbender, J. Heuveldop, G. Enriquez and H. Folster. 1985. Sistemas agroforestales de café (*Coffea arabica*) con laurel (*Cordia alliodora*) y con poró (*Erythrina poeppigiana*) en Turrialba, Costa Rica. I. Biomasa y reservas nutritivas. *Turrialba* 35: 253-242.

Alpizar, O.L.A., H.W. Fassbender, J. Heuveldop, H. Folster and G. Enriquez. 1986. Modelling agroforestry systems of cacao (*Theobroma cacao*) with laurel (*Cordia alliodora*) and poró (*Erythrina poeppigiana*) in Costa Rica. I. Inventory of organic matter and nutrients. *Agroforestry Systems* 4: 175-189.

Alvarez, H. 1956. Estudio forestal del "jaúl" (*Alnus jorullensis* H.B.K.) en Costa Rica. Tesis Mag. Agr. Turrialba, Costa Rica, IICA.

Anderson, E. 1950. An Indian garden at Santa Lucía, Guatemala. *Ceiba* (Guatemala) 1: 97-103.

Babbar, A.L.I. 1983. Descomposición del follaje en ecosistemas sucesionales en Turrialba, Costa Rica. Tesis Mag. Sc. Turrialba: Programa UCR/CATIE.

Baggio, A.J. 1982. Establecimiento, manejo y utilización del sistema agroforestal cercos vivos de *Gliricidia sepium* (Jacq.) Steud., en Costa Rica. Tesis Mag. Sc. Turrialba: Programa UCR/CATIE.

Baggio, A.J. and J. Heuveldop. 1982. Initial performance of *Calliandra calothyrsus* Meissm. in live fences for the production of biomass. Turrialba: CATIE.

Barradas, V.L. and J. Fanjul. 1984. La importancia de la cobertura arbórea en la temperatura del agroecosistema cafetalero. *Biotica* (Mex.) 9: 415-421.

Beer, J. 1983. Arboles de sombra en cultivos perennes. In L. Babbar, (ed.), Curso corto intensivo: Prácticas Agroforestales con énfasis en la medición y evaluación de parámetros biológicos y socioeconómicos. Turrialba, Costa Rica, 1983. Contribuciones de los participantes. Turrialba: CATIE.

————. 1984a. Estudio y promoción de sistemas agroforestales tradicionales en Centro y Sur América. In J. Beer and E. Somarriba (eds.), *Investigación en técnicas agroforestales tradicionales*. Turrialba: CATIE.

————. 1984b. Introducción al establecimiento de parcelas permanentes en asociaciones agroforestales tradicionales. In J. Beer and E. Somarriba (eds.), *Investigación de técnicas agroforestales tradicionales*. Turrialba: CATIE.

————. 1985. Promotion of tree planting on small farms in the area of Acosta-Puriscal, Costa Rica. Turrialba: CATIE.

————. 1987. Ventajas, desventajas y características deseables de los árboles de sombra para café, cacao y te. Turrialba: CATIE.

Beer, J. and E. Somarriba (eds.). 1984. Investigación de técnicas agroforestales tradicionales. Turrialba: CATIE.

Beliard, C.A. 1984a. Producción de biomasa de *Gliricidia sepium* (Jacq.) Steud., en cercas vivas bajo tres frecuencias de poda (tres, seis y nueve meses). Tesis Mag. Sc. Turrialba: Programa UCR/CATIE.

————. 1984b. Tablas de rendimiento de rebrotes (leña y forraje) en cercos vivos de *Gliricidia sepium* en la zona de Siquirres, Costa Rica. Turrialba: CATIE.

Beliard, C.A. and E. Mora. 1984. Preliminary fresh weight tables for *Gliricidia sepium* branches of live fence posts in the "La Francia" farm, Guápiles, Costa Rica. Turrialba: CATIE.

Benavides, J.E. 1983. Utilización de forrajes de origen arboreo en la alimentación de rumiantes menores. In L. Babbar (ed.), Curso corto intensivo: Prácticas agroforestales con énfasis en la medición y evaluación de parámetros biológicos y socioeconómicos. Turrialba, Costa Rica, 1983. Contribuciones de los participantes. Turrialba: CATIE.

Berish, C.W. 1983. Roots, soil, litter and nutrient changes in simple and diverse tropical successional ecosystems. Ph.D. thesis. University of Florida, Gainesville, Florida.

Bertisch, F. 1983. Diagramación de sistemas, con énfasis en fincas y en sistemas agroforestales. In L. Babbar (ed.), Curso corto intensivo: Prácticas agroforestales con énfasis en la medición y evaluación de parámetros biológicos y socioeconómicos. Turrialba, Costa Rica. 1983. Contribuciones de los participantes. Turrialba: CATIE.

Borel, R. 1986. Potencial de utilización de árboles leguminosos para la alimentación animal. In R. Bressani (ed.), Simposio sobre necesidades actuales y futuras de alimentos básicos en Centro América y Panamá. Guatemala, 1985. Actas. Guatemala. INCAP.

Bornemisza, E. 1983, Manejo e investigación en suelos bajo sistemas agroforestales. In L. Babbar (ed.), Curso corto intensivo: Prácticas agroforestales con énfasis en la medición y evaluación de parámetros biológicos y socioeconómicos. Turrialba, C.R. 1981. Contribuciones de los participantes. Turrialba: CATIE.

Bronstein, G.E. 1984. Producción comparada de una pastura de *Cynodon plectostachyus* asociada con árboles de *Cordia alliodora,* con árboles de *Erythrina poeppigiana* y sin árboles. Tesis Mag. Sc. Turrialba: Programa UCR/CATIE.

Brownrigg, L. 1985. *Home gardening in international development. What the literature shows.* Washington, D.C.: The League for International Food Education.

Budowski, G. 1955. Sistemas de regeneración de los bosques de bajura en la América Tropical. *Caribbean Forester* 17: 53–75.

————. 1957. Quelques aspects de la situation forestière au Costa Rica. *Bois et Forêst des Tropiques* 55:3–8.

————. 1959. Forestry practices of interest to coffee growers. *Coffee* (Turrialba, Costa Rica) 1: 49–52.

————. 1977. Agroforestry in the humid tropics: a programme of work. Report submitted to IDRC. Turrialba: CATIE (mimeo).

————. 1981. Agroforestry in Central America. In J. Heuveldop and J. Lagermann (eds.), *Proceedings of the Agroforestry Seminar,* Turrialba, 1981. Turrialba: CATIE.

————. 1982a. Professional education in agroforestry in Central America. Turrialba: CATIE.

————. 1982b. The socioeconomic effects of forest management on the lives of people living in the area: the case of Central America and some Caribbean countries. In E.G. Hallsworth (ed.), *Socio-economic effects and constraints in tropical forest management.* London: John Wiley.

————. 1983a. An attempt to quantify some current agroforestry practices in Costa Rica. In P.A. Huxley (ed.), *Plant research and agroforestry.* Nairobi: ICRAF.

————. 1983b. Manejo de un bosque secundario proveniente de un potrero abandonado. In L. Babbar (ed.), Curso corto intensivo: Prácticas agroforestales con énfasis en la medición y evaluación de parámetros biológicos y socioeconómicos. Turrialba, Costa Rica. 1983. Contribuciones de los participantes. Turrialba: CATIE.

————. 1983b. *The taungya system and its applicability in tropical America.* Turrialba: CATIE.

————. 1985. Home gardens in tropical America: a review. Turrialba: CATIE.

————. 1987a. Características críticas de árboles en sistemas agroforestales. Proceedings of the IUFRO/CATIE Meeting on Agroforestry. Turrialba: CATIE (forthcoming).

————. 1987b. *Live fences, a promising agroforestry practice in Costa Rica.* Tokyo: United Nations University (in press).

Budowski, G., D.C.L. Kass and R.O. Russo. 1984. Leguminous trees for shade. *Pesquisa Agropec. Bras.* 19: 205–222.

Budowski, G., R.O. Russo and E. Mora. 1985. Productividad de una cerca viva de *Erythrina berteroana* Urban. en Turrialba, Costa Rica. *Turrialba* 35 (1): 83–86.

Campos Arce, J.J. 1982. Los sistemas agroforestales en ladera y la conservación del suelo. Turrialba: CATIE.

Castillo, G. and J. Beer. 1983. Utilización del bosque y de sistemas agroforestales en la región de Gardi, Kuna Yala (San Blas, Panamá). Turrialba: CATIE.

CATIE. 1982. Huerto casero como componente integral de fincas pequeñas: estudio inicial del huerto casero (Cuestionario Confidencial). 1982. Turrialba: CATIE.

Cedeño, S.O. and P.J. Chavelas. 1980. Agroforestry in the field experiment forest of San Felipe Bacalar, Quintana Roo. In IUFRO/MAB Conference: Research on multiple use of forest resources. Flagstaff, Arizona. 1980. Washington: USDA Forest Service.

Chavarria, M. (ed.). 1979. Simposio Internacional sobre las ciencias forestales y su contribución al desarrollo de la América Tropical. San José, Costa Rica. Conicit and Interciencia.

Clarkin, K. 1982. Usted también puede tener árboles en su finca. Turrialba: CATIE.

Combe, J. 1981a. Advantages and limitations of pasture management with agroforestry systems. In J. Heuveldop and J. Lagemann (eds.), *Proceedings of the Agroforestry Seminar,* Turrialba, 1981. Turrialba: CATIE.

————.1981b. Taungya reforestation at CATIE, Turrialba, Costa Rica: *Terminalia ivorensis* with annual and perennial crops. In J. Heuveldop and J. Lagemann (eds.), *Proceedings of the Agroforestry Seminar,* Turrialba, 1981. Turrialba: CATIE.

————.1982. Agroforestry techniques in tropical countries: potential and limitations. *Agroforestry Systems* 1: 13–27.

Combe, J. and G. Budowski. 1979. Classification of agroforestry techniques. In G. De las Salas (ed.), *Proceedings of the Workshop on Agroforestry Systems in Latin America.* Turrialba: CATIE.

Combe, J. and N. Gewald (eds.). 1979. Guía de Campo de los ensayos forestales del CATIE en Turrialba, Costa Rica CATIE, Programa de Recursos Naturales Renovables, Turrialba, Costa Rica.

Combe, J., L. Espinoza, R. Kastl and R. Vetter. 1981a. Coffee plantation with alders: *Coffee arabica-Alnus acuminata.* In J. Heuveldop and J. Lagemann (eds.), *Proceedings of the Agroforestry Seminar,* Turrialba, 1981. Turrialba: CATIE.

————. 1981b. Growth of laurel in cocoa plantations and in pastures in the Atlantic Zone of Costa Rica. In J. Heuveldop and J. Lagemann (eds.), *Proceedings of the Agroforestry Seminar,* Turrialba, 1981. Turrialba: CATIE.

Combe, J., H. Jiménez Saa and C. Monge. 1981. Agroforestry: a bibliography. CATIE, Bibliografia No 6., Serie Bibliotecología y Documentación No 6.

Cook, O.F. 1901. Shade in coffee culture. Washington, D.C.: U.S. Dept. of Agriculture, Division of Botany.

De las Salas, G. (ed.). 1979. *Proceedings of the Workshop on Agroforestry Systems in Latin America.* Turrialba: CATIE.

De las Salas, G. and H.W. Fassbender. 1981. The soil science basis of agroforestry production systems. In J. Heuveldop and J. Lagemann (eds.), *Proceedings of the Agroforestry Seminar,* Turrialba, 1981. Turrialba: CATIE.

Detlefsen, R.G., P. Leiva and H. Martinez. 1984. Comportamiento inicial de tres especies forestales para producción de leña con y sin asocio de maíz (*Zea mays* L.) en La Máquina, Suchitepequez, Guatemala. *Tikali* (Guatemala) 3 (1): 114–128.

Dulin, P. 1982. Agroforestry as an appropriate land use system in the American tropics. Turrialba: CATIE.

Enriquez, G. 1983. Breve resúmen de los resultados del experimento central de plantas perennes de La Montaña. In L. Babbar (ed.), Curso corto intensivo: Prácticas agroforestales con énfasis en la medición y evaluación de parámetros biológicos y socioeconómicos. Turrialba, Costa Rica, 1983. Contribuciones de los participantes. Turrialba: CATIE.

Espinosa, B. 1984. Caracterización nutritiva de la fracción nitrogenada del forraje de madero negro *Gliricidia sepium* y poró *Erythrina poeppigiana.* Tesis Mag. Sa. Turrialba, C.R., Programa UCR/CATIE.

————. 1985. Untersuchungen über die Bedeutung der Baumkomponente bei agro forstwirtschaftlichem Kaffeeanbau an Beispielen aus Costa Rica. Dokt. Diss. Göttingen, Alemania Federal, Georg-August-Universität.

————.1986a. El componente arbóreo en el sistema agroforestal "Cafetal arbolado" en Costa Rica. *El Chasqui* (CR) 12:17–22.

————.1986b. Investigaciones sobre la importancia del componente arbóreo en el sistema agroforestal "cafetal arbolado" basándose en ejemplos de Costa Rica. Turrialba: CATIE.

Ewel, J. 1969. Dynamics of litter accumulation under forest succession in eastern Guatemala. M.S. thesis, University of Florida, Gainesville, Florida.

————. 1971. Biomass changes in early tropical succession. *Turrialba* 21:110–112.

————. 1976. Litterfall and leaf decomposition in a tropical forest succession in eastern Guatemala. *Journal of Ecology* 64:293–308.

Ewel, J. and L. Babar. 1981. La sucesión natural como modelo para el diseño de nuevos agroecosistemas tropicales. Turrialba: CATIE.

Ewel, J., C. Berish, B. Brown, N. Price and J. Raich. 1981. Slash and burn impacts on a Costa Rican wet forest site. *Ecology* 62:815–829.

Ewel, J., F. Benedict, C. Berish, B. Brown, S. Gliessman, M. Amador, R. Bermudez, A. Martinez, R. Miranda and N. Price. 1982. Leaf area, light transmission, roots and leaf damage in nine tropical communities. *Agroecosystems* 7:305–326.

Fassbender, H.W. 1984. Bases edafológicas de los sistemas de producción agroforestales. (Serie materiales de enseñanza No 21.) Turrialba: CATIE.

Fassbender, H.W., L. Alpizar, J. Heuveldop, G. Enriquez and H. Folster. 1985. Sistemas agrofoestales de caf. (*Coffea arabica*) y con poró (*Erythrina poeppigiana*) en Turrialba, Costa Rica. III. Modelos de la materia orgánica y los elementos nutritivos *Turrialba* 35 (4): 403–413.

Fassbender, H.W., L. Alpizar, J. Heuveldop, H. Folster and G. Enriquez. 1987. Modelling agroforestry systems of cacao (*Theobroma cacao*) with laurel (*Cordia alliodora*) and poró (*Erythrina poeppigiana*) in Costa Rica. III. Cycles of organic matter and nutrients. *Agroforestry Systems* (in press).

Fernandez, V.S. 1978. Comportamiento inicial de *Gmelina arborea* Roxb. asociado con maíz (*Zea mays* L.) y frijol (*Phaseolus vulgaris* L.) en dos espaciamientos en Turrialba, Costa Rica. Tesis Mag, Sc. Turrialba. C.R., Programa UCR/CATIE.

Fournier, O., L.A. 1981. Importancia de los sistemas agroforestales en Costa Rica. *Agronomía Costarricense* (1–2): 141–147.

Garriguez, R. 1983. Sistemas silvopastoriles en Puriscal. In J. Heuveldop and L. Espinoza (eds.), *El componente arbóreo en Acosta y Puriscal*. Turrialba: CATIE.

Gliessman, S.R. 1981. Los sistemas agroforestales como sistemas agroforestales en el trópico húmedo en México. In J.W. Beer and E. Somarriba (eds.), *Investigación de técnicas agroforestales tradicionales*. Boletin Técnico No 12. CATIE, Turrialba, Costa Rica.

Glover, N. and J. Beer. 1984. Spatial and temporal fluctuations of litter fall in the agroforestry associations *Coffea arabica* var. *caturra* - *Erythrina poeppigiana* and *C. arabica* var. *caturra* - *E. poeppigiana* - *Cordia alliodora*. Turrialba: CATIE.

————. 1986. Nutrient cycling in two traditional Central American agroforestry systems. *Agroforestry Systems* 4(2): 77–87.

Gonzalez de Moya, M. 1955. Ordenación de un bosque subtropical de crecimiento secundario en Costa Rica. Tesis Mag. Agr. Turrialba, Costa Rica, IICA.

Gross, L.S. 1983. Cercos vivos en la Península de Nicoya; establecimiento y manejo. Turrialba: CATIE.

Harcombe, P.A. 1977. The influence of fertilization on some aspects of succession in a humid tropical forest. *Ecology* 58:1375–1388.

Heuveldop, J. and L. Espinoza. 1984. El uso de encuestas en la investigación de técnicas agroforestales tradicionales. In J. Beer and E. Somarriba (eds.), *Investigación de prácticas agroforestales tradicionales*. Turrialba: CATIE.

Heuveldop, J., H.W. Fassbender, L. Alpizar, G. Enriquez and H. Folster. 1987. Modelling agroforestry systems of cacao (*Theobroma cacao*) with laurel (*Cordia alliodora*) and poró (*Erythrina poeppigiana*) in Costa Rica. II. Cacao and wood production, litter production and decomposition. *Agroforestry Systems* (in press).

Holdridge, L.R. 1951. The alder *Alnus acuminata* as a farm timber tree in Costa Rica. *Caribbean Forester* 12 (2): 47–53.

Jomenez, O.F. 1986. Balance hídrico con énfasis en percolación de dos sistemas agroforestales: café-poró y café-laurel, en Turrialba, Costa Rica. Tesis Mag. Sc. Turrialba, C.R. Programa UCR/CATIE.

Jones, J. 1983. La aplicación de la metodología de agroecosistemas a la agroforestería. In L. Babbar (ed.), Curso corto intensivo: Prácticas agroforestales con énfasis en la medición y evaluacion de parámetros biológicos y socioeconómicos. Turrialba, Costa Rica, 1983. Contribuciones de los participantes, Turrialba: CATIE.

Jones, J. and J. Campos. 1983. Actitudes hacia la reforestacion entre los agricultores de Piedades Norte, Costa Rica. In L. Babbar (ed.), Curso corto intensivo: Prácticas agroforestales con énfasis en la medición y evaluación de parámetros biológicos y socioeconómicos. Turrialba, Costa Rica, 1983. Contribuciones de los participantes. Turrialba: CATIE.

Kass, D.C.L., R.O. Russo and M.M. Quinlan. 1983. Leguminous trees as a nitrogen source for annual crops. *Agronomy Abstracts* 45.

Lagemann, J. 1981. Problems of agricultural production in the humid tropic lowlands. In J. Heuveldop and J. Lagemann (eds.), *Proceedings of the Agroforestry Seminar*, Turrialba, 1981. Turrialba: CATIE.

Lagemann, J. and J. Heuveldop. 1983. Characterization and evaluation of agroforestry systems: the case of Acosta-Puriscal, Costa Rica. *Agroforestry Systems* 1 (2): 101–115.

Leiva, P.J.M. 1982. Crecimiento inicial de *Cordia alliodora* Oken en plantación a campo abierto y bajo dos tipos de cubierta de bosque secundario tropical, en Siquirres, Costa Rica. Tesis Mag. Sc. Turrialba, C.R., Programa UCR/CATIE.

Leiva, P.J.M. and J. Lopez. 1985. Los sistemas agrofrestales de la cuenca del Rio Polochic composición y características. *Tikali* (Guatemala) (1-2) : 47–84.

Lindblad, P. and R.O. Russo. 1986. C_2H_2-reduction by *Erythrina poeppigiana* in Costa Rican coffee plantations. *Agroforestry Systems* 4 (1): 33–37.

Lozano, O.R. 1962. Postes vivos para cercos. *Turrialba* 12 (3): 150–152.

Maffioli, R. A. and M. Holle. 1981. Caracterización del huerto casero tropical en los cantones de Orotina y San Mateo; Alajuela (Costa Rica). Turrialba: CATIE.

Major, M., G. Budowski and R. Borel. 1985. *Manual of teaching methods for use in agroforestry intensive short courses.* Turrialba: CATIE.

Maldonado, U.T. 1986. La colonización del área de Taque-Taque, el uso de la tierra y los sistemas agroforestales. Reserva Forestal Rió Macho, Costa Rica. Análisis y perspectivas. Tesis Mag, Sc. Turrialba, C.R., Programa UCR/CATIE.

Marinero, R.M. 1962. Influencia del *Melinis minutiflora* Beauv. en el crecimiento de *Cordia alliodora* (R. & P.) Cham. Tesis Mag. Agr. Turrialba, Costa Rica. IICA.

Martinez, H.A. 1982. El uso del componente arbóreo en fincas de Guatemala. In Curso sobre metodología de investigación y técnicas de producción de leña. Guatemala. 1982. Actas. Guatemala, CATIE/INAFOR.

Maydell, H.-J. von. 1981. Criteria for planning and evaluation of agroforestry projects. In J. Heuveldop and J. Lagemann (eds.), *Proceedings of the Agroforestry Seminar*, Turrialba, 1981. Turrialba: CATIE.

Nations, J.D. 1981. The rainforest farmers. *Pacific Discovery* 34 (1): 1–9.

Nations, J.D. and R.B. Nigh. 1980. The evolutionary potential of Lacandon Maya sustained-yield tropical forest agriculture. *Journal of Anthropological Research* (EE.UU) 36 (1): 1–30.

OTS/CATIE. 1986. *Sistemas agroforestales: principios y applicaciones en los tropicos.* San José: OTS/CATIE.

Perez, C.A. 1954. Estudio forestal del laurel, *Cordia alliodora* (R. & P.) Cham., en Costa Rica. Tesis Mag. Agr. Turrialba, Costa Rica, IICA.

Pineda, M.O.J. 1986. Utilización del follaje de poró (*Erythrina poeppigiana*) en la alimentación de terneros de lechería. Tesis Mag. Sc. Turrialba, C.R., Programa UCR/CATIE.

Price, N. 1983. El huerto mixto tropical: un componente agroforestal de la finca pequeña. In L. Babbar (ed.), Curso corto intensivo: Prácticas agroforestales con énfasis en la medición y evaluación de parámetros biológicos y socioeconómicos. Turrialba, Costa Rica. 1983. Contribuciones de los participantes. Turrialba: CATIE.

Reiche, C. 1983. Implicaciones económicas del componente agroforestal. In L. Babbar (ed.), Curso corto intensivo: Prácticas agroforestales con énfasis en la medición y evaluación de parámetros biológicos y socioeconómicos. Turrialba, Costa Rica, 1983. Contribucione de los participantes. Turrialba: CATIE.

Rodbell, P. 1986. Manual de prácticas agroforestales, Tegucigalpa, Cuerpo de Paz.

Rodriguez, F.R.A. 1985. Producción de biomasa de poró gigante (*Erythrina poeppigiana* (Walpers). O.F. Cook) y King grass (*Pennisetum purpureum* x *P. thyphoides*) intercalados, en función de la densidad de siembra y la frecuencia de poda del poró. Tesis Mag. Sc. Turrialba, C.R., Programa UCR/CATIE.

Rodriguez, Q.J.E., M.R. Jimenez. and G.B. Canet. 1983. Actividades agrosilvopastoriles en Guatemala, Honduras y Panamá. San José, C.R., DGF/FAO.

Rose, D. and R. Salazar. 1983. Lineamientos generales para la evaluación de producción de biomasa y leña en cercas vivas viejas de *Gliricidia sepium*. Turrialba: CATIE.

Roskoski, J.P. 1982. Nitrogen fixation in a Mexican coffee plantation. *Plant and Soil* 67: 283–291.

Russo, R.O. 1983a. Efecto de la poda de *Erythrina poeppigiana* (Walpers) O.F. Cook (poró), sobre la nodulación, producción de biomasa y contenido de nitrógeno en el suelo en un sistema agroforestal "café-poró". Tesis Mag. Sc. Turrialba, C.R., Programa UCR/CATIE.

Russo, R.O. 1983b. Mediciones de biomasa en sistemas agroforestales. Turrialba: CATIE.
————. 1984. Arboles con pasto: justificación y descripción de un estudio de caso en Costa Rica.
    In J. Beer and E. Somarriba (eds.), *Investigación de técnicas agroforestales tradicionales.*
    Turrialba: CATIE.
Russo, R.O. and G. Budowski. 1986. Effect of pollarding frequence on biomass of *Erythrina
    poeppigiana* as a coffee shade tree. *Agroforestry Systems* 4: 145–162.
Ruiz, M. 1983. Avances en la investigación en sistemas silvopastoriles. In L. Babbar (ed.), Curso
    corto intensivo: Prácticas agroforestales con énfasis en la medición y evaluación de
    parámetros biológicos y socioeconómicos. Turrialba: CATIE.
Sabogal, M.C. 1982. Observaciones sobre la combinacion agroforestal tradicional de *Cedrela
    odorata* con *Coffea arabica* o pastos en Tabarcia-Palmichal (Cantón de Puriscal, Costa
    Rica.) Turrialba: CATIE.
————. 1983. Observaciones sobre la combinación de *Cedrela odorata* con café en Tabarcia-
    Palmichal (Canton Puriscal). In J. Heuveldop and L. Espinoza (eds.), El componente
    arbóreo en Acosta y Puriscal, Costa Rica. Turrialba: CATIE.
Samur, R.C. 1984. Producción de leche de cabras alimentadas con King grass (*Pennisetum
    purpureum*) y poró (*Erythrina poeppigiana*), suplementadas con fruto de bamano (*Musa*
    sp. cv. "Cavendish"). Tesis Mag. Sc. Turrialba, C.R., Programa UCR/CATIE.
Sauer, J.D. 1979. Living fences in Costa Rican agriculture. *Turrialba* 29 (4): 255–261.
Sequeira, W. 1984. Agroforestry systems and rural development in Costa Rica. M.S. thesis,
    University of Wales, Bangor.
Somarriba, E. 1982. Guayabo (*Psidium guajava* L.) asociado con pastos; método de análisis de
    producción de leña. Turrialba: CATIE.
————. 1985a. Arboles de guayaba (*Psidium guajava* L.) en pastizales. I. Produccion de fruta y
    potencial de dispersión de semillas. *Turrialba* 35 (3): 289–295.
————. 1985b. Arboles de guayaba (*Psidium guajava* L.) en pastizales. II. Consumo de fruta y
    dispersión de semillas. *Turrialba* 35 (4): 329–332.
————. 1986. Effects of livestock on seed germination of guava (*Psidium guajava* L.).
    Agroforestry Systems 4: 233–238.
Somarriba, E. and J. Beer. 1985. Arboles de guayaba (*Psidium guajava* L.) en pastizales. III.
    Producción de leña. *Turrialba* 35 (4): 333–338.
————. 1986. Dimensiones, volúmenes y crecimiento de *Cordia alliodora* en sistemas agrofores-
    tales. Boletín Técnico No 16, Turrialba: CATIE.
Somarriba, E., L.E. Vega, G. Detlefsen, H. Patino and K. Twum-Ampofo. 1986. Pastoreo bajo
    plantaciones de *Pisus caribaea* en Pavones, Turrialba, Costa Rica. *El Chasqui* (CR) 11:
    5–8.
Spiegeler, C.A. 1981. Comportamiento inicial de *Pinus oocarpa* Schiede. Asociado con cultivos
    anuales. Tesis Ing. Agr. Guatemala, Universidad de San Carlos.
Sumberg, J.E. 1985. Collection and initial evaluation of *Gliricidia sepium* from Costa Rica.
    Agroforestry Systems 3: 357–361.
Villatoro, P.R.M. 1986. Caracterización del sistema agroforestal café-especies arbóreas en la cuenca
    del Río Achiguate, Guatemala. Tesis Ing. Agr. Guatemala, Universidad de San Carlos.
Wilken, G.C. 1977. Integrating forest and small-scale farm systems in Middle America. *Agro-
    ecosystems* 3: 291–302.

7

Agroforestry in the dry zones of Africa: past, present and future

H.-J. von Maydell

Professor
Institute for World Forestry and Ecology
Federal Research Centre for Forestry and Forest Products
Leuschnerstr. 91, 2050 Hamburg 80, F.R.G.

Contents

Introduction

The challenge to develop agroforestry to solve apparent problems in tropical rural areas led to the establishment of ICRAF in 1977. At that time it was known that a great variety of integrated land-use systems had already been practised throughout semi-arid Africa since times immemorial. They had been empirically evolved and adapted to different environments as well as various ethnic groups. Very little attention was given to these forms of subsistence land use as they were seen to have little or no relevance to colonial and post-colonial, export-oriented, modern agriculture. Moreover, no need was felt to deal with them because no specific problems with regard to natural resources and social developments were recognized. For decades, rural development policies had concentrated on increasing yields from extensive cash-crop plantations, which had high external inputs in terms of technology and capital and aimed at outputs for overseas markets to earn foreign exchange rather than to supply products to African rural and urban populations.

Starting from the late 1970s, this attitude gradually changed. World markets for agricultural commodities had shifted. Traditional trade links were losing their previous importance after many states had gained their independence. Growing gaps in the terms of trade forced African governments to re-orientate their land-use policies as well as their general development strategies. With "Africanization", the importance of the hitherto-overlooked values of the African way of life were rediscovered. But, simultaneously, new problems of local food supply, environmental destruction, etc., emerged and conflicts between rural development and industrialization had to be solved. Finally, the severe drought throughout the Sahelian zone and other parts of Africa in the early 1970s had caused a shock among African politicans, agricultural services and the international community, indicating that something would have to be drastically changed in rural development in order to prevent future disaster. Thus, as a first step, at the United Nations Conference on Desertification in Nairobi, Kenya (1977) a "Plan of Action" was formulated which highlighted the role of sustainable-resources management and the till-then-underestimated importance of forestry within integrated land use.

The beneficial effects of "non-commercial" trees, shrubs and other woody perennials of the drylands that were recently brought to general awareness unfortunately very soon resulted in extremely high expectations as to what could be achieved by afforestation, reforestation or tree planting. Many of these expectations with regard to dry Africa soon proved to be unrealistic. In reality, multipurpose woody plants, either alone or in small groups and lines rather than closed forests or extensive greenbelts, and local species rather than introduced fast-growing and timber-producing "exotics", turned out to be much better adapted to various sites and demands. The forest services with their limited capacities, however, were not yet organized to change from their standardized programmes of conservation and timber production to the new issue of forestry as a component of integrated land use and "forestry for people". Suddenly, and often under severe time-pressure, they were expected to deal with a wide range of non-forestry subjects, i.e., the co-management of agricultural and pastoral lands. Farmers and herdsmen, as well as government agricultural and livestock services, were simultaneously expected to incorporate forestry components and functions into their management.

At all levels, therefore, it was felt that the common bond — integrating, harmonizing and optimizing land-use practices — was missing. This called for immediate interdisciplinary co-operation which became a common term, easy to spell but often difficult to realize in practice.

Research and development institutions, therefore, started to add to their previously dominant but isolated topics of commodity, technology and socio-economic research the "new" field of farming systems (Simmonds, 1985). This was a decisive step forward at all levels (national and international, including the Consultative Group on International Agricultural Research, CGIAR). Agroforestry-systems research, as initiated by ICRAF, has played a leading and stimulating role in this context. Some of the results originating from a special programme, namely, the Agroforestry Systems Inventory, have been published in *Agroforestry Systems*.

The past

The environment

"Dry Africa" includes all parts of the continent with less than about 1,500 mm of annual rainfall. Most of these lands lie within the tropical zone but they do extend beyond the

Tropics of Cancer and Capricorn. Dry Africa thus includes a great variety of climates and landscapes and they are inhabited by very different peoples. Therefore all attempts to generalize are of limited value and may lead to controversial discussions.

One common feature of these African regions, however, is a more or less pronounced water deficiency. The deficiency is either permanent, seasonal or sporadic with regard to horizontal or vertical distribution, quality and utilization. Rainfall is very variable in most parts. If rains are below average we call it a drought, although drought in its proper sense refers only to rainfall insufficient to support life and satisfy human demand.

Plants, animals and people have adapted to the harsh environment of dry Africa (Louw and Seely, 1982). Plants have developed specific mechanisms for drought escape and drought tolerance (e.g., groups of plants such as ephemerals, succulents, xerophytes, halophytes, psammophytes, plants employing the so-called C_4-photosynthetic pathway or CAM-plants that can select the C_3 or C_4 mode of life according to the prevailing conditions, etc.). Adaptation to stress implies also concentration or spatial restriction to small favourable sites and short periods of growth, as well as a high degree of resilience as compared with persistence. Animals and men have developed similar strategies of rather short-term flexible adaptation to changes in environmental conditions.

It is of paramount importance to acknowledge that the carrying capacity of arid and semi-arid lands in Africa is limited. By over-exploitation and/or misuse of natural resources, therefore, critical limits are rapidly reached, and once exceeded very soon result in an irreversible breakdown of the productive potential of a given site.

From the vegetation aspect, most ecosystems in dry Africa (excluding deserts and semi-deserts) can be classified under the term "savanna". These are characterized by grasslands with trees, palms and shrubs of varying density from open woodland to single thorny shrubs (see Le Houérou, this volume). Very few sites bear a climax vegetation of dense stands of woody perennials, and these sites are generally preferred for different forms of land use because of their better soils and water supplies. Wherever they remain unoccupied, they form islets important for the survival of both "refugees" and "nomads" among plants, animals and people. Typical sites of this kind are oases and forest islands. The role of such sites, which often support high population concentrations, has to be highlighted for dry Africa. They represent a network within an "ocean" of marginal lands, and their mutual feedbacks and interrelationships are a pre-condition for maintaining and further developing the natural and the socio-economic environment.

The role of woody perennials on range and farmlands

There is very little archaeological and historic evidence on agricultural — and thus agroforestry — systems of ancient Africa, except for Egypt and parts of north Africa which belonged to the Roman Empire. Most evidence refers to irrigation schemes, to specific crops and their utilization and to sites preferred for cultivation (Hall *et al.,* 1979). The first traces of agriculture in Egypt date back to about 12,500 BC with indications of plant cultivation and animal husbandry, and to 6000 BC for date-palm management. There is hardly any information available on tree cultivation and on land-use systems of these early times. What is known today has mainly been recorded during times of exploration of the continent, especially from the nineteenth century on and during the colonial period.

Trees, shrubs and palms have always played an important role in ecology as well as in human culture and economy throughout semi-arid Africa. However, the value of these woody perennials was not primarily derived from closed forests, but rather, and even predominantly, from individual specimens which were components of pastoral-, farm- or village-settlement lands. This has impeded understanding between temperate-zone-educated foresters, who think and act in terms of sustainable management of closed-forest systems, and African rural people, who are used to distinguishing specific single trees from which they wish to benefit.

Land tenure in these open bush- and woodlands was considered a tribal/community concern in traditional African land use (with private forest ownership being practically non-existent), whereas single fruit trees could well be individually owned and even be inheritable. Forests and forest products were used in a communal or collective way, generally free of charge and with very few restrictions. Forests were generally protected in colonial and modern national legislation and thus did not constitute an accessible resource. Scarcity of supplies, and more often dire need, provoked people to invade these forests in order to continue traditional uses, now classified as illegal, or to convert these forests into non-forest lands. These are some of the reasons why rural people in dry Africa are more closely affiliated with single trees outside the forest than with forest-land management.

The affiliation of trees outside the forests with people is also known from many examples of veneration of trees in African history. In ancient Egypt, for example, sycamore trees (*Ficus capensis* syn. *F. sycomorus*) were worshipped as symbols of fertility, suppliers of food for those alive and for the souls of the deceased. Other sacred trees were *Balanites aegyptiaca, Olea europaea* and *Phoenix dactylifera* (the date palm). All over semi-arid Africa, various tree, palm and shrub species offering food, medicine, gums and other exudants, or providing shade or an aesthetic sight, were venerated or at least cared for. Most of them grew in open woodlands and savannas, along riverbanks or in some oases, and in parks and gardens. Wood production for fuel, construction, etc., was given less importance because supplies were generally secured from other sources, mostly dead wood.

From Egypt, throughout the Sahara to the Sahel and the Sudan savannas, the *miombo* of East Africa, and open woodlands, savannas and semi-deserts in southern Africa, man's relation to trees was similar. Widely scattered trees formed an essential component of the natural vegetation in man-made environments and in daily life. The mighty baobab (*Adansonia digitata*) (Figure 1) may be considered a symbol of what a tree could mean to people. *Acacia albida* is another commonly protected and highly venerated non-forest tree of large areas in semi-arid Africa.

Those trees as essential elements of the landscape were used in three different ways:

1. Exploitation

Selective exploitation of natural resources took place among "mobile" (nomadic or transhumant people) as well as in sedentary farming societies. For example, the nomads, when migrating over vast areas, considered the woody vegetation as common property of their tribe or implied everyman's right to utilize what God had provided. No specific limitations had to be observed, since what was not used by them would most likely be used by others. This attitude still prevails with many people in the region and obviously constitutes a serious constraint to maintaining or regenerating a desirable density of woody vegetation on pastoral lands. Sedentary people generally enjoyed free access to communal/tribal resources within daily walking distances, i.e., 5–15 km. Collection of fuelwood, poles, and other products of trees and shrubs was unrestricted and considered a common right. As long as population densities remained low these practices could prevail without destructive effects on the environment or on the preferred species' survival.

Figure 1   *Adansonia digitata,* the baobab tree

2. Maintenance

Maintenance of selected, naturally grown trees took place with more intensive land use, especially in dry farming. The previous natural vegetation was changed or even removed in favour of field crops and permanent settlements. However, specific trees, palms and shrubs were deliberately left and eventually maintained and carefully protected against livestock and illegal use. Some of these trees were *Acacia albida, Butyrospermum parkii, Parkia biglobosa, Tamarindus indica, Ziziphus mauritiana,* and there were many others, depending on site, land use and tradition. Some of these trees became private property and could be handed down from generation to generation. The Emirs of Zinder, Niger, at times even passed death sentences upon people who felled *Acacia albida* trees.

3. Cultivation

Cultivation of trees was rarely practised in the past. Egypt was an exception. Parks and tree gardens, such as that in Amun (Nile delta), where 8,000 slaves were employed to maintain the trees, or Matoria, where the Holy Family was reported to have stayed for a while, were established and intensively managed like urban and individual ornamental plantations. Last but not least, the oases with their date palm and other tree or shrub plantations are outstanding examples of early selection, breeding, culture and integrated management of trees in rural environments. It is interesting to note that, whereas the date palm was intensively improved and cultivated, no effort has been made to genetically improve *Balanites,* originally an equivalent to the date palm and apparently much better adapted to the climate and sites.

In other parts of semi-arid Africa (outside oases) tree planting and breeding remained generally unknown until recent times when non-Africans started to propagate the so-called exotics such as cashew (*Anacardium occidentale*), neem (*Azadirachta indica*), eucalypts, various fruit trees (e.g., mango, *Citrus* spp.) and a few ornamental trees such as *Khaya senegalensis* and *Cassia* spp.

In essence, however, extensive exploitation of natural resources by migrating and "landless" people on the one hand and intensive use by sedentary farming communities or urban people on the other prevailed over many centuries. Both had an important influence on the development of early agroforestry-type land uses. As in other parts of the tropics, these traditional forms developed into three basic categories: silvopastoral, agrosilvopastoral and agrisilvicultural systems (Nair, 1985).

However, dry Africa is extremely heterogeneous. It is not a "khaki-coloured" environment with simple and uniform land uses. The local climate, for instance, may differ even within a few kilometres, especially with regard to quantity and distribution of rainfall. Soils show a remarkable variety, not only in their chemical components and physical structure but also in the micro-organisms essential for plant growth. The number of plant species that exist even under marginal conditions is remarkably high as long as degradation by repeated burning, overgrazing or monoculture cropping has not occurred. Man has adapted his life and his land-use patterns to the heterogeneity of the environment by developing a variety of practices. Consequently, hundreds of land-use systems have developed and many of them are still of great importance.

Silvopastoral systems

For thousands of years the regions of Africa with less than about 500 mm annual rainfall have been used for migratory animal husbandry. However, only parts of these vast grazing lands were utilized in a way which would allow their classification as silvopastoral agroforestry systems. But browse made up a varying proportion of livestock diet almost everywhere (Le Houérou, 1980, and this volume; IBPGR, 1984). Whether or not traditional forms of pastoral land use in Africa should be considered as agroforestry in its proper sense is a matter of definition and is subject to individual judgement because intentional cultivation or maintenance of fodder trees did not take place in these systems (Figure 2).

As a rule, these African dry lands can be classified as marginal, fragile ecosystems, of low carrying capacity and thus limited socio-economic development prospects. The ruling climatic constraint is a single, short rainy season with pronounced variation in quantity and spatial distribution of rainfall, especially in the more arid parts. Quantity and nutritional value of the plant biomass, as well as drinking water available for the animals, are thus subject to unpredictable but foreseeable fluctuations. As a consequence of the climate, the vegetation consists mainly of annual grasses with some perennial species and a varying number of shrubs and trees, attaining between 5 and 20 percent of the area depending on soil type and ground-water availability. The growth period of grasses varies between one and three months. Many, if not most, woody species are deciduous, i.e., leafless over the long dry season.

Over thousands of years, animal husbandry in the traditional forms of nomadism and semi-nomadism (transhumance) have proved to be optimally adapted to the carrying capacity of these lands and the rainfall-dependent variation in fodder production (Jahnke, 1982; Lusigi and Glaser, 1984; Galaty *et al.,* 1981). Recent problems have all been caused by unprecedented population growth (the resulting pressure on natural resources exceeding carrying capacities), and by external influences which have increasingly destroyed well-established regional structures.

According to Ruthenberg (1980), in nomadism the animal owners do not have a

Figure 2 Silvopastoral systems in the highlands of Ethiopia

permanent place of residence; they do not practise regular cultivation and their families move with their herds. Nomadism in this pure form was and still is mainly practised in desert and semi-desert environments north and south of the Sahara with less than 300 mm of annual rainfall and in limited parts of eastern and southern Africa. But these systems are less important than the semi-nomadic or transhumant systems in which livestock owners have a permanent place of residence which they maintain over several years. Whereas part of the family travels with the herds over long periods of the year and over long distances, women, children and older men may stay behind on lands which they cultivate over the rainy season in order to improve the family's food reserves and income.

Semi-nomadism is practised on a wide range of lands, extending into the arid zone with annual rainfall less than 300 mm and into 1,000–1,500 mm zones, thus overlapping with nomadism and with forms of sedentary animal husbandry which belong to agrosilvopastoral systems. Semi-nomadism or transhumance derives its importance from the fact that it benefits from high-quality, low-disease pastures in dry zones during the short rainy season and from the opportunity of spending the dry season where larger, although less nutritious, quantities of forage and adequate water supplies are available and where eventually agricultural crop residues can be utilized. Thus, during the rainy season, from about July to October, large herds migrate over hundreds of kilometres from the southern parts of the Sahel and the Sudan savannas to the northern Sahel and return to the south at the end of the rainy season.

From the point of view of agroforestry systems, competition, complementarity and mutual dependence of animals, woody perennials and grass/herbs deserve special attention. As long as stocking rates were low, detrimental impacts remained negligible and could easily be compensated for by alternating with more favourable sites. With increasing numbers of livestock over recent decades, however, problems have emerged. Since livestock, except goats, feed selectively on specific plants or parts thereof, the more palatable plants were continuously reduced and weeds invaded. This situation was seriously aggravated by the effects of seasonal grass/bush fires over large areas (Figure 3). The structure of herds also played a role. Many herdsmen prefer to specialize in cattle, sheep, goats, or camels, which leads to a one-sided exploitation of the resources. Mixed herds with cattle feeding on long grass, sheep on short grass and goats and camels browsing shrubs would improve overall carrying capacity.

Figure 3   Savanna rangeland severely degraded by frequent bush fire

In general, browsing is considered a necessity and a traditional right by herdsmen, but foresters consider it as an illegal practice causing severe damage. The presence of many villages and deep-wells in the Sahel gives evidence that the foresters' concerns are justified: on previously open woodland with scattered trees, often only stunted thorny shrubs survive (Figure 4).

On the other hand, livestock, like game, have always played an important role in promoting the growth of woody perennials by reducing competition from grass for nutrients and water, and by breaking seed dormancy through digestion and subsequent distribution of seeds with their droppings.

Thus, over long periods and vast areas in dry Africa, as long as population densities remained low, silvopastoral systems proved to be well adapted to the environmental conditions and to people's demands. Social structures and cultural life were based on nomadism and transhumance. This has changed within the past few decades. Therefore, recent moves to convert these "mobile" forms of "archaic" land use into sedentary, modern management systems is not only a question of technical and economic innovation but requires major socio-cultural change. For most African herdsmen their animals are more

Figure 4   Degraded *Balanites aegyptiaca* savanna 5 km off a deep-well in the Sahel

than mere farm commodities and production units; they are an integral part of their individual, family and ethnic identity.

Agrosilvopastoral systems

There has never been a distinct borderline between mobile, silvopastoral land-use systems in the more arid parts and sedentary agrisilvicultural systems in the more humid parts of Africa. Overlapping and transitional zones developed due to periodic variability of the climate and migration of different ethnic groups. Therefore, for most parts of the region, traditional agrosilvopastoral land-use systems were a characteristic feature.

They were based on agricultural-crop production during the rainy season and pastoral use of the same land during the dry period of the year. In the rainy season livestock were either brought to remote grazing lands by transhumance or, more commonly, driven daily to nearby open pastures next to the villages, the herds returning to the compounds or corrals at night, or even being tethered or kept in stables. Specific trees, shrubs and palms were left standing, deliberately maintained or even planted on the cultivated lands which formed an essential component of the system. The agricultural crop plants included cereals (millet, sorghum, barley, maize), cowpeas, groundnuts and melons in the dryer parts and a greater variety of plants on sites with better water supply and soils. These plants were cultivated mainly for food, their residues being used as fodder in the dry season. Cowpea is an outstanding example, the "straw" in some parts of the Sahel yielding even higher income than the beans. Many, if not most, of these crops were planted year after year until soil fertility was exhausted and the farmer had to shift. A fallow system was then applied for many years, the fallow land, including regrowth of woody plants from root suckers, coppice or invading seeds, being used as pasture the year round.

On extensive lands, however, trees and livestock provided for a sufficient recycling of nutrients to allow long-term or even sustainable cultivation. Most of the distinctive agrosilvopastoral systems were on such lands. They are still efficient and offer good prospects for future development.

Much more than the annual crops, the livestock and trees of these systems are typical "multipurpose" components. Livestock contribute to the system by providing food (mainly meat and milk) and other products such as hides, skins, hair, horn and sinew. Some supply draught power for cultivation and transport or are used for riding. In parts of Africa, manure is intensively applied to improve soil fertility, both in homegardens and within the fields of the agrosilvopastoral systems. Some dried manure is used as fuel for heating and cooking, e.g., in the highlands of Ethiopia. Livestock represented an important capital asset, as a source of income in addition to the agricultural crops, as an object of value in exchange for goods and services (including during marriages, etc.), and as a "living savings bank". Livestock represented wealth and security of subsistence and, in many societies, prestige.

Socially, keeping livestock in agroforestry systems provided employment, especially for unsalaried women and children, and specific income and property rights for them. Many of the highly complicated socio-cultural webs and networks were based on groups within the families or villages specializing in the management and use of animals (e.g., men — camels, transport trade; women — goats, food, subsistence).

Last but not least, livestock were important in using crop by-products and residues which otherwise would have been of little or no value, and in improving the productivity of marginal lands (i.e., poor sites within the system, as well as the croplands during the unproductive dry season) or fallow lands.

Trees, shrubs and palms which originally occurred on part of the land and occasionally regenerated during the fallow periods or survived on unutilized spots, were soon discovered to have multiple uses, many of them as food. Prominent species with edible fruit or seeds, leaves, shoots, flowers and gums are *Acacia senegal, Adansonia digitata, Annona senegalensis, Balanites aegyptiaca, Borassus aethiopum, Butyrospermum parkii, Cordyla pinnata, Detarium senegalense, Dialium guineense, Ficus* spp., *Hyphaene thebaica, Lannea* spp., *Parinari* spp., *Parkia biglobosa, Phoenix dactylifera, Sclerocarya birrea, Sterculia setigera, Tamarindus indica* and *Ziziphus mauritiana.* In addition, other species were used to provide leaves for teas, spices and condiments. Usually these trees, shrubs and palms were carefully protected but rarely actively regenerated and maintained. This was different with "imported" species from other regions of the world, many of which quickly became fully integrated. Species such as *Anacardium occidentale, Artocarpus* spp., *Citrus* spp., *Cocos nucifera, Mangifera indica, Moringa oleifera* and *Persea americana* were planted, bred, selected and grafted, and sometimes fertilized, irrigated and pruned. That is, a more or less pronounced arboriculture was practised even in early times within the agrosilvopastoral systems.

Forage-yielding species were given less priority. This was certainly because almost all wild-growing species proved to be palatable, although to varying degrees, and there were more than 100 species providing leaves, pods and bark as fodder. As browsing, mainly during the dry season, was commonly practised by free-roaming livestock, no specific attempts were developed to maintain or regenerate fodder trees. The management of *Acacia albida* for pods (a storable forage) can be considered an exception.

Trees, shrubs and palms within the system also provided fuel, wood for construction, furniture, fencing (including thorny branches), agricultural and household utensils, transport, etc. The long list of other uses included drugs for human and veterinary medicine, and also tanning materials, dyes, fibres, poisons and repellents.

The environmental benefits were only partially recognized. Shade was the main benefit sought, but trees were also used for shelter, windbreaks and soil improvement. From the socio-economic and cultural aspects, some species were maintained and utilized as cash crops, e.g., gum arabic and other gums and resins, for amenity (flowering and evergreen species), or as focal points for assemblies, cultural events or various forms of worship.

The outstanding tree species of these agrosilvopastoral systems has been, and still is, *Acacia albida,* which grows in most parts of dry Africa (Miehe, 1986). The species is almost perfect as a component of agroforestry systems. The large thorny tree keeps its leaves through the dry season and sheds them at the beginning of the wet season. Thus, under the trees, there is no competition for light during the cropping period while during the hot and dry months there is shelter for livestock, which also improve the soil through their droppings. Moreover, the roots penetrate into deep soil layers. They do not compete for water and nutrients with agricultural plants, but they intercept minerals at otherwise inaccessible depths and deposit them on the soil through the leaves which are shed just in time to fertilize the topsoil at the start of the cultivation season. It is estimated that 20–40 trees per hectare can provide nutrients equivalent to the normally required quantity of fertilizer (Figure 5). This was known to most farmers of the zone, and *Acacia albida* was carefully protected and maintained because the better crop yields underneath were strikingly apparent. The tree also produces large quantities of pods which mature and drop during the dry season. These pods provide a second "dry-season crop" and thus permit one to keep livestock within the system throughout the year. The tree itself has further uses, ranging from firewood and thorny fencing material to tannin, bee forage and a wide range of pharmaceuticals.

Figure 5 *Acacia albida,* planted on crop lands in Niger for optimal agrosilvopastoral land use

Although *Acacia albida* is the prominent agrosilvopastoral tree species in dry Africa, some others deserve to be mentioned as well because they have played an important role in integrated land use in the past and still have promising potential. These species include *Butyrospermum parkii* (the sheabutter or karit tree), *Borassus aethiopum* (the rhun palm), other palm species with a more local, site-specific importance, and *Parkia biglobosa*.

Agrisilvicultural systems

The trees, palms and shrubs just cited were also employed in agrisilvicultural systems, usually located in the wetter parts of the region and on suitable soils. The difference from agrosilvopastoral systems is to be seen in the fact that livestock were not a predominant component and were sometimes completely excluded. Trees and agricultural crop plants were grown together deliberately where intensive sedentary farming and permanent forms of individual land tenure prevailed. However, large numbers of livestock, and especially of cattle and camels, were absent due to health problems (e.g., the tsetse fly). Once more, *Acacia albida* and various fruit trees were preferred. A great variety of practices was applied (Steiner, 1982), especially in the slightly more humid regions (FAO/SIDA, 1981). For the dry zones, sorghum, millet and maize-based cropping systems were most important, but other combinations of crop plants, e.g. rice with trees, shrubs and palms, were also used (Figure 6).

Figure 6   Rice and oil palm in The Gambia

In summary, all traditional forms of agroforestry showed great variety, subject to the sites and shaped by the people involved. They had, however, some basic characteristics in common:

1.   Multi-purpose components of the systems, i.e., tree, animal and crop-plant species all

yielding various products and performing different functions such as recycling of nutrients and energy and helping to achieve self-sufficiency, stability and sustainability of the systems;

2. Site-adaptation, i.e., a tendency to optimize use of environmental conditions and natural resources over space and time, including small plots of land and applying a multistorey structure (seasonal complementarity);

3. Climatic adaptation, i.e., response to seasonality, uncertainty of rainfall, etc., by a high degree of resilience and by using drought-resistent species and practices;

4. Demand orientation, i.e., provision of benefits immediately to the people involved (the so called target-groups), covering most of the apparent requirements in the fields of food, fodder, energy, wood, various raw materials, medicine, environmental protection and improvement, and socio-economic and cultural priorities;

5. Emphasis on the use of locally available natural and human resources based on highly developed and balanced networks of work division and mutual exchange within the families, villages, etc., and even over large regions; and

6. Very little was done to adapt the systems to rapidly changing socio-economic and political structures and to technical progress. They thus remained "dormant" — oriented towards the past rather than the future.

The present

If the traditional forms of silvopastoral, agrosilvopastoral and agrisilvicultural agroforestry in Africa have been, and partly still are, successful, one may ask why and how the present rural problems have arisen. The answer, evidently, lies in the exponential population growth during the last few decades. Obviously, a simple return to previous subsistence economics after the relatively short period of colonial and post-colonial "errors", cannot solve the existing supply problems of urban and rural areas.

Regional development

If, in the past, local concerns formed the focus for most activities in rural dry Africa, a regional approach has become dominant in our times. Problems were shifted from the individual family or community level to specific zones within countries or even beyond national borders. The Sahel is one outstanding example. The expansion of decision levels over larger areas, including a variety of heterogeneous environments and different peoples, appears to be one common trend. This implies a tendency to generalize and standardize, which is, in some ways, in contrast to the characteristics of agroforestry and African tradition.

Political changes have resulted in the establishment of new states which do not necessarily correspond with natural boundaries and ethnic groups. In many instances this has interrupted basic interrelationships, mutual links and feedbacks. Free migration, essential for nomads and semi-nomads as well as other groups of the population, has been hampered, resulting in problems of supply, employment, and exchange of goods and services.

Modern technology has had an enormous impact, changing infrastructures, including transport and communications, reshaping landscapes, e.g., by extensive irrigation and/or hydroelectric schemes, and resulting in the large-scale application of mechanization and chemicals in agriculture.

Most important, medical care for both man and animals has made unprecedented progress. This, in turn, has contributed to the well-known and intensively discussed problems of excessive population growth, growth of animal herds, and the consequent escalating pressure on land resources. For the first time in African history, the rate of population growth exceeded that of agricultural production during the past two decades. In many countries this has led to almost permanent dependence on food imports and/or foreign aid. Large parts of dry Africa are now among the so-called least-developed countries of the world, suffering from a situation described by a continuously falling per capita income.

> On present trends, much of Africa — especially the Sahel, the dry belt in the south of the continent, the Horn of Africa and East Africa — appears to be heading towards human and ecological tragedy, unless far greater priority is given to agricultural development, conservation, and programmes to reduce population fertility than has hitherto been the case (FAO, 1984).

In such an emergency situation, the influence of external partners such as international organizations, bilateral aid, and non-governmental agencies, must necessarily increase and come to exert a strong influence on rural development in Africa.

Loss of potential productivity

As an immediate consequence of rapid population growth, large areas of potentially productive savanna range and woodlands are degraded by burning, inappropriate land use and exploitation beyond their carrying capacity. Desertification (IUCN, 1986; Baumer, 1987; Rapp et al., 1976) stands for a whole range of detrimental and sometimes irreversible processes which include loss of fertile topsoil by erosion, impairment of soil morphological properties by compaction, salinization of irrigated lands, the extinction of plant and animal species and loss of genetic diversity, and imbalances of water and energy regimes.

Pressure on Africa's croplands has escalated in recent years, resulting not only in accelerated deterioration but also in socially explosive competition, e.g., between herdsmen and sedentary farmers and/or urban and rural people in many parts of the continent. Pressure on croplands, in addition, is likely to result in falling yields if, for instance, fallow periods are shortened, cultivation is extended on to land with less fertile soils, marginal climatic or topographic conditions, or when animal dung is burnt for energy for lack of fuelwood. According to FAO (1984) only 20 per cent of all the utilizable lands of Africa were cultivated in 1975. The situation in the arid zones, however, was completely different. In 22 countries classified as severely endangered, 80–100 percent of the land was already occupied. Expansion of crop and rangelands has, therefore, little chance of improving the situation. In the year 2000, 29 countries and a population of more than 250 million people will face catastrophe if nothing is changed basically (FAO, 1984).

This is where the challenge for agroforestry begins. Can agroforestry really increase the area of productive cropland by replacing long fallow periods with sustainable, permanent production systems — i.e., by introducing multistorey production on the same piece of land? This could be decisive. To answer these questions, agroforestry research in dry Africa must be given high priority at all levels. Remarkable progress has been made recently in the form of inventories of promising traditional agroforestry systems compiled by ICRAF (Nair, 1987b) and other organizations and institutes, and in analyses of the systems. Much, however, remains to be done in extrapolating the existing agroforestry systems to ecologically similar conditions elsewhere, giving due consideration to changing situations and long-term sustainability. The AFRENA programme of ICRAF is expected to fill many gaps and to strengthen existing national research facilities.

Moreover, agroforestry should be developed immediately to intensify production per unit area by recycling of nutrients and nitrogen fixation. Improved application of manure and mulch should be promoted. Lack of nutrients, mainly phosphorus and nitrogen but also of trace elements, is a major constraint. FAO figures (FAO, 1983) indicating an average use of 19.5 kg per hectare or 7.4 kg per capita per year of mineral fertilizer in Africa as compared to 78.5 kg and 25.5 kg as a world average and 114.8 kg and 65.5 kg in industrialized countries, respectively, are significant and alarming.

Inevitably increasing demand

Within the past 25 years Africa's population has almost doubled, having now reached nearly 500 million people. Within another generation of only 17–25 years there will be 1,000 million Africans. The overwhelming problem of providing adequate food and energy for this population has to be solved. Social changes, such as growing urbanization, the tendency towards smaller household units (as compared with traditional rural entities), technical development and individual or group aspirations for improvement will also have consequences. Demand, however, does not only refer to food but to many other sectors of the rural economy. It is important to note that agroforestry can contribute a variety of goods and services simultaneously, and that agroforestry is suitable for site-specific diversification. This is in remarkable contrast to the above-mentioned tendency towards regional standardization and simplification and indicates the potential for regional solutions and alternatives.

At a meeting of some of the International Agricultural Research Centres (see Steppler, this volume) and ICRAF on agroforestry research in Africa, held in Nairobi, Kenya on 22–24 September 1986, the following were identified as issues of paramount importance in the context of further agroforestry development in Africa:

1. Lack of an institutional "niche" for agroforestry, both at national R & D and donor levels, thereby very small resources available;
2. Few trained scientists and research institutions with the multidisciplinary capability required for agroforestry R & D;
3. Psychological barriers rooted in the fact that farmers often have been more involved in tree cutting than tree planting;
4. Land-tenure issues;
5. Soil conservation/restoration is not usually perceived by the farmer as a problem requiring direct interventions;
6. Problems of inadequate supply of certified MPT seeds, and a great variation in species, compounded by lack of relevant information;
7. Extension services not normally dealing with trees.

The meeting agreed that all these problems could be alleviated, although it will require time, resources and convincing technologies. Other important problem areas discussed included the lack of knowledge on the proper management procedures for agroforestry technologies which limits the adoption of such technologies by farmers. It was stressed that in all cases farmers' perceptions of the problem were of paramount importance when designing research programmes. This also pointed to the need for more economic studies and proper follow-up of newly introduced agroforestry technologies.

African agriculture is 70–90 percent a subsistence economy, and for agroforestry in the dry regions the share may be even higher. A subsistence economy does not promote investment (due to lack of finance) or market-oriented production, however. Although there appears to be a concentration on the development of small-scale subsistence farming, agroforestry should use all possible means to incorporate cash crops with high yields, low risks and a promising market. In parts of East Africa, for instance, the income from tobacco exports would compensate for about 10 times the quantity of maize that could have been harvested from the same area. Similar calculations are available for coffee, tea and other cash crops. Keeping in mind the development aspirations of the farmers, the more recent condemnation of such crops should be thought over and new, "typically African", crops promoted.

Nevertheless, agroforestry, in its various forms of traditional systems and practices, is still widespread throughout dry Africa and thus forms a foundation to build upon. Its contribution to a strong subsistence economy is obviously great, although it is generally underestimated due to lack of statistical records.

Food
At least in dry Africa, food production is at present the main target of agroforestry and this is likely to continue. Local food production has become the main problem for rural development and urban supplies. The situation escalated to catastrophic dimensions during recent drought periods and/or political stress situations and has affected large parts of the continent. However, the increasing frequency of these catastrophes is just an indication of the ever-increasing instability and weakness of the rural human-ecological systems. Late, but hopefully not too late, it is now recognized that a certain degree of sustainable self-sufficiency in food supplies is essential for the overall development of the African countries. Agroforestry is expected to help in solving almost overwhelming problems. These expectations exist at all levels, but require support and co-operation of all parties involved from international and governmental agencies to the "grassroots", i.e., the individual land user.

As mentioned before, the contribution of agroforestry to food production is generally not statistically recorded for a variety of reasons. For example, the significance of the phrase "food from trees, shrubs and palms" is not fully understood. These sources are important in quantity as well as in quality, in being in seasonal complementarity with agricultural crops and animal products, and as emergency supplies in times of drought, as well as part of the regular diet. The latter becomes evident from what is offered at local markets, although this is only a small fraction of what is consumed overall. Moreover, most of these foodstuffs are obtained from "no-man's land", i.e. public lands, free of charge and so far require but a small labour input, mainly by women and children. This social aspect of "tree food" must not be overlooked.

Potential food production from trees and shrubs in northern Senegal (Ferlo) and northern Kenya (Turkana, Samburu) has been analysed by Becker (1984). Her results can be summarized as follows: the annual harvestable production of leaves and fruits amounts to about 150 kg per ha in the Saharo-Sahel, 300 kg per ha in the typical Sahel and 600 kg per ha in the Sudano-Sahel. This corresponds with the rule of thumb, derived from various observations in the West African Sahel, that in "normal" ecosystems the annual increment of non-woody biomass from trees, shrubs and palms in kg per hectare roughly equals the average annual rainfall in mm. Results from East and West Africa indicate that about 15 percent of that biomass can be classified as edible. Thus, in the above three ecological zones

23, 45 and 90 kg, respectively, of edible material would be available per hectare annually. Correlating these figures with an average population density of 1 per sq km, and assuming a ratio of 4:1 for leaves and fruits, 450–1,800 kg of edible tree/shrub fruit could be utilized per caput, which corresponds to 1.25–5.0 kg fruit per adult human being daily.

In the Ferlo/Senegal, an inventory of baobab (*Adansonia digitata*) was made in a typical area in a German-Senegalese rural development project. The results showed 6,611 baobab trees and 1,200 people lived on 20 sq km, i.e., 5.5 trees per caput. The leaves of these trees are intensively used as a green vegetable and dried as a powder and have a remarkable nutrient content: 100 g of fresh leaves contain 23 g dry matter, 3.8 g crude protein, 700 mg calcium and 50 mg ascorbic acid. They provide 69 calories. Even more valuable is the fruit pulp which is rich in vitamin B_1 and C, and the flour produced from the seeds contains up to 48 percent protein and 2 percent vitamin B_1 on a dry-weight basis.

This is only one example demonstrating the potential food production of woody perennials growing on agricultural and/or pastoral lands in the Sahel. From the many species available, merely a fraction are used for food, and, in normal times, people tend to be very selective about species, varieties, and parts consumed. Plants used by one group of people in one region may not be accepted by other groups in the same region or the same ethnic group in a different region. Moreover, there are pronounced differences in the quality of food, depending on varieties or provenances. *Adansonia digitata* may again be cited as an example. There are trees with soft and tasty leaves and others with fibrous and bitter leaves. As no cultivation has been practised with these local species, selection and breeding have not yet started but may offer a hitherto unknown potential for improvement. This should be tried with at least 30 food-tree species in the various parts of dry Africa.

Food production in agroforestry systems, of course, implies much more than producing and utilizing edible parts of woody perennials. Optimizing agricultural crop and animal production is the main concern and can be achieved as indicated for traditional systems such as those involving *Acacia albida*. It is interesting to note that, according to different observations in the Sahel and in other regions, and even outside Africa, a combination of trees/shrubs and annual plants (grass, cereals, herbs) is known to result not only in a considerable increase (about 300 percent) of total annual biomass production but specifically in higher yields of the annual plants, both in quantity and in quality. This has been recorded also from millet/sorghum fields with *Acacia albida* and from pasture lands with about 10 percent of the area covered by a light canopy of savanna trees (the author's unpublished observation). This indicates that *Acacia albida* agroforestry is a promising form of land use, but more research is needed to further develop the potential based on traditional systems and to clarify the underlying feedback cycles.

Forage:
Animal husbandry, as stated above, is the main land-use practice in dry Africa and is indispensable for food supply and the functioning of many agroforestry systems. Obviously, successful animal husbandry depends on the availability of sufficient forage. In traditional silvopastoral and agrosilvopastoral systems, a balance existed between stocking rates, natural vegetation and drinking-water supply. Where the carrying capacity was exceeded, people and herds could shift to more favourable sites. Recently this has become difficult due to increased numbers of people and animals and due to the extension of cultivation to lands formerly reserved for pasture. Moreover, desertification and degradation of vast areas are rapidly reducing the productivity of pasture lands.

The drought periods of the 1970s and between 1982 and 1985 have clearly indicated that more than relief programmes are needed to ensure the survival of millions of people and their livestock. These relief programmes have largely failed because of a number of misleading assumptions.

One of these assumptions is that the productivity of Sahelian pastures is primarily limited by rainfall. Although rainfall undoubtedly has a strong influence on annual phytomass increment, the deficiency of soil nutrients quite often plays an even greater role, especially the availability of phosphorus in the first part of the growing season and of nitrogen in the second part (Breman and de Wit 1983; Penning de Vries and Djitye, 1982). The northern (arid) parts of the Sahel are better endowed with nutrients and thus produce forage with a higher content of digestible protein, vitamins, trace elements, etc. However, the availability of water relative to phosphorus and nitrogen determines quality and quantity of the forage produced. Thus low water availability results in small amounts of nutritious phytomass, while higher water availability gives more phytomass but of increasingly inferior quality.

The north of the Sahel is, therefore, more suitable for animal husbandry than the south with regard to forage quality, but the carrying capacity is low and grazing is restricted to between two and four months a year. In addition, drinking water for the animals is usually not available beyond the rainy season. Livestock performance in the south Sahel and in the more humid savanna lands is poor, however, even if in the rainy season and thereafter relatively large quantities of forage are available, because the quality of that forage is low (Breman and de Wit, 1983).

This can be partially compensated for by the availability of higher-quality forage from trees and shrubs, either in the form of leaves, bark or fruit. Thus, browse in animal husbandry is less important in the northern than in the southern, generally more humid parts of Africa (Le Houérou, 1980; IBPGR 1984). The availability of browse plays an important part in silvopastoral and agrosilvopastoral systems and, thus in the agroforestry that is practised in most parts of the African dry lands (see Le Houérou, this volume).

A second misleading assumption is that livestock herds are generally too large because the nomadic and transhumant herdsmen maintain them for prestige. It is, however, accepted that the mean herd size for minimum subsistence must be 4–5 tropical livestock units (of 250 kg) per member of the family (Breman and de Wit, 1983). On average, not only is this figure not reached, but it has dropped to about 50 per cent during the recent drought periods. To compensate for the risk of losing half of their herds, the nomads and semi-nomads aim at keeping higher numbers of animals than required for subsistence in normal years.

As livestock also form one of the few possibilities for investment and thus serve as a "mobile savings bank", there is a tendency to keep large herds. The tendency is limited, however, by the availability of animals, fodder, and, partially, labour, as well as low animal fertility. Moreover, a clear distinction has to be made as to the animal species concerned. Camels and cattle generally mean more than economic assets to the people, whereas goats and sheep are used for subsistence or cash income.

A reason for the large numbers of animals lies also in their limited productivity. The average daily amount of milk that is available for human consumption ranges between 0.5 and 0.75 l per person. The mean annual increase in liveweight may reach 20 percent (or 50 kg) per TLU and 30–40 percent for sheep and goats, respectively. This corresponds to a production of 0.3–0.5 kg of animal protein per hectare annually for the nomadic and for the sedentary livestock systems in the Sahel and 0.6–3.2 kg for the transhumant systems, which is even better than under comparable climatic conditions in Australia and North America

(Breman and de Wit, 1983). Nevertheless, in parts of dry Africa the number of animals kept already exceeds the carrying capacities of rangeland ecosystems. This results in environmental degradation, low productivity of livestock, and critical supply situations (Figure 7).

Much of the present degradation/desertification of former productive rangelands in dry Africa is primarily due to lack of good management rather than overstocking. Moreover, reduction of stock numbers is usually unpopular and difficult, at least as long as no adequate compensation can be offered. Compensation in one sense could be the introduction or further development of complementary cash-crop production in transhumant agrosilvopastoral systems.

Thus, once more, agroforestry, mainly with its forms of agrosilvopastoral land use, is expected to improve the situation. However, there are constraints outside the range of technical development. First of all, there is an urgent need for an overall land-use policy in the countries concerned in order to regulate the allocation of lands to pastoral uses and to crop cultivation, respectively. Secondly, tenural rights may need revision, as would market policies for animal and agricultural crop products. Technically, the one-sided support of the veterinary services will have to be changed. Breeding and selection, diversification of herds and better marketing facilities can help to improve yields and reduce risks. Two bottlenecks will have to be overcome: the lack of soil nutrients (phosphorus and nitrogen) and of water to increase forage production. Digging wells other than those that are part of an integrated rural development programme has led to failures, while sustainable water management within agrosilvopastoral systems appears to be promising. In these systems the application of fertilizers may also prove to be economically feasible (which is not the case in pure pastoral management) if combined with optimal use of manure and leguminous woody perennials for soil improvement. Agrosilvopastoral systems could also help to avoid bushfires which are detrimental to soils and plant production. Sustainable management of

Figure 7 Large herds concentrating around wells in the arid zones during the dry season

Figure 8   Combined management of *Acacia nilotica* and *A. senegal* with fodder grass

improved pasture should be the target, a target that has already been achieved in some model projects such as the Ferlo of northern Senegal, and in other parts of the Sahel, North Africa, and East and southern Africa. Obviously, such sustainability depends on optimum grass and herb mixtures with woody perennials. This is one of the challenges to agroforestry in dry Africa. The introduction of more productive fodder shrubs and trees can supplement the presently available edible biomass and thus improve animal nutrition and animal productivity (Figure 8).

Energy

The energy crisis in arid and semi-arid Africa has been intensively discussed in many meetings and publications. Most households and many local small industries and handicraft concerns depend entirely or mainly on fuelwood or charcoal supplies. The crisis, however, is less evident in rural areas, where it is not yet fully recognized, while it has reached a critical stage in urban areas.

Most of the fuelwood is free and is collected by women and children. The average minimum demand for cooking appears to be about 1 kg of air-dry fuelwood per person per day, i.e., 360 kg, or about 0.7 m^3 of wood annually. In contrast to traditional forms of supply by free fuelwood collection, the increased population has led to the clearing of natural woody vegetation for crop and rangelands, and the "carrying capacity" of natural fuelwood resources has reached, or even surpassed, critical limits, resulting in severe fuelwood deficiencies on the one hand and in environmental degradation on the other. This is a recent development which has to be urgently controlled in order to avoid catastrophe. The great majority of fuelwood plantations in dry Africa (generally sponsored by foreign aid), have proved to be uneconomic in financial terms, the overall cost of production (including high overheads and marketing) far exceeding prevailing market prices, although ecologically they temporarily relieve the immediate pressure on near-by resources. Therefore, producing firewood in agroforestry systems by the landowners themselves is one way to solve the problem (Nair, 1987a). This can be done by managing the natural woody vegetation. The biomass increment of these trees and shrubs is, however, very low. One of the reasons is that, by tradition or by law, only dead wood is harvested. If the rotation period is shortened according to the maximum increment, much more could be harvested on a sustainable basis and presumably in smaller sizes which would yield more energy than (unsplit) thick pieces. Besides, the thick pieces are usually less appreciated because of the type of heat they offer and sometimes higher moisture content.

However, even forms of intensive firewood cropping with fast-growing species with a high coppicing potential may well prove to be financially attractive if combined with agriculture, e.g., in forms of alley cropping, or with range management. At present there is very little such activity because free exploitation of natural resources on the one hand and government/external project fuelwood plantations on the other are still believed to be adequate — which they are not. If pure fuelwood plantations by relief projects cannot yield a net revenue, this may be entirely different in agroforestry systems, where fuelwood is only one among a number of crops. This, in combination with energy-saving stoves and techniques, and with the development of alternative sources of energy, should be further promoted. In addition, biogas in agroforestry systems has not yet gained the necessary acceptance in Africa. Experience from Asian countries is available and could be transferred to African farmers.

Renewable raw materials

Most parts of dry Africa are deficient in exploitable mineral resources, and due to scarcity of foreign exchange, imports are restricted. The significance of renewable resources that are locally available or can be grown is evident. Agroforestry, providing a wide range of diverse products, deserves special attention.

Fuelwood has just been mentioned. Wood for construction, fencing, tools and implements, and many other products including handicrafts can also be produced. Again, people are very selective in the choice of species, and, as the most desirable ones have largely disappeared from the natural vegetation, the conservation of remaining single specimens on agroforestry lands and, in future, plantations of specific trees within agroforestry systems, can help to overcome foreseeable shortages. There are, however, a number of constraints which do not make timber production an attractive proposition within agroforestry. One is the problem of tenural rights, another the legal restrictions in the use of standing trees according to existing forest laws. Furthermore, only a limited number of exotic species are available that promise fast growth and high yields. However, these may not be compatible with other land uses and do not provide for a multitude of desired products and benefits. Therefore more should be done to incorporate multipurpose local species and to offer the landowner certified seeds and plant material and advise on appropriate cultivation techniques.

Vegetative propagation (by cuttings) of *Euphorbia balsamifera* and of *Commiphora africana,* as well as planting of hedges and live fences using various thorny species, could be further developed. One promising example is *Bauhinia rufescens* which can be used for thick-set hedges in nurseries, around compounds, etc., as tried in Burkina Faso (Figure 9).

Renewable raw materials go beyond wood. They include all the agricultural commodities in addition to food and forage, such as fibres, oils, and insecticidal and

Figure 9   Hedge of *Bauhinia rufescens* in a nursery three years after establishment

repellent extractives. The various gums and resins collected from trees and shrubs should be mentioned — gum arabic from Kordofan, Sudan and parts of the western Sahelian countries, various tanning and dyeing materials and fibres. A variety of pharmacological products that are essential for both human and veterinary medicine are obtained from wild-growing local trees and shrubs. The abundance of these plants has been drastically reduced by monoculture cropping but they are maintained as a free and always available "pharmacy" in most agroforestry systems. Again no statistical data are available, but without doubt agroforestry contributes substantially to health in dry Africa.

Environment

Under harsh or marginal conditions there is a general tendency, both ecologically and from the managerial aspect, to reduce heterogeneity of plant and animal species. Having fewer species, however, increases the instability of the system. This is evident from most monoculture crops in dry-zone Africa which can be successfully grown only with the support of substantial external inputs in the form of mineral fertilizers, agrochemicals, irrigation, etc. Agroforestry, by contrast, aims at a natural stability by diversification, multiple use and flexibility. Recently this has turned out to be more important for most rural people than higher yields alone. Combining different species, moreover, reduces the pressure on such limited resources as water and soil nutrients. The function of trees and shrubs as "nutrient pumps", transporting nutrients from deeper soil layers to the soil surface, is increasingly recognized in addition to their role in nitrogen fixation, erosion control, wind protection, and the provision of shade. The role of non-leguminous species in this context is still underestimated and deserves more attention. This includes mulch production and insect repellent effects in some species.

Not all trees, shrubs and palms should be incorporated into agroforestry systems uncritically. Species with an extensive lateral root system may be useful for controlling erosion and fixing moving sands, but they will most likely compete with crop plants. Species with a deep-reaching taproot may be excellent "nutrient pumps" but in turn may not stabilize the topsoil. Some species may attract birds, rodents, insects or other animals which can seriously damage agricultural crops or animals (e.g., the tsetse fly depends on woody habitats). Such aspects must be carefully considered. They are generally known to the local rural population, but questions concerning the compatibility of woody species with crops and livestock have to be answered soon in order to avoid failures and disappointment which may lead to a general rejection of agroforestry concepts by rural communities.

Socio-economic development

This phrase has been somewhat over-used and overloaded with significance. However, many essential aspects of the concept have been overlooked. Quite simply, shade in agroforestry systems results in socio-economic benefits. It is hard to work in the open fields in hot weather, and shade available in or around fields improves working conditions for the farmer. The main argument for planting trees in parts of western dry Africa is to provide shade in and around compounds, villages and fields. A very hot climate also reduces the productivity of animals.

More important, however, and as yet largely unsolved, is the problem of how to create more and better jobs in rural areas. At present, many of the most active and skilled people leave their rural habitats and migrate to urban centres, expecting better chances for earning money, better educational opportunities, social security, health care, etc. Old people, women and children stay behind and are faced with the problem not only of surviving but of producing a surplus to feed the city dwellers.

Agroforestry is labour-intensive. Does it simply increase the work load for those already burdened, or does it really provide more adequately paid jobs? Oases are one example. Their management is extremely labour-intensive. Many, if not most, oases at present suffer from non-availability of labour, and many of them have been abandoned or are in the process of collapse because the traditional social systems have changed. There are now more attractive job alternatives outside. The problems of manpower, jobs and income have to be solved first and at all levels if agroforestry in the dry zones of Africa is to have a chance in the future.

A solution would be better markets, i.e., guaranteed marketing possibilities and attractive prices, and physical and legal market access for agroforestry products. Why should a farmer grow fruit, cereals, vegetables, fuelwood, etc., and produce meat and milk if, when trying to sell this produce on the market, he can find no buyers. This happens frequently due to government policies which favour urban populations for political reasons (e.g., by extremely low food prices) and also because of food-aid programmes. Commonly preference is given to imported food (e.g. wheat instead of locally grown millet and sorghum).

These are the most serious socio-economic problems presently retarding the further development of agroforestry in the dry zones of Africa. In addition, the conflicts between traditional and government-land tenurial systems will have to be solved. People who do not own the land they cultivate, or cannot expect an extension of the lease over the period necessary for the rotation of woody perennials, can hardly be expected to invest in agroforestry. Tree planting is even actively prevented by some societies in order to avoid "sedentarization" and land acquisition by private people or by government agencies — because, according to tribal customs, property rights over cropland may be obtained by planting of trees.

The future

At present, in spite of many encouraging projects and remarkable research efforts, agroforestry in the dry zones of Africa is stagnant or on the verge of a standstill. Wherever progress is seen, the impulse primarily comes from outside, i.e., from international organizations, bilateral co-operative agreements or national governments. Very little initiative comes from inside the rural societies, i.e., from the target groups concerned. With their participation the maintenance of age-old practices may continue, but effective innovations are hardly forthcoming.

Why is this so? Is there a general resignation within rural societies in Africa or is there no future to be expected for agroforestry? Are there specific constraints to be overcome?

Resignation may play a role in some regions. Resignation can be traced back to a variety of reasons. One of them is "familism", i.e., the individual has to subordinate his concerns to the interests of his family/clan. Although providing a certain social security, this hampers innovation, because most decisions are made by elders. Closely linked is "parochialism" or "ethnocentrism", which means orientation towards the interior of a given society and a certain isolation from the outside. Fatalism and superstition are widespread and often in contrast to development efforts, and "status factors", e.g., with regard to specific types of labour (labour division between men and women) as well as socio-cultural restrictions or commitments have to be observed.

But people in the dry zones of Africa cannot afford resignation or inactivity in the future. Since the population will double within the next 17–25 years, no time can be lost in creating the conditions for their survival. Remaining at the present levels of extremely low agricultural production on the one hand and increasing environmental degradation on the other will inevitably lead to incredible hardship, if not the starvation of millions of people, within the next generation. An East African proverb clearly defines the challenge: "We have not inherited our environment from our fathers, we have loaned it from our children". Since there is no more free land left to migrate to if the region's own resources are exhausted, people will have to change from the previously highly developed strategies of mobile resilience, however well adapted to the environment, towards a more permanent strategy. Agroforestry in its various forms lends itself to sustainable land use with increased permanence, although it maintains a high degree of flexibility once a site has been chosen.

The fact that more people will have to be fed in the future will increase the demand for agroforestry products of all kinds, especially for food, forage, fuelwood and materials.

Population growth in other regions may even open additional markets for African rural products, e.g., in southern Asia. To meet these demands, however, a mere exploitation of the natural resources of African dry zones is not sufficient. As in all other sectors of the economy, investments will be needed for agroforestry.

Here, however, real bottlenecks exist. First, by tradition and lack of facilities, the majority of people in the dry zones of Africa were not accustomed to invest, except in children and livestock, and there are few resources available to be invested except labour. Moreover, confidence in investment has not increased in recent times.

Secondly, and perhaps more seriously, there is escalating competition for power, power based on the disposal of basic investment parameters. Previously, power was mainly attached to control over lands, i.e., over natural resources which are basic for investment. Private land ownership is hardly to be found in the dry zones of Africa. Traditionally the local societies allocated lands to individuals for a limited period and under specific conditions. At present, governments claim that right for themselves, partially overlapping with traditional patterns. Thus the individual land user has very little influence on long-term decisions and, therefore, little interest in long-term investment. Those disposing of the land, however, often do not have the resources and institutional structures for overall rural development. "When land is controlled by government or by a powerful few, the majority has little motivation or possibility to assure continued soil fertility or to maintain perennials" (FAO, editorial of *Unasylva,* No. 4, 1986).

Capital is the second asset for investment. People or groups disposing of money were, and still are, believed to rule others. Rural people in dry Africa certainly do not have the financial resources they would need for investment, and government or even foreign funds are more than limited. The alarming dependence on money supplies, either in the form of grants or credits, will have to be solved in a solid and effective way. Labour, number three of the investment "goods", is generally assumed to be abundant, but this is not true, as indicated above. It is not true with regard to seasonal distribution (only short cultivation periods), and labour is not available at sufficiently low cost. Mechanization as a possible solution would also require money and would remove the decision levels to donors outside the rural systems. Labour, however, includes all human skills, and these depend on both experience and education.

As stated, a wealth of experience, especially in resource management of agrosilvopastoral and silvopastoral systems, exists in the dry zones of Africa, specifically among farmers and herdsmen. This is the reason why it is so important to "rediscover" the existing knowledge and to make it available to all Africans; it is theirs. Education must be restructured and should be based on available knowledge, facts, local experience and applied research. Much progress has already been achieved by international and national institutes engaged in improving land use. ICRAF is contributing substantially to agroforestry development in Africa by its strategic research programme, by its world-wide information services and courses held. But the search for knowledge must also start at the grassroots level, in primary schools, in simple experiments and observations. The question behind all these efforts should be "What can be improved?" Research must be future-oriented. Our understanding of the feedback between ecological, socio-economic and socio-cultural processes is still so scant that up-grading of our knowledge must be a permanent challenge.

This is important for agroforestry. Agroforestry must not be an attempt to return to stone-age practices or to keep people at low levels of development and dependent on external inputs. On the contrary, it has to apply the most efficient technologies available in order to serve people and to maintain the potential of natural resources on a sustainable basis.

Agroforestry will have to be opened to dynamic technical progress, and is to be made a land-use concept suitable for spearheading development by its holistic systems approach. In future, for instance, biotechnology (particularly gene technology) can be expected to play a major role. This may range as far as sustainable and large-scale soil improvement by bacteria and breeding of high-yielding multipurpose plants and animals. Efforts should focus on reducing waste and on quality improvement since there are definite limits to quantitative growth. In future it will be unacceptable to produce more waste than target products unless waste can be adequately recycled in agroforestry enterprises.

Biotechnology will have to be applied to replace former practices of breeding plants and animals, specifically to shorten the cycles of tree production and reproduction. Modern techniques of cellular and subcellular engineering (such as gene splicing and recombinant DNA, cloning, hybridomas and monoclonal antibodies, protein engineering and nitrogen fixation) will open up hitherto unexploited resources. Simultaneously, however, their potential for misuse will have to be carefully watched.

> Tissue culture is one of the most important components of biotechnology. The research in agriculture and forestry stands at the threshold where a large number of plants produced through tissue culture are being sold commercially, and it is already a multi-million dollar industry. Some of the most important aspects of the use of tissue culture are micropropagation for biomass energy production, production of disease-free and disease-resistant plants, induction and selection of mutants resistant to pests, pathogens, adverse soil conditions, drought, temperature, herbicides, etc., production of haploids through anther culture, wide hybridization through embryo rescue, somatic hybrids and cybrids through the fusion of protoplasts, transformation through uptake of foreign genome, nitrogen fixation, cryopreservation of germplasm, etc. Moreover, for the improvement of forest and horticultural trees, biotechnology enables faster multiplication of desease-free elite and rare stocks by micropropagation (Bajaj, 1986).

Many people are searching for a paradise which in reality never existed on earth. Remarkably, however, paradise was not described as open fields and grazing lands or closed forest. Paradise was a highly diversified garden, in some way a perfectly functioning "agroforestry" system, in which valuable trees played an essential role. Misuse of fruit trees finally led to the expulsion of man from that paradise.

It is worthwhile to reflect from time to time upon the meaning of old symbolic wisdom. It is not only worthwhile but absolutely necessary to know more about the wisdom of rural people in the arid zones of Africa in order to further develop agroforestry to the year 2000 and beyond. But likewise it is important to elaborate new strategies by combining knowledge from inside the target groups involved and from external research organizations. This is the type of investment which is likely to yield the highest rates of interest in material and intellectual terms.

For the future of agroforestry, however, more than research, technical development and funds will have to be mobilized. An overall national land-use strategy is required for all countries, based on realistic facts and targets, interdisciplinary and co-ordinated efforts towards a holistic systems approach, the harmonization of human needs and the environmental carrying capacity. The African CILSS-countries have taken some very important steps into this direction. Moreover, these strategies will have to be effectively applied in practice. The responsibility for the future, therefore, lies not only with the farmers, scientists and technicians but first of all with the policy makers. This has to be stated quite clearly.

Above all, agroforestry is not a panacea of land-use methods. It may be useful or even optimal under specific conditions but indifferent or even unsuitable elsewhere. Agroforestry may not be acceptable and practicable for all people, and some groups may be more active and successful in developing agroforestry than others. The challenge is thus to concentrate on the most promising sites and systems and to develop these like islands within other forms of land use. The islands, hovever, should be mutually linked to form a network in order to increase the region's carrying capacity — a web requiring minimum material inputs and providing, above its traditional elasticity, sustainability and the development potential that is needed for the livelihood of future generations.

REFERENCES

Bajaj, Y.P.S. (ed.). 1986. *Biotechnology in Agriculture and Forestry — I. Trees.* Berlin: Springer Verlag.

Baumer, M. 1987. *The potential role of agroforestry in combating desertification and degradation of the environment.* Wageningen Technical Centre for Agricultural and Rural Co-operation (CTA) (ACP-EEC Lomé Convention).

Becker, B. 1984. Wildpflanzen in der Ernaehrung der Bevolkerung afrikanischer Trockengebiete: Drei Fallstudien aus Kenia und Senegal. Inst.f.Pflanzenbau u.Tierhyg.i.d.Trop. u.Subtrop, Goettingen: *Goettinger Beitrge zur Land- und Forstwirtschaft in den Tropen und Subtropen,* Heft 6.

Breman, H. and C.T. de Wit. 1983. Rangeland productivity and exploitation in the Sahel. *Science* 221: 1341–1347.

FAO. 1983. *Fertilizer yearbook,* Vol. 32. Rome: FAO.

————. 1984. *Land, food and people.* Rome: FAO.

FAO/SIDA. 1981. Agroforesterie Africaine. Une étude préparée par la Faculté des Sciences Agronomiques de l'Etat Section forestierè des pays chauds. Gembloux, Belgique. Rome: FAO.

Galaty, G.J., D. Aronson and P.C. Zalzman. 1981. *The future of pastoral peoples.* Proceedings of a conference held in Nairobi, Kenya, 4–8 August 1980. Ottawa: IDRC–175e.

Hall, A.E., G.H. Cannell and H.W. Lawton. (eds.). 1979. *Agriculture in semi-arid environments.* Berlin: Springer Verlag.

Le Houérou, H.N. 1980. *Browse in Africa: The current state of knowledge.* Addis Ababa: International Livestock Centre for Africa (ILCA).

IBPGR. 1984. Forage and browse plants for arid and semi-arid Africa. Kew, U.K.: Royal Botanical Gardens, and International Board for Plant Genetic Resources (IBPGR).

IUCN. 1986. *The IUCN Sahel report: A long-term strategy for environmental rehabilitation.* Gland, Switzerland: International Union for Conservation of Nature and Natural Resources (IUCN).

Jahnke, H.E. 1982. *Livestock production systems and livestock development in tropical Africa.* Kiel, FRG: Kieler Wissenschaftsverlag.

Louw, G. and Seely, M. 1982. *Ecology of desert organisms.* London/New York: Longman.

Lusigi, W.J. and G. Glaser. 1984. Désertification et nomadisme: une étude pilote en Afrique orientale. *Nature et Resources* 20 (1): 21–31.

Miehe, S. 1986. *Acacia albida* and other multipurpose trees on the Fur farmlands in the Jebel Marra highlands, Western Darfur, Sudan. *Agroforestry Systems* 4: 89–119.

Nair, P.K.R. 1985. Classification of agroforestry systems. *Agroforestry Systems* 3: 97–128.
————. 1987a. Agroforestry and fuelwood production. In D.O. Hall and R.P. Overend (eds.), *Biomass: Renewable energy*. Chichester: John Wiley.
————. 1987b. Agroforestry systems inventory. *Agroforestry Systems* 5: 301–318.
Penning de Vries, F.W.T. and M.A. Djitéye (eds.). 1982. *La productivité des pâturages sahéliens: Une étude des sols, des végétations et de l'exploitation de cette ressource naturelle*. Agricultural Research Reports 918. Wageningen: Centre for Agricultural Publishing and Documentation.
Rapp, A., H.N. Le Houérou and R. Lundholm (eds.). 1976. Can desert encroachment be stopped? A study with emphasis on Africa. *Ecological Bulletin* No. 24. Stockholm: NFR/UNEP/SIES.
Ruthenberg, H. 1980. *Farming systems in the tropics*. 3rd ed. London: Oxford University Press.
Simmonds, N.W. 1985. *Farming systems research: A review*. World Bank Technical Paper 43, Washington: World Bank.
Steiner, K.G. 1982. *Intercropping in tropical smallholder agriculture with special reference to West Africa* Schriftenreihe der GTZ. No. 137. Eschborn: GTZ.
United Nations. 1977. Round-up, Plan of Action and Resolutions, United Nations Conference on Desertification, 29 August–9 September 1977. New York: United Nations.

8

Agroforestry in the Indian subcontinent: past, present and future

G.B. Singh

Assistant Director-General
Indian Council of Agricultural Research
Krishi Bhavan, New Delhi 110001, India

Contents

Introduction

India, Pakistan, Bangladesh, Nepal, Bhutan and Sri Lanka constitute the "Indian subcontinent". The subcontinent has a total population of about 1,000 million people and a total land area of 4.13 million sq km. Only 20 percent of the total land area is under forest, ranging from 3.8 percent in Pakistan to about 70 percent in Bhutan. It is estimated that the actual area under vegetal cover may be only half of that reported. Besides the need for increasing food production to feed the increasing population of the subcontinent, the urgency of meeting fuel, fodder and timber requirements and preserving the ecological and environmental balance cannot be understated.

Agroforestry has a long tradition in the Indian subcontinent. The socio-religious fabric of the people of the subcontinent is interwoven to a very great extent with raising, caring for and respecting trees. Trees are integrated extensively in the crop- and livestock-production systems of the region according to the agroclimatic and other local conditions.

While the multi-tier tree-crop combinations in the homegardens of the humid lowlands meet cash and household requirements, the *kherji* (*Prosopis cineraria*) and agricultural-crop combination in the hot arid region meets needs for fodder, small timber and food. The *Alnus nepalensis* and *Amomum subulatum* combination in the humid sub-temperate regions of Nepal and Bhutan and Sikkim State of India is an excellent example of a commercial but traditional agroforestry system. Deliberate growing of trees on field bunds, their sporadic distribution in agricultural fields, and the systematic retention of shade trees in tea and coffee plantations are other common examples of prevalent agroforestry practices. Similarly, it is a common practice to utilize the open interspaces in the newly planted orchards and forests for cultivating crops for 2–3 years and to interplant shade tolerant crops such as turmeric and ginger later.

In India alone it is estimated that the deficit of fuelwood by the year 1990 will be of the order of 100 million m³ per year, and presently only a third of the fodder that is required to feed the country's livestock population is available. About 44 percent of the geographical area of India is subjected to serious erosion hazards, causing an estimated loss of 5,334 million tonnes of soil annually, 1,572 million tonnes of which find their way into the sea. In order to realize the goal of producing 225 million tonnes of food by the year AD 2000 in India, an estimated 20 million tonnes of plant nutrients would be required, a substantial part of which will have to be contributed by sources other than chemical fertilizers.

To cope with some of these challenges and tackle these problems of food and environmental security, the potentials of agroforestry need to be fully exploited. Presently some initiatives are being undertaken by the governments, farmers, non-government organizations and industries in the subcontinent to develop appropriate agroforestry systems and popularize them. Matchbox and paper industries have introduced cultivation of poplar trees in the areas about 30°N. Farmers have taken to planting *Eucalyptus, Casuarina* and *Acacia* trees either on field bunds or as block plantations. Cultivation of fruit trees has become popular in marginal lands in the arid regions. However, these efforts are only the beginning. Further sustained effort is required to develop agroforestry systems suitable for each agro-ecological region of the Indian subcontinent.

Agro-ecological characteristics of the Indian subcontinent

The monsoon, the seasonal reversal of winds and associated rains, develops most prominently over the Indian subcontinent and adjacent seas. The average annual rainfall of about 100 cm over the Indo-Gangetic plains is high for a region lying mostly in the subtropical high-pressure belt. Contrasts in climate are very striking: Cherrapunji, with the heaviest rainfall in the world, and the Thar Desert, with as little as 8 cm of rain per year, are almost on the same latitude (26°N). Almost every type of climate, from equatorial to alpine, is found in one part or the other of the subcontinent. The alpine climate occurs in the Himalayas within 400 km of the Tropic of Cancer. The northern plains experience higher summer temperatures than other land areas on the same latitude, and while over a large area relative humidity is over 80 percent during the south-west monsoon, the air is very dry at the height of summer in the north-west.

India has been classified into eight broad agro-ecological regions. Since the neighbouring countries have almost the same agroclimate as in the adjoining parts of India, an extrapolation of these agro-ecological zones to the countries of the Indian subcontinent gives a practical understanding of the entire subcontinent. These agro-ecological regions are shown in Figure 1, and their major features are described below.

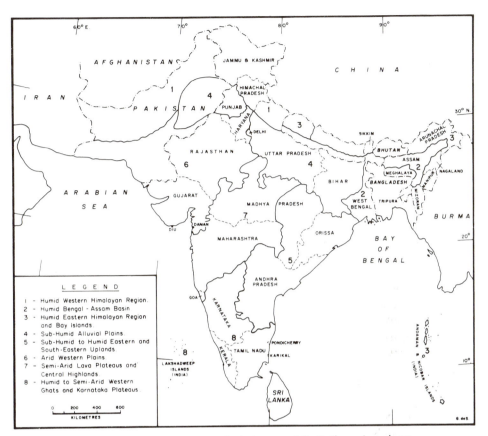

Figure 1   The major agro-ecological zones of the Indian subcontinent

1.   Humid western Himalayan region

The climate in this region varies from hot and subhumid tropical in the southern low tracts to temperate cold alpine and cold arid in the northern high mountains. Annual precipitation ranges from 8 to 350 cm. Three major groups of soils, namely mountain meadow soils, submontane meadow soils and brown hill soils (Cryosorolls, Crychepts, Hapludalfs, Palehumults) occur. Only about 5–17 per cent area is under cultivation. Orchards provide temperate and subtropical fruits. The region has great potential for further development of horticulture and orchards. Indiscriminate felling, overgrazing, soil erosion and soil acidity are the main liabilities. Rich forest wealth and alpine grasslands most suitable for sheep rearing and fruit gardening are the assets. The vast glaciers, after thawing, provide ample water for irrigation in the plains. Crop and animal husbandry can complement the regional economy and allied industries can provide adequate employment opportunities.

2.   Humid Bengal-Assam basin

This region represents the Ganga-Brahmaputra alluvial plain of continental alluvium and deltaic deposits. The climate is hot and humid with average annual precipitation varying from 220 to 400 cm. Floods are a common feature from mid-July to October. The predominant soil groups are alluvial, red, brown hill and coastal (Udifluvents, Haplaquents, Hapludalfs, Paledalfs, Palehumults, Udipsamments) with extensive patches of saline and alkali soils in the deltaic tracts. Tropical deciduous forests in the west and humid tropical forests in the south constitute the diversified vegetation. In the Assam valley and the submontane tracts, bamboo and cane forests are common. *Sal (Shorea robusta)* forest abounds on high lands. In Bengal, about 75 percent of the total area is under cultivation, whereas in the Assam valley, cultivated land is limited because of floods. Rice is the major crop of the region.

3.   Humid eastern Himalayan region and Bay islands

The eastern Himalayan ranges, with a widely varying elevation, contribute to the rugged physiography of the region. About 90 percent of the total precipitation (200–400 cm) is received during the monsoon. The predominant soil groups are brown hill, red loamy, red and yellow, alluvial and laterite (Hapludalfs, Hapludults, Paleudalfs, Plinthudults and Udifluvents). In the foothills, alluvial soils are loam to sandy loam in texture and slightly acidic to neutral in reaction. The soils contain a fairly high percentage of organic matter. The region is endowed with rich forests. Tropical evergreen forests are found in the foothills and plains, and temperate evergreen forests in the middle Himalayas and valleys. The greater Himalayan ranges are covered by conifers.

In the Bay islands, mangrove forest is present along the coast. Bamboo is extensively grown in the tropical evergreen belt. The main feature of the farming system is shifting cultivation practised in the north-eastern hills. Horticulture, animal husbandry and aquaculture have great potential.

4.   Subhumid Sutlej-Ganga alluvial plains

But for the Terai-Bhabar submontane strip, the entire region is level and built up by the mighty rivers originating in the Himalayas. Extremes of climate prevail in the region. The average annual rainfall varies from 30 to 200 cm in the plains and ground frost is common in December and January. The predominant soil groups are calcareous sierozem in the south-west, reddish chestnut in the submontane zone, alluvial soils and patches of saline alkali soils (Calciorthids, Hapludalfs, Calciorthents, Ustochrepts and Salorthids). Cultivated land is 65–85 percent of the total area. The major part of the region consists of introduced vegetation of *sal,* teak and bamboo. In Terai soils, tall grasses are found which serve as good raw material for the paper industry.

5.   Subhumid to humid eastern and south-eastern uplands

The region is characterized by undulating topography, denuded hills, plateaus, mature river valleys, highlands of the Eastern Ghats and wide basins. Elevation ranges from 150 to 1,500 m. The climate is tropical monsoonal and subhumid to humid from west to east. Rainfall ranges from 100 to 180 cm. Predominant soil groups are mixed red and black, red and yellow, red sandy, laterite, black, riverine alluvial and coastal sandy alluvial (Pellusterts, Chromustent Haplustults, Ochragults, Haplustalfs, Hapluquants). Rice is the predominant crop in the high-rainfall areas, coastal plains and deltaic tracts. In the drylands, sorghum is the major crop. In the eastern part of Madhya Pradesh and Orissa States of India, forests occupy 40–45 percent of the total land. Teak and *sal* are the dominant plantation species. Soil salinity and floods in the coastal districts are the two major impediments to agricultural production.

6.   Arid western plains

The region is an extensive alluvial plain dotted with sand dunes, sandy plains, saline depressions and granite hills intersected by the Aravalli range. Annual precipitation is scanty and erratic, ranging from 10 cm in the extreme west to about 65 cm in the extreme east. Most of the rains are contributed by the south-west monsoon during July-September. The deficiency of moisture varies from 40 to 100 percent from east to west.

Predominant soil groups are alluvial, grey brown alluvial, black desert, saline and alkaline (lithic Entisols, Calciorthids, Salorthids and Ustochrepts). Frequent dry spells make the region drought prone and incapable of sustaining plant growth. The region is devoid of trees. Scrub and scattered vegetation are further depleted by cutting and browsing. Natural grasslands have been considerably degraded. In the arid regions pearl millet and sorghum are the major crops, whereas in the semi-arid environment and irrigated areas a variety of crops could be grown.

7.   Semi-arid lava plateau and central highlands

The region is a plateau ranging from 300 to 1,400 m of horizontally bedded sedimentaries and basalt. Between the highlands, there are wide alluvial plains. The climate is semi-arid with extremes of temperature and variable rainfall. The annual precipitation, confined to June-October, varies from 70 to 125 cm except in the Western Ghats where it varies from 335 to 743 cm. The major soil groups of the region are alluvial black and lateritic, mixed red and black and yellowish brown (Paleustalfs, Rhodustalfs, Haplustalps, Chromusterts, Plinthudults, and Haplustults). The dominant forest types include moist and dry deciduous, semi-evergreen and thorny and scrubby vegetation. Semi-evergreen forests are confined to the Western Ghats and the rest of the plateau is covered with deciduous types. Low and uncertain rainfall, undulating topography and inadequate irrigation impede agricultural activities.

8.   Humid to semi-arid Western Ghats and Karnataka Plateau

The region can be divided into the Western Ghats Plateau, river valleys, undulating rocky plains and coastal plains. Rainfall varies from 60 to 300 cm. Important soil groups are black, red, lateritic and alluvial (Pellusterts, Chromusterts, Ustifluvents). The western slopes of the Eastern Ghats and the Nilgiris have tropical evergreen forests and the eastern slopes and uplands grow deciduous forests with sandalwood, eucalyptus, *sal* and teak. The cultivated area varies from 10 percent in the Western Ghats to 54 percent in the Karnataka plateau. About 10 percent of the land is irrigated. Dry farming is practised throughout the region. Forest land and plantation areas have the potential for growing fodder trees, legumes and grasses.

Some important traditional agroforestry systems

Mixed farming systems have been a traditional way of life for the farmers of the Indian subcontinent. In every village there are combinations of tree, crop and animal-husbandry activities according to the local requirements. Some of these combinations have stood the test of time and are practised extensively. The most important of these agroforestry systems are described below.

Shifting cultivation (*jhum*)

This sequential agroforestry system, believed to have originated in the Neolithic period around 7000 BC (Sharma, 1976), is still extensively practised in the North Eastern Hill (NEH) Region and some other humid and hilly parts of the Indian subcontinent. FAO (1981) estimates that about 4,500,000 people practise shifting cultivation annually over an area of 750,000 ha in this region. The reasons for its continuance are linked to ecological, socio-economic and cultural factors, including lack of communication because of remoteness. Details of the tenurial system, field operations, crop mixtures and socio-economic considerations involved in the practice have been well documented by Borthakur *et al.* (1981) and FAO (1981). It has been indicated that due to increasing pressure of population the length of forest fallow has decreased alarmingly from the original 30–40 years to as low as 2–3 years at present. This gradual shrinkage in the cycle of *jhuming* is the sign of an imminent breakdown of the system because of the very short time that is available for the land to restore its fertility.

Studies conducted under humid tropical conditions in India have reported a loss of 40.9 t ha^{-1} of soil in the first year of *jhuming,* resulting in a heavy drain of plant nutrients from the *jhum* plots (FAO, 1981). Since no fertilizer input is given to these fields, the resultant productivity of the crops is low. The gross value of the products from one hectare of *jhum* was only Rs 505–655* with a maximum return of Rs 2.48 per man day and an input-output ratio of 1:1.28.

Concerned by the continuous degradation of the environment of the tribal families dependent on shifting cultivation, the Government of India, in collaboration with the local state governments, launched a number of initiatives to control the problem. The main approach of these schemes was to provide some terraced and afforested land planted with fruit trees to each family so that the farmers could be attracted towards settled agriculture. However, surveys revealed that the project did not meet with much success.

This failure was attributed to:

(i)   Non-acceptance of an abrupt change in the socio-economic traditions;
(ii) Production level in the first year being low in terraces; and
(iii) Lack of follow-up action by the Government agencies in terms of technical advice, input availability and development of an appropriate marketing infrastructure.

The approach now being tried is to improve the existing *jhum* system by providing suitable management packages based on local resources. The abandoned *jhum* lands are planted with desirable tree species and legumes so that fertility can be restored as quickly as possible. Research and development efforts towards an appropriate alternative farming system are also in progress. However, the acceptance of these new components in the system depends on provision of an adequate infrastructure for marketing and an assured supply of essential inputs for the *jhum* farmers.

The taungya system

The taungya system, a method of establishing forest species in temporary combination with field crops, was attempted in the Indian subcontinent soon after its first introduction by Brandis in Burma in 1856. However, regular taungya cultivation was not taken up until

---

* US $ 1 = Indian rupees (Rs) 12.70 (1987).

1911 when it was used for raising plantations of *Shorea robusta* and *Tectona grandis* in 1912. Later, it became a standard practice for regenerating and/or establishing forest plantations in a number of places (see King, this volume). The system, which is initiated and executed by the Forest Department, allows the cultivators to raise agricultural crops in the reserve forest area allocated for new plantations. Input and care given to the agricultural crops in the interspaces improve the growth of the associated trees. The major forest species raised in taungya cultivation are *Shorea robusta* and *Tectona grandis*. A number of other fast-growing tree species were also taken up under this system in order to exploit the advantages of the soil enriching benefits associated with the growing of agricultural crops in the inter-row spaces. The intercrops varied according to the agro-ecological situation. Some of the main crops are rice, millet, maize, gram, mustard, sugarcane, cassava, cotton and potatoes. The system is practised in India and Sri Lanka and the various steps being taken to make it more effective have been described in detail (FAO, 1981).

Studies to evaluate changes in fertility of forest soil after the harvest of inter-cultivated crops have registered a definite increase in organic carbon, phosphorus and potassium (FAO, 1981). In poplar plantations, the yield of interplanted rice, maize and wheat was 4,000, 3,000 and 3,000 kg per hectare, respectively. The height of *Dalbergia sissoo* in combination with peanuts increased up to 23 percent compared to control plots in the initial stages (58.8 cm against 47 cm). Intercropping of turmeric (*Curcuma longa*) in the established plantation of *Tectona grandis, Shorea robusta* and other common forest species resulted in encouraging responses. A yield of 1,200 kg per hectare of turmeric was obtained when the plantations were two years old. With increasing age of plantations, however, the yield of turmeric decreased.

Under the taungya system as followed in Sri Lanka, the emphasis is on reafforestation of land that is abandoned by non-resident cultivators within a period of three years. The main tree species planted are *Tectona grandis* and *Eucalyptus camaldulensis* and intercropping of a variety of agricultural crops is done. Prominent among such intercrops are rice, maize, plantain, chilli and mustard. The area under this system in Sri Lanka is decreasing drastically due to the lack of farmers interested in the practice. The high labour requirement is also a major deterrent.

Although the taungya system helps the initial establishment and growth of forest species, minimizes the cost of maintenance of the trees, and in some cases earns revenue for the forest department on the one hand and provides employment to the rural poor, the interest of the forest department in this system is decreasing. The difficulty the forest officers have in getting the area vacated is said to be the major deterrent factor. Sometimes the farmers cause damage to the growing trees and adjacent forest areas. However, considering the ecological and socio-economic benefits of the system, there is a strong case for forest departments to pursue the approach further.

Prosopis cineraria (khejri)-based system

Historically, *Prosopis cineraria,* locally known as *khejri,* has played a significant role in the rural economy of the arid north-west region of the Indian subcontinent. This tree is an important constituent of the vegetation system and is a source of animal feed, fuel and timber. Its pods are used as a vegetable. It improves the fertility of the soil beneath it, is well adapted to arid conditions and stands up well to the vagaries of climate and browsing by animals. The rural communities encourage the growth of *khejri* in their agricultural fields, pastures and village community lands, as depicted in Figure 2. Through experience,

Figure 2 *Prosopis cineraria* (*khejri*) trees in the agricultural fields in western Rajasthan, India (Photo: Central Arid Zone Research Institute, Jodhpur, Rajasthan).

farmers have realized its usefulness and learnt that it does not adversely affect crop yields but actually improves grain yield and forage biomass production. However, one of the limitations of *khejri* is its rather slow growth. Extensive studies carried out by the Central Arid Zone Research Institute (CAZRI) in Jodhpur, India on the growth pattern, influence on associated agricultural crops and biomass/economic productivity have been summarized in a CAZRI monograph (Mann and Saxena, 1980).

Prosopis cineraria has a very deep taproot system and hence it does not generally compete for moisture with the associated crops. The tree is ready for lopping during the eighth year of its life in the 350–400 mm rainfall zone. The anticipated annual yields of fuelwood, dry leaves and pods from a mature *khejri* tree in the 350–500 mm rainfall zone are 40–70 kg, 20–30 kg and 5 kg per tree, respectively. Singh and Lal (1969) reported an increase in soil silt and clay content to 120 cm under *Prosopis cineraria* while in the open field it was only down to 90 cm. Soil-moisture studies conducted under different desertic tree species (Aggarwal *et al.,* 1976; Gupta and Saxena, 1978) have demonstrated higher moisture content under *khejri* trees than in open fields. Average moisture depletion rates of 1.8 mm per day and 2.1 mm per day day, respectively, were observed from the soil under *Prosopis cineraria* and *Prosopis juliflora*. Available plant nutrients were also recorded to be higher under the *Prosopis cineraria* compared to the open field. The improved physical soil conditions, coupled with higher availability of nutrients under the *khejri* canopy, explain the better growth of the crops associated with it.

Agroforestry involving the *khejri* tree is an age-old practice in arid and desertic conditions. The farmers in the region know the value of these trees and ordinarily will not cut them for fuel. The density of *khejri* trees varies from a few trees to 120 per ha depending on the soil type and rainfall. Alluvial plains with sandy loam soils invariably support more trees. In a well-developed tree stand on alluvial soils all trees are lopped for foliage (*loong*) in a systematic manner during November and December each year. When winter sowing is to be done early, the lopping is completed by the end of October, particularly when wheat is to be sown. The lopped trees remain dormant up to the middle of February. The plant sprouts well with several new branches which are again cut during May-June for livestock feeding. This management of the crown during dry periods keeps down the amount of foliage in the crown. In June-July (the rainy season) the undersown crops and the canopy cover of the tree develop simultaneously. The crop and tree do not compete with each other. Annual crops draw their moisture and nutrients from the top 50–60 cm of soil, whereas the effective root system of the tree is below this depth.

The crops which are normally raised along with the *khejri* tree are millets, especially *Pennisetum glaucum,* and a variety of legumes. The *khejri* tree, though initially slow growing, provides a continuous supply of fodder, fuel, timber and vegetables even in drought years. There is a need for careful selection of comparatively fast-growing *khejri* germplasm and to supply these seeds to other areas.

Homestead agroforestry system

A homestead (or homegarden) is an operational farm unit in which a number of tree species are raised along with livestock, poultry and/or fish mainly for the purpose of satisfying the farmer's basic needs. Such a farming system is traditional to the eastern and southern parts of the Indian subcontinent. Homestead agroforestry practices have been described by Khaleque (1987) from Bangladesh, Nair and Sreedharan (1986) from Kerala, India, and Liyanage *et al.* (1985) from Sri Lanka. A typical homestead with a multitude of crops presents a multi-tier canopy configuration (Soemarwoto, this volume). The leaf canopies of the components are arranged in such a way that they occupy different vertical layers with the tallest component having foliage tolerant of strong light and high evaporation demand and shorter components having foliage requiring or tolerating shade and high humidity. The major portion of the upper canopy goes to coconut, which is followed by other crops such as black pepper, cacao and tree species. The lower storey of the harvesting plane is occupied by banana and cassava and other tuber crops. At the floor level, pineapple, vegetables and other herbaceous crops are grown. However, wide variation in the intensity of tree cropping is noticeable among homegardens in different places. This is generally attributed to the differences in socio-economic conditions of the household and their response to externally determined changes, particularly prices of inputs and products, dependence on land and tenurial conditions, etc. A schematic representation of the activity of an intensive homegarden, as described by Nair and Sreedharan (1986), is presented in Figure 3.

A study conducted by a voluntary organization in the homesteads of Kerala reported a net income of Rs 1,550, 3,848 and 3,950 from a 0.12 ha plot in the first, second and third year, respectively (Nair and Sreedharan, 1986). The components of the system consisted of a three-tier planting of fodder, cassava, vegetables, a number of bananas, cloves and coconut with *Leucaena* being planted all around. A cow, which formed a part of the scheme, not only provided milk for consumption and sale but also manure. Chemical fertilizers were not used at all. It is evident from this case study that farmers with ingenuity

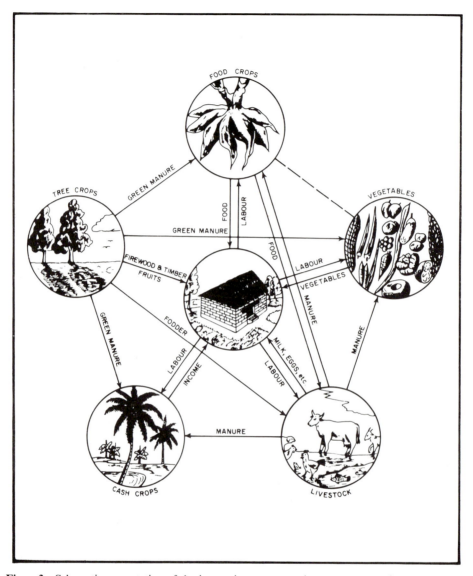

Figure 3   Schematic presentation of the interaction among major components of the homestead agroforestry system of Kerala, India (Source: Nair and Sreedharan, 1986).

have been able to develop a self-sustaining organic farming system capable of yielding a high income without much input other than the family labour.

The coconut-based farming system has been so popular and productive in Kerala, India that farmers are converting their rich rice lands into coconut lands. The system, besides satisfying the primary needs of the farmer, also helps to conserve the fertility by nutrient cycling in spite of the high intensity of cropping (Nair, 1984). The increased microbial activity in the rhizosphere, high labour utilization and risk-minimization are the other benefits associated with the system (Nair, 1979). A mixed farming practice causes substantial improvements in the physical and biological characteristics of the soil (Nelliat and Shamabhat, 1979). The use of waste materials for feeding the cattle, poultry and fish results in efficient recycling and the homestead soil is very well covered to protect it from the beating action of the torrential rain experienced in these areas.

However, there is a need to look into the best possible arrangements of different components in the system and also to explore the inputs required to maintain the productivity of the system. The Central Plantation Crops Research Institute, Kasaragod, Kerala, India along with the Kerala Agricultural University, have on-going research on the various management aspects of high-density coconut-based multiple cropping systems.

Multipurpose tree planting in agricultural fields

A less systematic form of tree growing on farm lands is traditionally practised almost everywhere in the Indian subcontinent. The approach and uses for tree planting in unirrigated semi-arid and arid areas are different to those in the irrigated and humid areas.

Unirrigated arid and semi-arid areas

In these areas the systems have a strong pastoral bias, permitting the combination of animal husbandry with agriculture since free grazing of cattle is rarely allowed. The systems are agrisilvicultural rather than silvopastoral. *Prosopis cineraria, Acacia nilotica, Acacia cupressiformis, Pongamia pinnata,* and *Gliricidia sepium* are some of the tree species which are present in farmlands for meeting fodder requirements. The partial shade cast by the trees reduces insolation and soil temperature which is beneficial to the associated field crops. It is a common sight in certain parts of Uttar Pradesh, India to see individual fields surrounded by very closely planted rows of *Dalbergia sissoo* and *Syzygium cuminii* which are normally raised on high bunds along the boundaries. Similarly *Azadirachta indica, Melia azedarach,* and *Albizia lebbek* have been traditionally planted on the field borders in the semi-arid areas. Under dryland cropping systems, deficiency of soil moisture is the greatest limitation for crop production. Normally, these tree combinations are selected in such a way that the different species do not extract much moisture from the same soil layer.

Irrigated and subhumid or humid areas

The prosperity generated by the "Green Revolution" in the irrigated areas of the Indian subcontinent made the general scarcity of wood all the more troublesome and farmers therefore took the progressive step of planting fast-growing trees around their homesteads, along field boundaries and irrigation channels and also within the fields. The preferred tree species are *Eucalyptus tereticornis, Poplus* spp. and *Dalbergia sissoo* which cast only light shade (Figure 4). Eucalyptus grows fast to produce a straight cylindrical bole which is in much demand as round and sawn timber and as pulpwood. Lops and tops are used as fuel. *Dalbergia sisoo* provides high-quality furniture wood, fuel and fodder.

This kind of planting, although a traditional practice, has quite recently became a prominent feature of the rural landscape, especially in Punjab, India (see Figure 5). The cash value realized from the sale of trees is very high and more than compensates for any marginal adverse effects that the planting of such trees may have on the crop yields. The trees do not directly compete with the shallow-rooted agricultural crops either for the irrigation water or for fertilizers, nor do the species selected cast enough shade to be detrimental to the crop growth. Trees do attract birds, but perhaps the loss of grain is more than compensated for by the elimination of insect pests (FAO, 1981).

Figure 4   Eucalyptus trees planted on the boundary of an irrigated wheat field in northern India

Figure 5   Onion (*Allium* sp.) as an intercrop with poplar trees in Punjab, India

Environmentally and economically the trees are a boon to the farmer and it is no wonder that eucalyptus planting has been taken up with much enthusiasm. It has not only revolutionalized tree culture but agriculture as well and its contribution to the prosperity of the region is conspicuous.

Present status of agroforestry

The Indian subcontinent with 995 million people and 540 million livestock is one of the most densely populated regions in the world (240 persons and 130 livestock per sq km; FAO, 1985). Demographic projections for India alone estimate a population of about 1,000 million people by the year AD 2000. To support such a large population the amount of food, fodder, fibre, timber and energy required will be enormous. Fuelwood, which is still the major source of rural non-commercial energy (40–50 per cent) is in great shortage. As against the estimated requirement for India of 133 million tonnes of fuelwood per annum, all the present and projected sources produce only about 39 million tonnes (Government of India, 1982). Similarly, in Pakistan, against a requirement of 18.9 million m^3 of fuelwood, state forests produce only 1.9 million m^3 (Sheikh, 1987). The situation in Bangladesh, Nepal and Sri Lanka is no better.

One of the worst repercussions of fuelwood scarcity is the burning of considerable quantities of cowdung and crop residues which otherwise could be used in restoring soil fertility. The Committee on Fodder and Grasses appointed by the Government of India in 1985 estimated the availability of only 224.08 million tonnes of green fodder against a requirement of 611.99 million tonnes for supporting the present livestock population. Since a decline in soil productivity due to intensive-cropping and improper management of soil is a general feature in the subcontinent, and chemical fertilizers are very expensive, the use of biological alternatives for enriching the soil is one of the major concerns in these countries. Present land-use patterns presented in Table 1 show that only 24.3 percent of the subcontinent is under forest, woodland and permanent pastures. Apart from the forest area being precariously small, it is also extremely degraded and hardly 50 percent of it may be under good vegetal cover. Village woodlands are seriously depleted. Tree planting on unused lands, categorized as "waste lands and barren lands", and in conjunction with agricultural crops on arable land, seems to be the main hope for providing food and

Table 1   Land use in the countries of the Indian subcontinent (10^3 ha)

	India	Pakistan	Bangla-desh	Nepal	Sri Lanka	Bhutan	Total
Total area	328,759	80,394	14,400	14,080	6,561	4,700	448,894
Land area	297,319	77,872	13,391	13,680	6,474	4,700	413,436
Arable land	164,850	19,911	8,891	2,290	1,080	94	197,116
Permanent crops	3,500	369	220	29	1,122	6	5,246
Permanent pastures	11,900	5,000	600	1,786	439	218	19,943
Forest and woodland	67,420	3,050	2,106	2,308	2,383	3,285	80,552
Other land	49,649	49,542	1,574	7,267	1,450	1,097	40,579
Irrigated	39,700	15,320	1,920	640	550	-	58,130

Source: FAO, 1985.

environmental and economic security in the subcontinent.

Government-run social forestry schemes have made a good impact in some areas. Recently, some non-governmental organizations (NGOs) have also been active in this area. But the major thrust in integrating trees with agriculture will have to come from the farmers themselves. The motivation may be diversification of production for meeting their own requirements, cash generation, or even the scope for sustainable production in marginal and arid areas compared to the field crops. Some farmers have already taken the initiative in this direction, and to meet the need for an appropriate management strategy for tree-crop integration the national governments have started strengthening the research and development infrastructure. A brief account of some such efforts is given below.

India

Research work on fuel and fodder production systems by the Central Soil and Water Conservation Research and Training Institute, Dehra Dun; on silvopastoral systems by the Indian Grassland and Fodder Research Institute, Jhansi; on shelterbelts and agrisilviculture systems by the Central Arid Zone Research Institute, Jodhpur; and on a mixed system consisting of agricultural, horticultural and silvicultural components as an alternative to shifting cultivation by the Indian Council of Agricultural Research (ICAR)'s research complex for the North-Eastern Hill region was initiated during the late 1960s and 1970s (Singh and Randhawa, 1983). Even prior to the above initiatives by the ICAR, the Forestry Department of the Government of India had taken up elaborate studies on taungya cultivation.

Vishwanathan and Joshie (1980) successfully utilized the bouldery riverbeds in Dehra Dun by raising *Dalbergia sissoo* and *Acacia catechu* trees in conjunction with *Chrysopogon fulvus* and *Eulaliopsis binata* grasses. The yield from such a system over a period of 19 years is given in Table 2.

Table 2   Yield from a silvopastoral system on a bouldery river bed

Treatment	Yield
Dalbergia sissoo (9.15 x 9.15 m) + *Chrysopogon fulvus*	64.0 t wood ha^{-1} 5.5 t grass ha^{-1} yr^{-1} (average of 17 years)
Acacia catechu (4.55 x 4.55 m) + *Eulaliopsis binata*	71.3 t wood ha^{-1} 5.3 t grass ha^{-1} yr^{-1} (average of 15 years)

Source: Vishwanathan and Joshie, 1980.

Under similar conditions, Singh *et al.* (1982) tried four fodder-tree species, namely *Albizia lebbek, Grewia optiva, Bauhinia purpurea* and *Leucaena leucocephala* in association with *Eulaliopsis binata* and *Chrysopogon fulvus*. The results are given in Figure 6. The *Grewia optiva* and *Eulaliopsis binata* system proved to be the most rewarding in terms of total return on such degraded lands under rainfed conditions.

Figure 6   *Eulaliopsis binata,* a good fodder grass, under *Grewia optiva*

While evaluating the effect of trees planted on farm boundaries, Khybri *et al.* (1983) reported a more pronounced effect of boundary trees on winter wheat than on a rainy-season crop of rice, perhaps due to the limitation of moisture availability during the winter season. Srivastava and Narain (1980) studied the effect of a 20 year-old *Eucalyptus tereticornis* tree line planted on the farm boundary on crops under rain-fed semi-arid conditions. Serious reductions in crop yields were noticed of up to 10 m in the case of green gram, up to 5 m in black gram and only 2 m in the case of sorghum. At the same site (Kota, India), Prasad and Verma (1983) studied the effect of *Leucaena* alleys on associated crops of sorghum, sorghum and pigeon pea, and sorghum and black gram. The results, presented in Table 3, indicate some reduction in the yields of pigeon pea and black gram grown in association with *Leucaena* compared to pure crop, but the additional biomass yield from *Leucaena* makes the combination more paying.

In order to study the interference effect of tree roots on associated agricultural crops, Singh and Dayal (1974) dug a 15 m x 1.0 m trench along the field boundary of a 15-year-old plantation of *Dalbergia sissoo* and *Acacia arabica* planted on peripheral bunds of cotton and tobacco fields and evaluated the crop yields. In this semi-arid situation elimination of *D. sissoo* roots from the crop-root zone increased the yield of cotton from 956 kg ha^{-1} to 1,056 kg ha^{-1}; the yield of tobacco increased from 1,206 kg ha^{-1} to 1,859 kg ha^{-1} by pruning the roots of the *Acacia arabica*. However, Dadhwal and Narain (1984) did not find any significant effect from trenching along *Eucalyptus* rows on maize under high-rainfall conditions.

Studies conducted on agroforestry systems in arid regions of India have been described by Shankarnarayan *et al.* (1987). The introduction of *Acacia tortilis,* a comparatively fast-growing tree, in the arid zone has helped in the stabilization of sand dunes and early establishment of shelterbelts. *Acacia tortilis* with the grass *Cenchrus ciliaris* in a silvopastoral system earned maximum returns among the different combinations tried.

Table 3   Yield of leucaena and intercrops*

Leucaena (alone)	Leucaena with intercrops		Crops (alone)	
3.75 m x 25 cm–3,700	(i) *Leucaena* (3.75 m x 25 cm)	1,800	Pigeon pea	1,650
75 cm x 25 cm–14,000	+			
	Pigeon pea	1,280	Sorghum	370
	+			
	Sorghum	380		
	(ii) *Leucaena* (3.75 m x 25 cm)	1,400	Pigeon pea	1,690
	+			
	Pigeon pea	1,370	Black gram	350
	+			
	Black gram	208		
	(iii) *Leucaena* (3.75 m x 25 cm)	1,200	Sorghum	660
	+			
	Sorghum	460		

* Air dried biomass of leucaena and grain yield of intercrops in kg ha^{-1}
Source: Prasad and Verma, 1983.

Lopping of eight-year-old *Holoptilia integrifolia* caused a favourable effect on the associated crops of mung bean (*Vigna mungo*) and cluster bean (*guar*) (*Cyamopsis tetragonoloba*), as shown in Table 4.

Table 4   Grain yield under lopped and unlopped eight-year-old *Holoptilia integrifolia* (kg/ha)

Species	Unlopped tree	Lopped tree	Crop without tree
Vigna mungo (mung bean)	201.75	252.06	246.21
Cyamopsis tetragonoloba (guar)	93.90	244.14	253.14

In another experiment, the growth rates of *Prosopis cineraria* and *Acacia albida* and their influence on the associated crops of mung bean and *guar* were studied. *Acacia albida* was found to grow much faster with a height increment of 110 cm in three years compared to only 15 cm in the case of *P. cineraria*. The results of this study are presented in Table 5.

It is interesting to observe that, in general, the association of trees with crops improved the crop yield in these experiments. During 1985, when the general level of crop yield was very low due to moisture stress, the yield of *guar* improved due to the association of *Prosopis cineraria*.

In the semi-arid region at Jhansi the performance of *Leucaena leucocephala* was evaluated in a silvopastoral combination (Gill and Patil, 1983). It was found that the tree was compatible with a number of forage-production systems. Total biomass production in association with *Leucaena* was invariably higher compared to monocrop systems. Borthakur *et al.* (1981) suggested a "silvi-agri-horti" system as an alternative to shifting cultivation in the North-Eastern Hill Region of India. The system is more productive than shifting cultivation and other indigenous systems and reduces the risk of soil and water loss

Table 5 Grain yield of mung bean (*Vigna mungo*) and guar (*Cyamopsis tetragonoloba*) under tree species (1983–1985)

Treatments	Mung (10^2 kg ha^{-1})			*Guar* (10^2 kg ha^{-1})		
	1983	1984	1985	1983	1984	1985
Prosopis cineraria at 5 x 5 m	7.85	8.50	0.68	3.60	7.15	1.16
P. cineraria at 5 x 10 m	7.44	.6.08	0.75	4.45	5.14	1.28
P. cineraria at 10 x 10 m	8.68	7.90	1.11	3.70	6.53	1.34
Acacia albida at 5 x 5 m	8.08	9.30	0.81	3.32	6.11	0.65
A. albida at 5 x 10 m	7.08	6.70	0.61	5.08	5.37	0.76
A. albida at 10 x 10 m	7.92	7.51	0.74	0.34	6.34	1.28
Control (crops without trees)	6.40	7.30	1.06	3.20	6.41	0.95

to a great extent. Singh and Singh (1987), while summarizing the work on alternative farming systems for the drylands of semi-arid tropics of India, highlighted the effect of the addition of *Leucaena* loppings to the soil in the improvement of sorghum grain yield (Table 6). It is interesting to note that the addition of biomass from other sites in combination with fertilizer nitrogen increased the yield of sorghum. However, the communication does not indicate the quantity of lopping added in the experiment.

Table 6 Effect of *Leucaena leucocephala* planting, lopping addition and levels of fertilizer nitrogen on the grain yield of *rabi* sorghum; Solapur, India

Treatments	Grain yield (10^2 kg ha^{-1})			Total N uptake (kg ha^{-1})		
	N levels			N levels		
	0	25	50	0	25	50
1. Control (without *Leucaena* planting and addition of lopping from outside)	6.2	12.1	16.8	16.6	22.3	45.0
2. No planting of *Leucaena* but addition of loppings from outside	10.7	15.8	17.4	26.33	43.2	48.0
3. *Leucaena* planting without addition of loppings	4.7	10.8	12.7	13.5	26.3	33.8
4. *Leucaena* planting with addition of loppings	9.3	13.9	16.6	24.3	36.4	44.5
C.D. (0.05)	1.83					

Besides the Institutes of the Indian Council of Agricultural Research (ICAR), certain state agricultural universities have also initiated systematic research work on agrogforestry systems. The pioneers among them are the Tamil Nadu Agricultural University, Coimbatore and the Agricultural University located at Dharwar.

On the recommendation of the task force constituted during the Agroforestry Seminar organized at Imphal, India in May 1979, ICAR launched an All India Coordinated Research Project on Agroforestry spread over 20 centres in the country in 1983. The core research activities at these centres are (i) conducting a survey of the existing agroforestry practices, (ii) collection and evaluation of the local agroforestry tree species, and (iii) development of location-specific agroforestry systems. With the increasing interest in agroforestry in the country, the programme of the All India Coordinated Research Project on Agroforestry has since been extended to 11 more centres covering all the 23 state agricultural universities, and it has been decided that a National Research Centre on Agroforestry be established during the seventh five-year plan of India (1985–1990).

Pakistan

Since the area under forest in Pakistan is very small (3.8 per cent), about 58 percent of timber and 90 percent of fuelwood requirements are being met from farm and wasteland plantations. If integration of trees with agricultural systems is needed anywhere in the world, the plains of Pakistan deserve the highest priority. Tree planting has not been properly organized and still a large majority of the rural community is not fully motivated to plant trees due to the belief that the trees compete with agricultural crops for water, nutrients, light and space and harbour birds which cause damage to the associated crops. However, the introduction of poplar, a multipurpose tree, has changed the rural scene in some areas. There is tremendous demand for poplar wood from the various wood-based industries in Pakistan. It has also been established that *Eucalyptus* can be successfully grown in marginal lands where average annual rainfall is about 300 mm.

Certain studies on the tree/crop interface were conducted in the country. Sheikh (1987) found that crop yields were affected by growing them in association with *Acacia nilotica* and *Dalbergia sissoo*, the maximum loss occurring in a 2-metre radius from the tree. Yields were poorer from the crop planted on the northern side of the tree.

Kermani (1980) reported the results of large-scale experiments to find the effect of growing *Acacia nilotica* and *Eucalyptus camaldulensis* in association with cotton, wheat, sesamum and sorghum. It was found that the yields of agricultural crops were higher when grown in association with *E. camaldulensis*. The eucalyptus and cotton combination was the best, giving higher yields of wood as well as better monetary returns from the agricultural crops. Khattak *et al.* (1980) found that the yield of wheat under *Dalbergia sissoo* was significantly higher than that of wheat under *Eucalyptus citridora, Populus deltoides* or *Bombax ceiba*. Sheikh and Haq (1986) summarized the results of tree/crop interface experiments conducted in the Thal Desert and observed an almost uniform response pattern. Wheat yield was depressed up to 5 metres from the tree rows, but total yield of wheat from a tree-associated plot was higher than that of the control plot. Trees (*D. sissoo, Tamarix aphylla* and *P. deltoides*) were not different in their effect on crop yield. The farmer was able to make good the crop loss through the sale of trees at fairly good prices. Sheikh and Haq (1986) recorded a depressing effect of poplar rows on the yield of sugarcane crop up to 10 m away from the trees.

The important findings emerging from the different studies conducted in Pakistan, as summarized by Sheikh (1987), are (i) trees in close proximity to agricultural crops depress the yield of the latter, (ii) the effects vary depending upon the species of trees and crops, and

(iii) farmers are prepared to plant trees provided the species are fast growing and have a good market value. The information available so far indicates that a great potential exists for agroforestry in the plains of Pakistan. However, development efforts need to be backstopped by necessary research to find solutions to the various problems encountered.

Bangladesh

The majority of people in Bangladesh obtain their fuel, fodder and timber from homestead forests. According to one estimate, homestead forests produce about 70 percent of sawlogs and about 90 percent of fuelwood and bamboo consumed in Bangladesh (Byron, 1984). A survey on the practice of homestead plantations indicated that farmers prefer fruit trees to other species, and fuelwood is seen as a by-product.However, as suggested by Khaleque (1987), there is a need to work out the homestead planting in such a way that most of the requirements of the household are met. Moreover, the development of a marketing infrastructure would be necessary to make fruit-tree planting profitable.

Sri Lanka

The country practises a number of agroforestry systems. The most prominent are *chena,* a form of shifting cultivation, some forms of taungya, intercropping under coconuts, homegardens, growing tea and coffee under the shade of trees and windbreaks/shelterbelts (Liyanage *et al.,* 1985). *Moringa oleifera* is commonly planted in hedges around homesteads throughout the dry zone. It is the normal practice to grow cowpea and other crops under the tree since the crown is light and the yields of crops are not decreased by the tree's shade. *Sesbania grandiflora,* a short-lived tree of medium height, is often grown in and near homesteads for its tender leaves and flowers that can be used as a vegetable.

Nepal and Bhutan

Growing of multipurpose tree species in the agricultural fields and around the homestead specifically for fodder and fuel is a common practice. Planting *Amomum subulatum* (large cardamom) in association with *Alnus nepalensis* is a major cash-generating agroforestry system in these countries. A comprehensive list of multipurpose tree species found on the farmlands in Nepal is given by Fonzen and Oberholzer (1985). The multipurpose tree species common to Bhutan and Sikkim have been described by Singh and Pazo (1981).

Conclusions

We can conclude that the role of agroforestry in meeting either present or future requirements of fuelwood, food, fodder and small timber and for environmental protection has been very well recognized in the Indian subcontinent. What is now required is the development of location-specific, need-oriented systems along with the necessary support systems so that farmers can get the required seedlings and other inputs easily and market the produce at competitive prices.

With a population growth rate of over 2 percent in most of the countries of the subcontinent, the challenge of meeting basic requirements of the population is a major concern. The exploitative tendency of modern agriculture, aimed at higher and higher production per unit area, is in fact rapidly degrading the basic production capacity of the ecosystem. Long-term experiments conducted in India have indicated very clearly that in high-production areas the soil is quickly depleted of essential plant nutrients. In certain regions where ground-water recharge is slow the rapid withdrawal of ground water has alarmingly lowered the watertable. Village woodlands and grazing grounds have vanished, resulting in acute shortages of fuel and fodder. Government-managed forests that are already denuded are heavily exploited for meeting timber requirements. Deforestation and cultivation on steep slopes have resulted in heavy soil losses and water runoff. These things have happened, perhaps, because agricultural research and development efforts have given greater emphasis to individual production sectors or components than to the whole system.

Farmers in the subcontinent basically practise a mixed-farming system. Their way of life is an integration of different components for optimum production without necessitating much external input. Agroforestry presents an excellent opportunity for such low-input socio-economic situations as it provides for the integration of various production factors for achieving need-based goals. Whether it is the problem of apple boxes in the hills, or fodder and fuel in the semi-arid/arid areas, or shelterbelts in deserts, or ground cover in high-rainfall areas, or cash crops in high input areas, trees can play a role and can be suitably integrated with the existing agricultural production system. It is for the scientists engaged in agroforestry research to identify trees for a specific role in a particular ecosystem and to synthesize and develop the agroforestry system based on existing components so that the production can be optimized without impairing the quality of the resources. Fortunately, in the Indian subcontinent the information on the component technology is not lacking, whether it is tree, crop, animal, soil, water or environment. In some countries of the subcontinent, work on the development of appropriate agroforestry systems has already started; others are in the process of developing the needed infrastructure. However, considering the urgency of such an approach for application to the field, there is no need to wait for elaborate infrastructural developments. What may be required is to develop multidisciplinary teams which are capable of synthesizing appropriate systems and have the capacity to manage the components for realizing the objectives of the systems.

Agroforestry research should have as a high priority the identification of appropriate tree species for their assigned role in the system in a particular environment. A wide range of indigenous and exotic species is available for selection. The ingenuity of the researcher lies in developing a management system which may integrate the selected tree species with on-going land-use activities in such a manner that the overall production gains are higher, without impairing the basic resources. It may be necessary to go for more "on farm" trials so that the synthesized systems can be tested and modified on farmers' fields before they are tried for large-scale extension. A number of systems failed to become popular with farmers because of the absence of the necessary input supply and marketing infrastructure. This may be crucial for agroforestry because overproduction of wood, fruit or fodder without any market can affect the farmers adversely.

The scope for tackling the problems of the rural areas through agroforestry interventions in the Indian subcontinent is very promising. There are a number of indigenous and exotic tree species which are already being grown on agricultural lands. Farmers are responsive to the idea and are prepared to adopt the alternative farming systems. The scarcity of fuelwood and fodder is widespread and is affecting seriously the living conditions of rural people. In such a situation an appropriate technology of tree integration with agriculture will be immediately accepted. However, a major effort awaits agroforestry scientists to synthesize site-specific technology and policy makers to develop a suitable infrastructure for the disposal of diversified products from such a system.

REFERENCES

Aggarwal, R.K., J.P. Gupta, S.K. Saxena and K.D. Muthana, 1976. Some physico-chemical and ecological changes of soil under permanent vegetation. *Indian Forester* 102: 863–872.

Borthakur, D.N., R.N. Prasad, S.P. Ghosh, R.P. Awasthi, R.N. Rai, A. Varma, H.H. Datta, J.N. Sachan and M.D. Singh. 1981. Agroforestry-based farming system as an alternative to *jhuming. Proceedings of the Agroforestry Seminar held at Imphal, India, 1979.* New Delhi: ICAR.

Byron, N. 1984. People's forestry: A novel perspective of forestry in Bangladesh. *ADAB News* (Dhaka) 11 (20): 31–37.

Dadhwal, K.S. and P. Narain. 1984. Root effects of boundary trees on the *rabi* crops can be reduced by trenching. *Soil Conservation Newsletter,* 3 (2), Central Soil and Water Conservation Research and Training Institute, Dehra Dun, India.

Fonzen, P.F. and E. Oberholzer. 1985. Use of multipurpose trees in hill farming systems in Western Nepal. *Agroforestry Systems* 2: 187–197.

FAO. 1981. *India and Sri Lanka: Agroforestry.* FAO/SIDA Forestry for Local Community Development Programme. FAO. Rome.

————. 1985. *Production yearbook,* 39. Rome: FAO.

Gill, A.S. and B.D. Patil. 1983. Subabul (*Leucaena leucocephala*) for forage production in agriculture in India. Bulletin of the Indian Grassland and Fodder Research Institute, Jhansi, India.

Government of India. 1982. *Report of the fuelwood study committee.* Planning Commission, New Delhi.

Gupta, G.P. 1975. Sedimentation production status report on data collection and utilization. *Soil Conservation Digest* 3(2): 10–21.

Gupta, S.P. and S.K. Saxena. 1978. Studies on the monitoring of the dynamics of moisture in the soil and the performance of ground flora under desertic communities of trees. *Indian Journal of Ecology* 5: 30–36.

Kermani, W.A. 1980. Developing multiple use silvicultural practices for forestry of arid regions. *Proceedings of IUFRO/ MAB Conference, Research on Multiple Use of Forest Resources,* Flag Staff, Arizona.

Khaleque, K. 1987. Homestead forestry practices in Bangladesh. *Proceedings, Workshop on Agroforestry for Rural Needs,* Vol. 1. New Delhi, India.

Khattak, G.M., M.I. Sheikh and A. Khaliq. 1980. Growing trees with agricultural crops. *Pakistan Journal of Forestry* 31: 95–97.

Khybri, M.L., R.K. Gupta and S. Ram. 1983. Effect of *Grewia optiva, Morus alba* and *Eucalyptus* hybrid on the yield of crops under rainfed conditions. Annual Report, Central Soil and Water Conservation Research and Training Institute, Dehra Dun, India.

Liyanage, M. de S., K.G. Tejwani and P.K.R. Nair. 1985. Intercropping under coconuts in Sri Lanka. *Agroforestry Systems* 2: 215–228.

Mann, H.S. and S.K. Saxena (eds.). 1980. *Khejri.* Monograph No. 11, Central Arid Zone Research Institute, Jodhpur, India.

Nair, P.K.R. 1979. *Intensive multiple cropping with coconuts in India.* Berlin/ Hamburg: Verlag Paul Parey.

————. 1984. *Soil productivity aspects of agroforestry.* Nairobi: ICRAF.

Nair, M.A. and C. Sreedharan. 1986. Agroforestry farming systems in the homesteads of Kerala, Southern India. *Agroforestry Systems* 4: 339–363.

Nelliat, E.V. and K. Shamabhat (eds.). 1979. *Multiple cropping in coconut and arecanut gardens.* Technical Bulletin 3, Central Plantation Crops Research Institute, Kasargod, India.

Prasad, S.N. and B. Verma. 1983. Intercropping of field crops with fodder crop of *subabul* under rainfed conditions. Annual Report, Central Soil and Water Conservation Research and Training Institute, Dehra Dun, India.

Shankarnarayan, K.A., L.N. Harsh and S. Kathju. 1987. Agroforestry in the arid zones of India. *Agroforestry Systems* 5: 69–88.

Sharma, T.C. 1976. The pre-historic background of shifting cultivation. In *Proceedings of Seminar on Shifting Cultivation in North East India.* New Delhi: Indian Council of Social Science Research.

Sheikh, M.I. 1987. Agroforestry in Pakistan. *Proceedings of a Workshop on Agroforestry for Rural Needs.* New Delhi, India, February 1987.

Sheikh, M.I. and R. Haq. 1986. Study of size, placement and composition of wind breaks for optimum production of annual crops and wood. Final Technical Report, PF1, Peshawar, Pakistan.

Singh, A. and R. Dayal. 1974. Peripheral trenching for eliminating root effect of forest trees bordering agricultural crops in ravine areas. Annual Report, Central Soil and Water Conservation Research and Training Institute, Dehra Dun, India.

Singh, G.B. 1980. Shifting cultivation and its control in the North Eastern Hills Region — A critical review. Paper presented in the National Symposium on Soil Conservation and Water Management, Dehra Dun, India (unpublished).

Singh, G.B. and P.O. Pazo. 1981. Agroforestry in the Eastern Himalayas. *Proceedings of the Agroforestry Seminar held at Imphal, India, 1979.* New Delhi: ICAR.

Singh, G.B. and N.S. Randhawa. 1983. Status paper on agroforestry. Presented during ICAR/ICRAF group meeting held at New Delhi, India (unpublished).

Singh, H., P. Joshi, M.K. Vishwanathan and M.L. Khybri. 1982. Stabilization and reclamation of eroded areas and wastelands utilization of bouldery riverbed lands for fodder, fibre and fuel. Annual Report, CSWCRTI, Dehra Dun, India

Singh, K.S. and P. Lal. 1969. Effect of *Khejri* (*Prosopis cineraria* Linn) and *Babool* (*Acacia arabica*) trees on the soil fertility and profile charateristics. *Ann. Arid Zone* 8: 33–36.

Singh, R.P. and S.P. Singh. 1987. Alternate farming systems for drylands of semi-arid tropics of India. Paper presented in the National Symposium on Alternate Farming Systems, IARI, New Delhi, February 1987 (unpublished).

Srivastava, A.K. and P. Narain. 1980. Competition studies on tree crops association with *Eucalyptus tereticornis* under farm forestry at Kota. Annual Report, Central Soil and Water Conservation Research and Training Institute, Dehra Dun, India.

Vishwanathan, M.K. and P. Joshie. 1980. Economic utilization of class V and VI lands for fuel and fodder plantation. Annual Report, Central Soil and Water Conservation Research and Training Institute, Dehra Dun, India.

9

Indigenous shrubs and trees in the silvopastoral systems of Africa

Henry N. Le Houérou

Centre National de la Recherche Scientifique
Centre L. Emberger
Route de Mende
B.P. 5051, 34033 Montpellier Cedex
France

Contents

Introduction

Silvopastoral and agrosilvopastoral systems in Africa

The role of shrubs and trees

Present trends in African silvopastoral systems

Conclusions

References

Introduction

Most of the natural grazing lands in intertropical Africa, i.e., those under summer rainfall regimes, can be called "silvopastoral", for most are found in either dry forest, woodland, bushland, wooded savanna or shrubland.* Purely herbaceous grazing lands, i.e., grass savanna or "grassland" *sensu stricto,* are rather rare and restricted to particular conditions of soil, climate and/or management: for instance on Vertisols, some types of shallow soils, or wet soils, or under very arid climatic conditions such as in the northern Sahel, or some

---

* Woodland — open dry forest with 50% or more canopy cover;
  Bushland — sparse trees *and* shrubs;
  Wooded savanna — grassland with a more or less continuous herbaceous layer (annual or perennial) with trees and shrubs having a canopy cover of less than 50 percent;
  Shrubland — no trees, shrubs making up about 50 percent of the overall canopy cover.

southern African velds, or under management practices that tend to eliminate woody species.

In Africa north of the Tropic of Cancer (Mediterranean Africa), under climatic conditions characterized by a winter rainy season, silvopastoral systems of livestock production are more or less confined to semi-arid to perhumid bioclimates. The North African arid zone is, to a very large extent, dominated by a dwarf-shrub type of steppe vegetation and therefore the livestock production systems developed in such grazing lands cannot be called silvopastoral. The African desert systems, characterized by contracted types of vegetation clustered along the drainage network, can, perhaps somewhat paradoxically, be called silvopastoral systems since they include both shrubs and trees, though these are very sparse.

These different silvopastoral systems represent about 80 percent of the African natural grazing lands (Le Houérou, 1977).

Shrubs and trees in the silvopastoral production systems constitute the basic feed resource of more than 500 out of the 660 million head of livestock (FAO, 1985), i.e., 165 and 218 million tropical livestock units (TLU)* respectively. A number of studies suggest that ligneous species represent an average 10–20 percent of the overall annual stock diet in these production systems in terms of dry-matter uptake, but much more in qualitative terms as one of the main sources of protein and minerals in the diet, particularly during the dry seasons.

In addition to their fodder value, trees and shrubs (sometimes collectively referred to in agroforestry jargon as "trubs") play an essential role, not only as fodder reserve for critical periods, but also as a multipurpose resource, fulfilling many basic needs of these pastoral populations. For example, they are a source of energy and fuel, fibres for clothing and handicrafts, wood for construction, tools, human food, medicines for man and animals, drugs and dyes, shelter for man and beast (zribas, bomas), maintenance of soil fertility via nutrient cycling, soil protection from erosion agents, etc.

In various revegetation projects, such as reforestation, establishment of forage reserves, watershed management, sand-dune stabilization and agroforestry, the question often arises: what species to select; exotics or native? Common sense would suggest avoiding any dogmatism in such matters, and the taking of a pragmatic view based on experience. But this does not always happen: advantages of both exotics and native species are advanced. This, in my judgement, is a false dilemma. One should select whatever has proved best adapted and most productive under the circumstances and for the objective that is being pursued without undue consideration to ideological principles! In some instances exotics may prove best suited while in other circumstances native species may be preferable for various reasons such as self-regeneration, the wishes of local users, etc.

Silvopastoral and agrosilvopastoral systems in Africa

Africa north of the Tropic of Capricorn, which is considered in this paper, is 96 percent of the continent's surface area, that is, 28.2 million km² including the islands. This area is extremely varied in terms of climate: Mediterranean, tropical, equatorial, subtropical, sub-equatorial, montane, and Afro-Alpine climates are found (Le Houérou and Popov, 1981). Most of the soil classes of the world are represented to a greater or lesser extent

---

* 1 TLU = approx. 250 kg liveweight of animal.

(FAO, 1974, 1978). The African flora comprise some 36,000 species of flowering plants, 30,000 of which belong to the intertropical zone (Brenan, 1979). About 9,000 of those are trees or shrubs (Wickens, 1980). About 75 percent of the latter (6,750) are browsed (Whyte, 1974), and over 4,500 have been recorded as being useful to man (Wickens, 1980).

Shrubland, woodland and savanna communities occupy around 10,000 km², i.e., 35 percent of the continent, the remainder being desert (36 percent), dense forest (22 percent) and cropland (7 percent) (Le Houérou, 1977).

Mediterranean Africa

Silvopastoral systems in the Mediterranean region of Africa are different in their biological composition, structure and physiognomy from those in intertropical systems. However, some similarities may be found with the highland silvopastoral systems of East Africa, with quite a number of common genera such as *Pistacia, Olea, Rhus, Osyris, Jasminum, Colutea, Buxus, Ceratonia, Asparagus, Juniperus, Argyrolobium*, and a physiognomy and structure that are strongly reminiscent of the Mediterranean *garrigue* and *maquis* with sclerophyllous shrubs and forests with acicular-leaved trees (*Juniperus, Podocarpus*) (Le Houérou, 1984). The area of the main types of silvopastoral systems of the Maghrib countries is shown in Table 1. The table shows that there is a total of about 8.8 million hectares of silvopastoral systems in northern Africa.

These silvopastoral systems contribute substantially to the nutrition of some 10 million head of sheep-equivalents in the region (Table 2). The silvopastoral systems are estimated to provide about 10 percent of the total feed consumption by the regional livestock population, the balance coming from dwarf-shrub steppes, stubble and fallows, fodder crops, crop residues, agro-industrial by-products and concentrated feed, including cereal grain (mostly barley). The annual productivity of these silvopastoral systems in the region is in the neighbourhood of 400 Scandinavian feed units (SFU)* per hectare (Le Houérou, 1976b, 1980b). Productivity varies widely from one system to another, with values ranging from 100 to 1,000 SFU ha^{-1}yr^{-1}. Similarly, the period of utilization varies greatly between systems; the highland systems (*Juniperus thurifera, Abies* sp., *Cedrus atlantica*) are mainly used as summer pastures within transhumant systems of animal husbandry. The lowland systems are used differently by various livestock species: cattle and sheep use them when stubble and fallow are not available from late fall to early summer, whilst goats use them all the year round.

The transhumant and nomadic systems of northern Africa are on the decline, except for very short seasonal transhumance of a few kilometres. Animal production in these silvopastoral systems represents 60–80 percent of their productivity in economic terms (Le Houérou, 1976b, 1980a), the balance being fuelwood, charcoal, honey, distillation of essential oils, cork and some timber.

These North African silvopastoral systems have been the subject of extensive study and review by the author (Le Houérou, 1973, 1976a, 1978, 1980a,b,c, 1981a,b, 1987a,b,c,d).

The forest and woodland, and therefore the silvopastoral systems, are on the decline in northern Africa. This decline has two aspects: the reduction of the total area and a continuous decline of productivity per unit area.

The reduction of area under wooded land (forest, woodland and shrubland) is

---

* 1 SFU = 3 kg dry matter, being equivalent to the net energy of 1 kg barley.

Table 1   Silvopastoral systems in North Africa (10^3 ha)

System	1	2	3	4	5	6	7	8	9	Total
Country										
Algeria	70	650	700	30	850	160	100	580	-	3,140
Egypt	-	-	-	-	-	-	-	-	-	-
Libya	-	-	10	-	-	-	50	150	-	220
Morocco	10	400	1,350	120	70	750	500	400	700	4,300
Tunisia	30	130	120	-	50	30	70	420		1,140
Total	110	1,180	2,180	150	1,270	940	720	1,550	700	8,800

System 1:   *Quercus faginea* forest and woodland; humid and perhumid zone.
System 2:   *Quercus suber* forest and woodland; subhumid and humid zones.
System 3:   *Quercus ilex* and *Q. coccifera* shrublands and woodlands; semi-arid and subhumid zones.
System 4:   *Cedrus atlantica* forest and woodland; cold subhumid and humid zones.
System 5:   *Pinus halepensis* forest and woodland, semi-arid to subhumid zones.
System 6:   *Tetraclinis* woodlands and shrublands; arid and semi-arid zones.
System 7:   *Olea europaea sylvestris* and *Pistacia lentiscus* shrublands; semi-arid to humid zones.
System 8:   *Juniperus phenicea* shrublands; arid and semi-arid zones.
System 9:   *Argania sideroxylon* parkland and shrubland; arid and desert zones of S. W. Morocco.

Lesser areas under other systems include:
System 10:   *Pinus pinaster mesogeensis,* high rainfall zones, acidic soils in Tunisia (2,000 ha), Algeria (1,200 ha), Morocco (15,000 ha).
System 11:   *Abies moroccana,* in high elevation: 5,500 ha in N. Morocco; *Abies numidica,* in high elevations: 1,000 ha in N. Algeria.
System 12:   *Juniperus thurifera,* at elevations above 2,500 m: 31,000 ha in Morocco and 1,000 ha in Algeria.
System 13:   *Cupressus atlantica:* 9,500 ha in Morocco; *Cupressus sempervirens:* 100 ha in Algeria and 100 ha in Tunisia.
System 14:   *Quercus toza* (deciduous oak): 14,000 ha in N. Morocco.
System 15:   *Pinus nigra:* 100 ha in N. Morocco.
System 16:   *Cupressus dupreziana,* a few hectares in a deep wadi bed in the Tassili n'Ajjer in the central Sahara of Algeria.

estimated to be of the order of 1 percent per annum (Le Houérou, 1981a; Bourbouze, 1982). There are two reasons for this:

1. The ever-increasing pressure of man due to high population growth (3.5 percent per annum in Libya, Morocco and Algeria; 2.8 percent in Egypt and Tunisia — that is, a doubling period of 20 and 25 years, respectively), resulting in clearance of forest land and grazing land for cultivation and increased pressure of wood cutting, charcoal making and browsing; and
2. Forest legislation and policies that are obsolete. Forest land and grazing lands belong to the state and the local communities have no say in their management.

Table 2   Livestock population in northern Africa (10^3 adult head)

Species	Cattle	Horses and mules	Donkeys	Sheep	Goats	Camels
Algeria	1,400	380	570	14,700	3,000	164
Egypt	1,825	11	1,780	1,450	1,500	54
Libya	200	14	105	4,800	1,500	135
Morocco	3,300	776	1,200	12,000	4,500	200
Tunisia	600	127	210	5,230	1,030	177
Total	7,375	1,308	3,865	38,180	11,530	730
Conversion rate	5	5	3	1	0.8	7
Sheep equivalents	36,875	6,540	11,595	38,180	9,224	5,110
Grand total:	107,524	sheep equivalents (SE)				

Annual feed requirement: 108 million SE x 300 SFU = 32.4 billion SFU. Production from silvopastoral systems = 3.0 billion SFU (from about 9 million ha).
Production per hectare = 330 SFU/ha = 1,000 kg DM ha^{-1} yr^{-1}.

Source: FAO *Production Yearbook*, 1984.

The protection of forest by state agents is a matter of:

1. Repression which can no longer be enforced because of the very high population pressure; and
2. A shift of policy and legislation giving management responsibility to the local communities of users which could perhaps improve the situation in this respect (Bourbouze, 1986).

The decline in productivity is a result of over-exploitation — overstocking, overgrazing, overbrowsing, excessive woodcutting. Trampling, runoff, erosion, loss of soil fertility and loss of productivity of the overall ecosystem ensue. The current level of productivity is often only a third of that in reasonably well managed ecosystems and as low as a tenth in extreme cases: rain-use efficiency (RUE) factors of 0.5–1.0 kg DM ha^{-1} yr^{-1} mm^{-1}, instead of the usual 4–6, are common (Le Houérou, 1984).

Intertropical Africa

General climate
Intertropical Africa is a large area of some 22 million km² encompassing a number of ecoclimatic zones (Le Houérou and Popov, 1981). Precipitation occurs mostly during the long-days season (summer), although there are monomodal and bimodal rainfall patterns. The number of rainy seasons depends usually on latitude: one rainy season in the 10–23° latitudes N and S and two seasons in the 0–10° latitudes on both sides of the equator. The monomodal regime is called "tropical" and the bimodal "equatorial". Both tropical and equatorial intertropical climates have dry (desert) and wet (rain forest) subtypes, which tend to become "amodal" (no rains or continuous rains) in extreme situations.

Frost may occur in higher elevations. The lower altitude of frost occurrence naturally depends on the latitude: about 1,000 m a.s.l. near the Tropics of Cancer and Capricorn and around 2,500 m a.s.l. at the equator — that is, an increase of about 63 m for each degree of decrease in latitude or 0.57 m per km towards the equator. The isoline of year-long night frost is about 1,500 m higher, i.e., 2,500 m at the Tropics of Cancer and Capricorn and 4,000 m at the equator, and the lapse rate with latitude is similar. Frost occurrence gives rise to subtropical and subequatorial climates according to the number of rainy seasons. These montane climates in Africa are ecologically quite peculiar: they are mainly located in eastern and southern Africa. The areas with year-long night frost are referred to as Afro-Alpine: they too have very peculiar types of vegetation.

The areas occupied by the main ecoclimatic zones in intertropical Africa are shown in Table 3.

Table 3   Ecoclimatic zonation of intertropical Africa (10^3 km²)

Main types	Tropical	Equatorial	Montane	Afro-Alpine	Total	%
Hyperarid	3,170	423	50	-	3,643	18
Arid	1,573	459	118	-	2,150	11
Semi-arid	1,084	541	277	-	1,902	9
Subhumid	1,676	415	425	10	2,526	12
Humid	2,648	283	237	40	3,208	16
Hyperhumid	2,500	4,200	150	20	6,870	34
Total	12,651	6,321	1,257	70	20,299	100
Percent	62	31	6	0.003	100	

Source: Le Houérou and Popov, 1981.

Desert silvopastoral systems

In the desert zone, silvopastoral systems are restricted to the drainage network in or along wadis which originate in montane areas such as the Ahaggar, Tibesti Air and Adrar. These are the only sites used by desert pastoralists. Because of episodic runoff and deep underflows, permanent vegetation is present with more-or-less scattered individual trees and shrubs such as *Acacia tortilis* var. *raddiana, A. ehrenbergiana, Balanites aegyptiaca, Maerua crassifolia, Capparis decidua, Grewia tenax, Salvadora persica, Ziziphus lotus saharae, Leptadenia pyrotechnica, Ochradenus baccatus, Aerva persica, Tamarix aphylla, Moringa peregrina, Nucularia perrini, Traganum nudatum, Cornulaca monocantha.* These are mixed with herbs, legumes and perennial grasses such as *Panicum turgidum, Lasiurus hirsutus, Cymbopogon proximus, Pennisetum dichotomum, Stipagrostis pungens* and a palatable perennial desert sedge, *Cyperus conglomeratus.*

East African deserts have a similarly large number of tree and shrub species in the silvopastoral vegetation that colonizes the wadis and depressions of the Danakil, Somali and Chalbi deserts: *Acacia raddiana, A. tortilis tortilis, A. t. spirocarpa, A. nubica, A. edgeworthii, A. reficiens, Balanites aegyptiaca, Hyphaene thebaica, Salvadora persica, Boscia* sp., *Cadaba* sp., *Dobera glabra* mixed with dwarf shrubs such as *Duospermum eremophilum, Indigofera spinosa,* and *Sericocomopsis hildebrandtii.* There are also perennial grasses such as *Panicum turgidum, Chrysopogon plumulosus, Cenchrus ciliaris,*

Andropogon spp. and a number of annual grasses of low value such as *Aristida adscensionis, A. mutabilis, Chloris pycnothrix,* and *Tetrapogon cenchriformis.* These systems are used in the same way as in the Sahara by nomadic pastoral tribes such as the Afar, Issa, Somali, Boran, Rendille and Turkana. These silvopastoral zones in the deserts are the very basis of these pastoral systems since there is virtually no perennial vegetation outside the silvopastoral depressions and wadis. More than any other, these desert silvopastoral systems are threatened by desertisation for the same reason of demographic growth as in northern Africa. But without these systems life cannot be sustained in the desert zones.

Arid and semi-arid silvopastoral systems
These systems cover large areas in the Sahel (3 million sq km), in East Africa (Ogaden, Somalia, northern Kenya, northern Tanzania), and Botswana, Kalahari and Namibia in southern Africa (nearly 4 million sq km). Arid and semi-arid silvopastoral systems cover about a third of intertropical Africa and sustain about 60 percent of the livestock population of the continent (Le Houérou, 1977).

These silvopastoral systems are essentially savannas with various densities of shrubs and trees from less than 100 to over 3,000 individual "trubs" per hectare, dotting a herbaceous layer of annual grasses and herbs. The canopy area of the ligneous stratum also varies greatly from less than 1 percent to over 30 percent. The arid zone (100–400 mm of annual rainfall) is dominated by species in the genus *Acacia* and related genera of the Mimosoideae subfamily of legumes. In East Africa, species in the genus *Commiphora* are often found with the acacias. The semi-arid zone (400–600 mm of annual rainfall) is more varied in terms of woody cover. The Combretaceae often play a major role (*Combretum* spp., *Terminalia* spp., *Anogeissus* sp., *Guiera* sp.), but there are still a few acacias, particularly *A. seyal,* on fine-textured soils, but also *A. senegal, A. mellifera* in the Sahel, *A. etbaica, A. drepanolobium* and *A. bussei* in East Africa, whilst agrosilvopastoral systems are dominated by *Acacia albida* (*Faidherbia albida*).

There are about 100 main species of browse in the silvopastoral systems of the Sahel and about as many in the arid and semi-arid zones of eastern and southern Africa. About 30 percent of the species are common to the three regions (Walker, 1980; Lamprey *et al.,* 1980; Le Houérou, 1978, 1980a,b; von Maydell, 1983, this volume). In terms of animal nutrition woody species play a major role in these systems because they are the only source of protein, carotene and phosphorus during the long dry season and therefore livestock and wildlife cannot survive without them. This fact was not fully realized by scientists until recently, but it is now well accepted among livestock scientists, wildlife specialists, ecologists and range scientists (McKell, *et al.,* 1972; McKell, 1987; Le Houérou, 1973, 1980a,b).

Among the woody species, Capparidaceae play a major role in arid zones because of their high content of nitrogen and minerals (15–25 percent protein and 8–16 percent of non-silica minerals in the dry matter). Particularly significant genera in this respect are *Maerua, Cadaba, Boscia, Crataeva,* and *Capparis* (Lamprey *et al.,* 1980; Le Houérou, 1980c).

In the present socio-economic context, the role of browse in silvopastoral systems is to provide stability and productivity of livestock production, which is the major source of livelihood and income in arid and semi-arid African zones. The role they play could, in principle, be substituted by various concentrated feeds such as non-protein nitrogen (urea), salt licks and molasses, as practised in climatically comparable situations in developed countries such as the USA and Australia, but, given the terms of trade between livestock products on the one hand and the cost of transport and industrial products and by-products on the other, solutions of this type will not be used in pastoral zones of intertropical Africa in the foreseeable future.

Silvopastoral systems in the subhumid and humid climates

Subhumid and humid silvopastoral systems cover very large areas in Africa: most of the Sudanian ecological zone from the Atlantic Ocean to Lake Victoria and the Indian Ocean, and large parts of eastern and southern Africa, including the *miombo,* with a total area of 6–7 million sq km, i.e., about a third of intertropical Africa. Some 20 percent of the African livestock population and a large, but unknown, proportion of its wild ungulate populations live in this area. The livestock population amounts to 120–130 million head or 40–50 million TLU.

These are wooded savannas of the Sudanian ecological region, and the homologous zone in the southern hemisphere is the *miombo.* It also includes the so-called derived savannas of the Guinea-Congolese rain forest ecological zone. The structure and function of this type of ecosystem are quite different from those of the arid and semi-arid zone, for various reasons. First, there are three very distinct sets of components: trees, shrubs and tall grasses. Capparidaceae and Mimosoideae do not play the major role they do in drier climates: the botanical composition of the woody layers is more diversified, with the exception of the *miombo.* Trees of the Leguminosae family, mainly Papilionoideae and Caesalpinioideae, are actually dominant in the *miombo.* Perennial tall grasses are dominant in the herbaceous layer — these include many Andropogonoideae, particularly the genera *Andropogon, Hyparrhenia* and related genera (Le Houérou, 1977). Secondly, bush fires play an essential role in the ecosystem, particularly as regards the grazing resource, since they induce regrowth of the perennial grasses in the dry season, thus easing the feed shortage at that period. Thirdly, soils are very different: large areas have ferruginous tropical soils and ferralitic hardpans, which practically prohibit cultivation on the interfluves. Fourthly, climate is more favourable with a rainy season lasting 5–9 months in the Sudanian and *miombo* zones and 10 or more months in the Guinea-Congo zone. Finally, livestock husbandry is impaired by trypanosomiasis since most of these silvopastoral zones are infested to a greater or lesser degree by tsetse flies (*Glossina* spp.), the vectors of the trypanosomes. As a consequence, animal husbandry is not of great importance in these zones as long as they remain silvopastoral in nature since the tsetse flies only live in areas with some tree or bush cover. Indeed, one of the methods of elimination of trypanosomiasis is destruction of the woody layer, hence of the silvopastoral system! However, there are trypanotolerant breeds of stock, such as, for cattle, the N'Dama breed originating in the mountains of Fouta Djalon and Nimba in Guinea and the West African shorthorns, a complex of breeds from Ghana, Nigeria and Ivory Coast. Smallstock also have trypanotolerant breeds such as the Djalonke sheep of Guinea and neighbouring countries, and the achondroplasic West African dwarf goat of Nigeria, Ghana, Ivory Coast and neighbouring countries.

These breeds lend themselves to some livestock husbandry in tsetse-infested areas. Some countries, such as Ivory Coast and Zaire, have long since initiated a policy of meat production based on these trypanotolerant breeds, particularly the N'Dama, rather than the "negative" policy adopted in other countries of destroying the tree layer in order to permit an animal-husbandry system based on exotic breeds, possibly in conjunction with the intensive use of chemicals. The policy of encouraging local trypanotolerant breeds seems a sound one because various tsetse-eradication schemes have rather poor records and trypanotolerant breeds allow the preservation of ecological diversity, including maintaining a tree layer.

In the humid and sub-humid climates, the silvopastoral systems do not play the major role that they do in arid and semi-arid situations. This is particularly so because there are viable alternatives to the use of browse in the relatively shorter dry seasons in these regions. For example, the most currently used alternative is burning. The burning of the tall *Andropogon* grass at the beginning of the dry season allows some regrowth a few weeks later which provides a small amount of fodder of high quality that would adequately complement the roughage available in the unburnt areas. A skilful use and timing of controlled burning may thus ensure an adequate diet for stock throughout the dry season. This method has been used for decades, particularly in large meat-production ranches based on the N'Dama breed at Mwesi in southern Zaire (Le Houérou, 1977). Nevertheless, in areas of mild tsetse infestation browse remains a major component of livestock diet during the dry season in the subhumid zones, especially between the isohytes of 600 and 1,200 mm of mean annual rainfall where both Zebu cattle and Taurine trypanotolerant cattle co-exist and interbreed (e.g., the West African Mere, the Sanga and the Nilotic Ankole). This remark is also valid for the dwarf goat in the humid zones of West Africa which consumes a large proportion of browse in its diet and plays an important role in the household economy of the small West African farmers (Mecha and Adegbola, 1980).

Highland (montane) silvopastoral systems

Montane silvopastoral systems do not represent very large areas in Africa, but they harbour a large livestock population. Together with the Afro-Alpine ecological zone, they cover some 50,000 sq km and support 120–150 million head (almost 50 million head for Ethiopia alone), or 20–30 percent of the intertropical African livestock population.

These montane silvopastoral systems play an essential role, particularly during the rainy season, unlike the systems in arid and semi-arid zones. This is because these montane areas are intensively cultivated so that stock have nowhere to go during the cropping season except to silvopastoral areas on rugged terrain. During the dry post-harvest season the stock mostly feed on crop residues. Moreover, the montane silvopastoral systems are also used as transhumant grazing land for neighbouring pastoral arid zones during dry seasons. Thus these areas are used by resident herds in the rainy season and by transhumant pastoralists during the dry season. They are thus heavily exploited all the year round, and, as a result, they are heavily depleted in terms of grazing value in spite of the favourable climatic conditions. They are therefore covered by herbaceous species of little value such as *Pennisetum schimperi, P. villosum,* and *Eleusine jaegeri,* the shrub and tree layers having been heavily chopped, lopped, overbrowsed, cut down, or eradicated.

The role of shrubs and trees in these montane silvopastoral systems is difficult to quantify or even evaluate. The only thing which is certain is that they are receding at a rapid pace. A study on a number of test zones in the Ethiopian highlands, based on the comparison of aerial photos taken 20 to 25 years apart, yielded frightening results (Haywood, personal communication).

The role of shrubs and trees

As mentioned above, the shrubs and trees in the silvopastoral systems of Africa have a multiplicity of roles, and the literature on them is extensive. Therefore the topic will be dealt with in very general terms with special emphasis on two subjects, namely, browse and the screen function.

Traditional uses

There are innumerable traditional uses of shrubs and trees. Some African languages, for instance, use the same word to mean tree and medicine (Dia, 1986). The traditional uses may be grouped into 13 main headings:

1. Food and drink for humans;
2. Medical treatments for humans and livestock;
3. Browse for livestock and wildlife;
4. Beekeeping and honey production;
5. Dye and tanning material;
6. Source of energy — firewood and charcoal;
7. Building, furniture, fencing material;
8. Tools for agriculture, cottage industry, musical instruments, etc.;
9. Religious objects, art, handicraft, witchcraft;
10. Fibre for textiles, ropes, mats and carpets, etc.;
11. Shade and shelter for plants (sciaphytes, cover-crops), livestock, wildlife and humans (palaver trees);
12. Protection against water and wind erosion, maintenance of soil fertility and productivity, nutrient recycling;
13. Water storage tanks (baobab).

Generally speaking almost all parts of trees are used in one way or another. To cite one example, *Acocanthera schimperi* in the East African montane zone produces delicious fruits for human consumption, is a good firewood species, and the bark of the roots is used, after some preparation, to make a lethal arrow poison!

Browse

The browse aspect of trees and shrubs has been reviewed in some detail by the author (Le Houérou, 1980a) for western and southern Africa, by Lamprey *et al.* (1980) for East Africa, by Walker (1980) for southern Africa, and by Lawton (1980) for the *miombo*.

There are about 200 main browse species in intertropical Africa. Special mention needs to be made of legumes (mainly Mimosoideae and Caesalpiniodeae), Capparidaceae, Combretaceae, Tiliaceae, Rubiaceae and Rhamnaceae, in decreasing order of importance, in animal diets. Average values of chemical composition and feed value, based on a review by the author (Le Houérou, 1980c) for some 850 samples and 150 species from East and West Africa are: crude protein 13 percent of the DM, crude fibre 24 percent, fat 4 percent, nitrogen-free extract 48 percent, ash 11 percent, phosphorus 0.18 percent and calcium 1.75 percent. Capparidaceae yield an average crude protein content of 20 percent, while legumes average 16 percent. Leaves of some Capparidaceae have an average crude protein content of 30 percent, which makes them similar in this respect to protein-rich concentrated feed! A figure of 44 percent crude protein has been reported in the leaves of *Justicia salvioides* a species of Acanthaceae from the *miombo* (Lawton, 1980).

Feed value has been estimated using various *in vivo* trials on digestibility and animal performance. For leaves and shoots, DM digestibility is, in general, 50 ± 5 percent. There are exceptions of very low digestibility (25–35 percent) as well as very high figures of 60–70 percent *in vivo* (Mabey and Rose-Innes, 1964, 1966).

Some specialists of animal production have questioned the validity of the proximal analyses data, as well as those concerning digestibility of energy or protein, due to the presence of tannins and other polyphenolic compounds that may inhibit the actual assimilation of nitrogen and sugars, particularly in the intestine. The author's studies on animal performance over long periods, using sheep feeding only on known quantities of selected browse, do not confirm the negative impact of browse on sheep performance. On the contrary, the experiments carried out continuously over nearly one year, showed that animals feeding on the 10 species of browse under experiment showed a performance in direct relation to the amount of feed consumed. Moreover, sheep feeding partly on range and receiving a browse complement in the pen performed significantly better than those feeding on a pure range diet, even under excellent range conditions.

The productivity of browse in arid and semi-arid silvopastoral systems was reviewed by the author (Le Houérou, 1980a). The order of magnitude of productivity is 1 kg DM ha^{-1} mm^{-1} of consumable biomass per year, where mm refers to average rainfall per year. That is about a third of the productivity of the herbaceous layer. This figure is, of course, liable to enormous variability depending on the ecosystem concerned and on its present and past management. The range is usually 0.1–5.0 kg DM ha^{-1} yr^{-1} mm^{-1}. Production of 5,000 kg DM ha^{-1} yr^{-1} has been reported from *Acacia hockii* in riverine conditions in East Africa (Pellew, 1980), a figure that is probably close to the maximum under natural conditions.

Average individual "trub" production is of the order of 1 kg DM yr^{-1}, with a range of 0.08–50.0 kg, most figures ranging from 0.5 to 5.0 kg (Bille, 1977; Poupon, 1980; Hiernaux, 1980; Dayton, 1978; Pellew, 1980; Kelly, 1973; Kennan, 1969; Rutherford, 1978; Cissé, 1980a,b, 1986).

Equally important are the availability of browse in critical seasons and its direct accessibility to livestock and wildlife. The maximum standing crop of browse occurs naturally during the rainy season, but the most important is its availability during the dry season and the so-called "pre-rainy" season. In the Sahel, for instance, the most critical seasons are the hot dry season from March to May and the pre-rainy season in June-July. During these seasons, not more than a third of the maximum standing crop of that present in October is available to animals, although some of the dried leaves that fall are gathered by stock. Thus the amount of browse available at the most critical periods does not represent more than 20–25 percent of annual production. The maximum standing crop in September-October in the Sahel represents about 80 percent of annual production. But this does not tell the whole story because part of the crop can be stored in the form of dried pods such as those of *Acacia albida, Acacia tortilis, Prosopis juliflora,* and *P. cineraria.* These pods are traded in local markets for a monetary value close to that of similar concentrates of similar feed value, which also shows that local farmers and stock owners are well aware of the actual value of these products.

The distribution and structure of available browse in space is another important issue. Smallstock can browse up to about 130 cm above the ground, cattle about 2 m, camels 3 to 4 m, and giraffe up to 5 m. The distribution of browse in these strata is thus an important point to consider in relation to the species of livestock which are kept (goats are known to be able to climb trees). Distribution of browse above the ground is extremely variable and depends on many factors: nature and botanical composition of the "trub" layer, kind of present and past management applied to it (trimming, lopping, etc.). In central Mali, for instance, Hiernaux and Cissé (1983) estimated that about 50 percent of the browse is located between the ground and a height of 2m, 30 percent between 2 and 3 m, and some 20 percent above 3 m. The horizontal distribution is also of relevance since the woody layer is usually not randomly distributed but follows a pattern more or less linked to runoff and drainage. This horizontal distribution, in turn, affects the nature of the species which are present and therefore affects livestock behaviour as well as the practices of the herder who will seek particular species in their preferred sites.

In the Sahel, browse species consist of groups with specific characteristics which determine the type of management required (Gillet, 1986). Thus one could differentiate those species having permanent or quasi-permanent green leaves such as *Maerua crassifolia, Balanites aegyptiaca, Boscia senegalensis, B. agustifolia, B. salicifolia, Cadaba glandulosa* and *Salvadora persica* (which all belong to the family Capparidaceae except *Balanites* and *Salvadora*).

Then there are those species which leaf early at the end of the dry season (precession of leafing) such as *Commiphora africana;* these are used only for a short period in June-July before green grass becomes available. *Commiphora africana* is, therefore, a most important species at this particular time of the year, which is the most critical in the annual cycle when there is little feed available elsewhere and animals are emaciated by nine months of dry season and malnutrition. Moreover, females that are usually in a late stage of pregnancy at this time need protein-rich diets which only *Commiphora* can supply at that time.

The winter flowering species also play a significant role as the flowers, shoots, young leaves and young fruits are consumed then. This is the case, for instance, with several Combretaceae (*Combretum, Terminalia*) and also with some of the Sudanian species such as *Gardenia erubescens* (Rubiaceae).

Species which set leaves right at the onset of the rains, such as *Combretum aculeatum* and *Feretia apodanthera,* also need to be differentiated. These are also keenly sought species at a time of scarcity, and they play a role somewhat similar to *Commiphora* but in a somewhat different way. They are browsed throughout the rainy season and the first half of the dry season, whereas *Commiphora* is ignored as soon as green grass is available, and sheds its leaves soon after the last September rain.

A fifth group is composed of species whose foliage is lopped and fed to penned or household animals, or prestige animals (household sheep — "moutons de case" — and horses). Among these, there are five main species in West Africa: *Pterocarpus lucens,* particularly in Mali and Niger, *Pterocarpus erinaceus* in southern Senegal, *Khaya senegalensis* in several countries, *Terminalia avicennoides* in Niger, and *Acacia albida* in many countries of Africa.

A last group includes the species usually found around termitaria which remain green longer in the dry season than in the surrounding bush; they are therefore important browse species, particularly *Grewia bicolor, Grewia villosa, Boscia salicifolia, Maerua oblongifolia, Feretia apodanthera.*

The screen function

The screen function is obvious from field observations: animals and people seek shade at midday when shade temperatures may reach 35–45° C. In depleted rangelands of the arid and semi-arid zone, grass-layer production is usually much higher (up to twice or more) under shade than in the open; and it remains green 4–6 weeks longer at the end of the rainy season. Measurements in the semi-arid zone of Botswana (25° S, 25° 50′ E; 550 mm; 1,000 m altitude) over several years showed that under the canopy of *Peltophorum africana, Acacia tortilis* and *Grewia flava,* solar radiation and windspeed were reduced by about 50 percent as compared to a nearby open test area. As a consequence, potential evapotranspiration was reduced by 70 percent under the canopy, while the continuous grass layer of *Panicum maximum* had a production 26 percent higher when grazed and 12 percent when ungrazed (Pratchell, personal communication). Bille (1977) reported a production of the grass layer twice as large in the shade as compared to the open in the Sahel of Senegal (combined production of grass and shrubs). The photosynthetic efficiency during the growing season was 1.4 percent, whereas it barely reached 0.3 percent in the surrounding open grassland (global radiation 180 kcal cm^{-2} yr^{-1} (752 kJ); incident radiation 90 kcal cm^{-2} yr^{-1} (376.2 kJ); 1 g DM = 4.2 kcal cm^{-2} yr^{-1} (17.6 kJ); NPP = 125 percent of maximum standing crop). The efficiency of the system was thus 4.6 times greater in the multistorey vegetation structure as compared to a monostratum grass layer. These facts, however, should not be overgeneralized as there are instances where shrub cover may reduce production from grassland understorey, particularly when the ecosystem is in a good dynamic status (Le Houérou, 1984).

Present trends in African silvopastoral systems

All monitoring studies in the Sahel and East Africa show a considerable regression of the woody cover over the past 30 years, caused by a combination of two phenomena: prolonged droughts from 1970 to 1985 and over-utilization resulting from the growth of human and stock populations which has been occurring at a rate of about 2.5 percent per annum. The comparison of aerial photographs taken at about 20-year intervals has given absolute proof of the substantial decline of the tree and shrub cover in Chad, Sudan, Niger, Mali, Burkina Faso, Senegal, Ethiopia, Somalia, and Kenya. The decline has been of the order of 20–35 percent between 1954 and 1975 and has accelerated since the early 1980s (Gaston, 1975; Haywood, 1981; De Wispelaere, 1980; De Wispelaere and Toutain, 1976; Gaston, 1981; Lamprey, 1975, 1983; Lusigi, 1981; Boudet, 1972, 1977; Le Houérou, 1977, 1980a, 1981b; Peyre de Fabrègues, 1985; Peyre de Fabrègues and De Wispelaere, 1984; Gillet, 1986; Le Houérou and Gillet, 1986). These results are confirmed by field surveys over the whole Sahel; for instance the mean mortality of trees and shrubs during the 1969–1973 drought has been estimated at 40–50 percent. Over some areas the die-off was total. Detailed research on the functioning of the Sahelian ecosystems, conducted at Fete Ole in northern Senegal over a period of 10 years, including the 1969–1973 drought, showed an average mortality of 20 percent under conditions of total protection (Bille, 1977, 1978; Poupon, 1980). Naturally, under current conditions of overexploitation the situation is much worse, with mortalities often reaching 50 percent. A study of the ligneous population in Fete Ole showed that 30 years of total protection would be required to offset the effects of the 1969–1973 drought and to bring the production back to its pre-drought level (Bille, 1978). But the subsequent 1980–1985 period was far worse (Tucker *et al.,* 1986; Le Houérou and Gillet, 1985; Peyre de Fabrègues, 1985). According to most ecologists familiar with the Sahel situation, the Sahel might never recover from these two consecutive droughts (Peyre de Fabrègues and De Wispelaere, 1986). The lack of regeneration after the droughts can be explained by three main factors:

1. A great many trees were badly pruned or lopped in order to provide fodder for stock during the drought. Many trees did not survive this treatment, mainly because cutting was done over the whole canopy or using methods which permanently damaged it. Even if the trees survived this treatment they were not able to withstand any subsequent bush fires (Piot, 1980);

2.   The small quantity of seeds and seedlings produced were immediately eaten out by hungry stock;

3.   Individual trees or shrubs in areas near the dry limit of their geographic distribution suffered most; whole stands died off and regeneration is thus most unlikely. This is the case, for instance, with *Sclerocarya birrea, Commiphora africana, Guiera senegalensis* and others in the Ferlo of Senegal (Piot and Diaite, 1983;).

4.   Firewood gathering has disastrous effects on browse and the development of populations of browse species. Wilson (1980) showed that the small town of Niono in Central Mali, with a population of 15,000 inhabitants (growing at 12 percent annually) consumed the equivalent of the browse production of 37,000 hectares of *Pterocarpus lucens* as firewood annually. This species is one of the best, if not *the* best browse species of the Sahel: its foliage is sold on the markets. If the present demographic growth of the town remains unchanged over the next 15 years, the firewood consumption will reach the equivalent of the production of 400,000 hectares of browse annually, thus reducing the stocking capacity of the neighbourhood by 60,000 head of cattle since there is little, if any; regeneration of *Pterocarpus* under the present-day conditions of exploitation.

Assuming that there will be three drought periods in the next hundred years — there have been four since the beginning of the present century — it is to be feared that the ligneous cover of the Sahel will be reduced to close to zero within the next 50 years (unless continuous regeneration could be ensured through good management). Such regeneration being virtually non-existent as environmental degradation worsens, the Sahel would become unexploitable by livestock outside the rainy season, unless animals are fed concentrates and/or urea and minerals in order to offset the shortage of browse. Of course, such complementary feeding is possible in principle as it is used in other arid regions, though under totally different socio-economic conditions. However, this solution is not likely to be a viable one in the Sahel or under East African socio-economic conditions in the foreseeable future because of cost/benefit considerations and the lack of adequate managerial skills. Given current meat producer prices and prevailing offtake rates in pastoral production systems, scarcely any investment is economically feasible (though slight differences exist between countries in this respect). In fact, one cannot see any way in which livestock could be supplemented for 8–9 months of the year with the present prices of meat and concentrates (De Montgolfier-Kouevi and Le Houérou, 1980; Le Houérou, 1987b).

The destruction of the ligneous cover of browse species would therefore lead to a dead-end for the livestock industry in the Sahel and in large parts of East Africa before the middle of next century, if not earlier, if population growth continues unabated. Generally speaking, the situation is not yet as desperate in East Africa as it is in the Sahel; but locally it may be just as bad, particularly in Somalia, south-east and north-east Ethiopia and northern Kenya. An unmistakable indicator of environmental deterioration is when traditional cattle and sheep pastoralists shift to rearing camels and goats, that is from grazers to browsers. Some ethnic groups such as the Rendille and Samburu of northern Kenya have already made this shift.

Conclusions

As indicated previously, for nutritional and economic reasons, browse should represent 20–25 percent of domestic ruminants' intake (30 percent in the dry season, 5 percent in the rainy season) in the arid and semi-arid silvopastoral areas of Africa. How could such a balance be ensured? Or, to put it in another way, how could sound management be established on the principle of sustained maximum output? Such an objective would require a careful and co-ordinated course of action.

At the conceptual and planning level, the philosophy and objectives ought to be clearly defined, and the strategy and means to attain the objective targets should be clearly spelt out at the outset and selected and described. This implies a dynamic livestock policy, including stratification of the industry, a marketing and a price policy giving a fair share to the producer, and also promotion of quality products.

At the technical execution level, in the field, the problem is theoretically simple — if not easy. It is, after all, a question of adapting stocking rates to the carrying capacity of the ecosystem, that is an offtake of the primary production of not more than 25–30 percent. In practice this would mean that human and animal densities would be controlled and that simple techniques such as deferred grazing, periodic enclosures and the adaptation of watering regimes to the density and seasonal occurrence of watering points would be applied. Of course, this kind of management implies choices and daily decisions, i.e., the notion of responsibility. Yet the present situation in dry Africa is characterized by a general lack of responsibility at resource-management level, in other words in communal or common ownership of land and water.

Water and pasture are actually common or public resources, whereas animals are privately owned, so it is in every user's interest to draw a maximum and immediate profit from the common resource without bothering about what may happen to it in the long run. Such a situation results in destruction of the common resources in the long run for the sake of the individual's own immediate benefit. This has been labelled the "looting strategy" (Le Houérou, 1977), or the "tragedy of the commons" (Hardin, 1968).

It is quite obvious that no rational system of any kind can be implemented without the concept of responsible management, whether by individuals or groups. Meeting these responsibilities involves fundamental land reforms in terms of land tenure, and of land and water usufruct. Such systems, on a collective basis, existed long ago in pre-colonial Africa but with a much lower density of human and animal population so that they resulted in a fairly steady balance between the resources and their utilization. This equilibrium, which was certainly not ideal as regards productivity, human dignity or social justice, is badly endangered today by severe overutilization. If drastic socio-political reforms are not implemented without delay in order to ensure the rational management of shrubland ecosystems, the arid and semi-arid zones of Africa will have to face a very deep crisis threatening their present main resource.

The situation in the subhumid and humid eco-climatic zones is certainly less gloomy, and development prospects are much better, at least on deep soils. But what will happen to the immense areas of shrublands and forests growing on shallow ferruginous or lateritic (ferralitic) hardpans? At present browse plays an important part in livestock production, especially during the second half of the dry seasons.

REFERENCES

Bille, J.C. 1977. Etude de la production primaire nette d'un écosystème sahélien. Trav. & Doc. de l'ORSTOM, No. 65, Paris.
——————. 1978. Woody forage species in the Sahel, their biology and use. *Proceedings of the First International Rangeland Congress,* Denver, Colorado.
Boudet, G.G. 1972. Désertification de l'Afrique tropicale sèche. *Adansonia,* ser. 2, 12 (4): 505–524.
——————. 1977. Désertification ou remontée biologique au Sahel? *Cah. ORSTOM, Ser. Biol.,* XII, 4: 293–300.
Bourbouze, A. 1982. L'élevage dans la montagne marocaine. Organisation de l'espace et utilisation des parcours par les éleveurs du Haut Atlas. Doct. Dissert. Inst. Nat. Agron., Paris-Grignon.
——————. 1986. Adaptation *à* différents milieux des systèmes de production des paysans du haut Atlas. *Techniques et Cultures* 7, 59–94.
Brenan, J.P. 1979. Some aspects of the phytogeography of tropical Africa. *Ann. Missouri Bot. Garden* 65: 437–478.
Cissé, M.I. 1980a. The forage production of some Sahelian shrubs: relations between maximum leaf biomass and various physical parameters. In H.N. Le Houérou (ed.), *Browse in Africa.* Addis Ababa: ILCA.
——————. 1980b. Effects of various stripping regimes on foliage production of some browse bushes of the Sudano-Sahelian zone. In H.N. Le Houérou (ed.), *Browse in Africa.* Addis Ababa: ILCA.
——————. 1986. Les parcours sahéliens pluviaux du Mali Central. Caractéristiques et principes techniques pour une amélioration de leur gestion dans le cadre des systèmes de production animale existants. In I.A. Touré, P.I. Dia and M. Maldague (eds.), *La problématique et les stratégies sylvo-pastorales au sahel.* CIEM, Univ. de Laval, Quebec and UNESCO, Paris.
Dayton, B.R. 1978. Standing crop of dominant *Combretum* species at three browsing levels in the Kruger National Park. *Koedoe* 21: 27–76.
De Montgolfier-Kouevi, C. and H.N. Le Houérou. 1980. Study on the economic viability of browse plantations in Africa. In H.N. Le Houérou (ed.), *Browse in Africa.* Addis Ababa: ILCA.
De Wispelaere, G. 1980. Les photographies aeriennes témoins de la dégradation du couvert ligneux dans un écosystème sahélien sénégalais. Influence de la proximité d'un forage. *Cah. ORSTOM, sér. Sces Hum.,* XVII, 3–4: 155–166.
De Wispelaere, G. and B.Toutain. 1976. Estimation de l'évolution du couvert végétal en vingt ans, consécutivement à la sécheresse dans le Sahel voltaïque. *Rev. Photointerpr.* 76: 3/2.
Dia, P.I. 1986. Espace, ressources et pasteurs du sahel sénégalais. In I.A. Touré, P.I. Dia and M. Maldague, *La problématique et les stratégies sylvo-pastorales au sahel.* CIEM, Univ. de Laval, Quebec and UNESCO, Paris.
FAO. 1984. *FAO Production Yearbook.* Rome: FAO.
——————. 1985. *FAO Production Yearbook.* Rome: FAO.
FAO/UNESCO. 1972/1975. *World soil map 1/5,000,000: Africa.* FAO, Rome and UNESCO, Paris.
Gaston, A. 1975. Etude des pâturages du Kanem après la sècheresse de 1973. Farcha, Tchad: IEMVT.
——————. 1981. La végétation du Tchad: évolutions récentes sous influences climatiques et humaines. Thèse Doct. Sces, University of Paris XII, IEMVT, Maisons-Alfort.
Gillet, H. 1986. Principaux arbres fourragers du sahel sénégalais. In I.A. Touré, P.I. Dia and M. Maldague (eds.), *La problématique et les stratégies sylvo-pastorales au sahel.* CIEM, University of Laval, Quebec and UNESCO, Paris.
Hardin, G. 1968. The tragedy of the commons. *Science* 162: 1243–1248.
Haywood, M. 1981. Evolution de l'utilisation des terres et de la végétation dans la zone soudano-sahélienne du projet CIPEA au Mali. Addis Ababa: CIPEA/ILCA.
Hiernaux, P. 1980. Inventory of the browse potential of bushes, trees and shrubs in an area of the Sahel of Mali: methods and initial results. In H.N. Le Houérou (ed.), *Browse in Africa.* Addis Ababa: ILCA.
Hiernaux, P. and M.I. Cissé. 1983. Les ressources naturelles des systémes de production animale du Gourma. Etude exploratoire. Doc. de Progr. AZ/92 F, CIPEA, Bamako.
Kelly, R.D. 1973. *A comparative study of primary production under different kinds of land use in south eastern Rhodesia.* Ph. D. dissertation, University of London.

Kennan, T.C.D. 1969. The significance of bush in grazing land in Rhodesia. *Rhod. Scient. News* 3: 331–336.

Lamprey, H.F. 1975. Report on the desert encroachment reconnaissance in the northern Sudan. Nairobi: UNEP (mimeo).

————. 1983. Pastoralism yesterday and today: the overgrazing problem. In F. Bourlière (ed.), *Tropical savanna. Ecosystems of the world*, Vol. 13. Amsterdam: Elsevier.

Lamprey, H.F., D.J. Herlocker and C.R. Field. 1980. Report on the state of knowledge on browse in East Africa in 1980. In H.N. Le Houérou (ed.), *Browse in Africa*. Addis Ababa: ILCA.

Lawton, R.M. 1980. Browse in the Miombo. In H.N. Le Houérou (ed.), *Browse in Africa*. Addis Ababa: ILCA.

Le Houérou, H.N. 1973. Ecologie, démographie et production agricole dans les pays méditerranéens du tiers-monde. *Options Méditerranéennes* 17: 53–61.

————. 1976a. Problèmes et potentialités des terres arides nord-africaines. *Options Méditerranéenes* 26: 17–35.

————. 1976b. The rangelands of North Africa: typology, yield, productivity and development. In *Proceedings of the symposium on evaluation and mapping of tropical African rangelands*. Addis Ababa: ILCA.

————. 1978. The role of shrubs and trees in the management of natural grazing lands (with particular reference to protein production). Position paper, item No. 10, VIIIth World Forestry Congress, Jakarta, Indonesia and FAO, Rome.

————. 1977. The grasslands of Africa: classification, production, evolution and development outlook. *Proceedings of XIIIth International Grassland Congress*, Vol. I. Berlin, DDR: Akademie Verlag.

————. 1980a. Browse in northern Africa. In H.N. Le Houérou (ed.), *Browse in Africa*. Addis Ababa: ILCA.

————. 1980b. The role of browse in the sahelian and sudanian zones. In H.N. Le Houérou (ed.), *Browse in Africa*. Addis Ababa, ILCA.

————. 1980c. Chemical composition and nutritive value of browse in West Africa. In H.N. Le Houérou (ed.), *Browse in Africa*. Addis Ababa: ILCA.

————. 1981a. The impact of man and his livestock on mediterranean vegetation. In F. di Castri, D.W. Goodall and R.L. Specht (eds.), *Mediterranean-type shrublands. Ecosystems of the world*, Vol. 11. Amsterdam: Elsevier.

————. 1981b. Impact of the goat on Mediterranean ecosystems. Proceedings of the 32nd Annual Meeting of the European Association for Animal Production, Sheep and Goats Commission, Zagreb, Yugoslavia.

————. 1984. Rain-use efficiency: a unifying concept in arid lands ecology. *J. Arid Envir.* 7: 213–247.

————. 1987a. The shrublands of Africa. In C.M. McKell (ed.), *Shrub biology and utilization*. N.Y.: Academic Press (in press).

————. 1987b. An assessment of the economic feasibility of fodder shrub plantation in Africa. In C.M. McKell (ed.), *Shrub biology and utilization*. N.Y.: Academic Press (in press).

————. 1987c. Inventory and monitoring of rangeland ecosystems in the Sahel. FAO, Rome and UNEP, Nairobi.

————. 1987d. The grazing-land ecosystems of the Sahel. In *Ecological studies*. Heidelberg: Springer Verlag.

Le Houérou, H.N. and H. Gillet. 1986. Conservation vs desertisation in the African arid lands. In M.E. Soulé (ed.), *Conservation biology, the science of scarcity and diversity*. Sunderland, Massachussetts: Sinauer Associates.

Le Houérou, H.N. and G.F. Popov. 1981. An ecoclimatic classification of intertropical Africa. Plant Production Paper No. 31, FAO, Rome.

Lusigi, W.J. 1981. Combatting desertification and rehabilitating degraded production systems in Northern Kenya. Technical Report No. A-4. Nairobi: IPAL, UNESCO.

Mabey, G.L. and R. Rose-Innes. 1964/66. Studies in browse plants in Ghana: *Emp. J. of Exper. Agric.*: XXXII, 126: 114–130; 127: 180–190; 128: 274–278; N. Ser., 2: 27–32, 113–117.

McKell, C.M. (ed.). 1987. *Shrub biology and utilization*. New York: Academic Press.

McKell, C.M., J.P. Blaisdell and R.J. Goodin (eds.). 1972. Wildland shrubs — their biology and use. *USDA Forest Service, General Technical Report INT-I*, Washington, D.C.

Mecha, I. and T.A. Adegbola. 1980. Chemical composition of some southern Nigeria forage eaten by goats. In H.N. Le Houérou (ed.), *Browse in Africa*. Addis Ababa: ILCA.

Pellew, R.A. 1980. The production and consumption of Acacia browse and its potential for animal protein production. In H.N. Le Houérou (ed.). *Browse in Africa*. Addis Ababa: ILCA.

Peyre de Fabrègues, B. 1985. Quel avenir pour l'élevage au Sahel? *Rev. Elev. Medec. Vet. Pays Tropic.* (38) 4: 500–508.

Peyre de Fabrègues, B. and G. De Wispelaere. 1984. Sahel: fin d'un monde pastoral? *Marchés Tropicaux* 2488–2491 (12 Oct.).

Piot, J. 1980. Management and utilization methods for ligneous forage: natural stands and plantations. In H.N. Le Houérou (ed.), *Browse in Africa*. Addis Ababa: ILCA Addis Ababa.

Piot, J. and I. Diaité. 1983. Systèmes de production de'élevage au Sénégal. Etude du couvert ligneux. CTFT, Nogent/Marne and LNERV/ISRA, Dakar.

Poupon, H. 1980. Structure et dynamique de la strate ligneuse d'une steppe sahélienne au Nord-Sénégal. Trav. & Doc. de l'ORSTOM, No. 115, Paris.

Rutherford, M.C. 1978. Primary production ecology in South Africa. In M.J.A. Werger, *Biogeography and ecology of Southern Africa*. The Hague: Junk.

Tucker, C.J., C.O. Justice and S.D. Prince 1986. Monitoring the grasslands of the Sahel. *J. of Remote Sensing* 7 (11): 1571–1581.

Tucker, C.J., C.L. Vanpraet, M.J. Sharman and G. Van Ittersum. 1985. Satellite remote sensing of total herbaceous biomass production in the senegalese Sahel: 1980–1984. *Remote Sensing of the Environment* 17: 233–249.

von Maydell, H.-J. 1983. Arbres et arbustes du sahel. Eschborn, F.R.G: GTZ. Walker, B.H. 1980. A review of browse and its role in livestock production in Southern Africa. In H.N. Le Houérou (ed.), *Browse in Africa*. Addis Ababa: ILCA.

Wickens, G.E. 1980. Alternative use of browse species. In H.N. Le Houérou (ed.), *Browse in Africa*. Addis Ababa: ILCA.

Wilson, R.T. 1980. Fuelwood in a Central Malian town and its effects on browse availability. In H.N. Le Houérou (ed.), *Browse in Africa*. Addis Ababa: ILCA.

Whyte, R.O. 1974. The use and misuse of shrubs and trees as fodder: Africa. Joint publication No. 10, Imperial Agricultural Bureaux, Bureau of Pasture, Field Crops Forestry and Nutrition, Aberystwyth, Wales.

10

Homegardens:
a traditional agroforestry system
with a promising future

Otto Soemarwoto

Director
Institute of Ecology, Padjadjaran University
Bandung, Indonesia

Contents .

Introduction

Homegardens may have originated in prehistoric times when hunters and gatherers deliberately or accidentally dispersed seeds of highly valued fruit trees in the vicinity of their camp sites (Hutterer, 1984). Brownrigg (1985) in his literature review, mentioned that homegardens in the Near Eastern region were documented in paintings, papyrus illustrations and texts dating to the third millenium BC. These ancient gardens may have originated as early as the seventh millenium BC. They were attached to temples, palaces, elite residences and the homes of the common people. The homegarden was mentioned in an old Javanese charter of AD 860 (Terra, 1954). Perhaps the first published report in modern times was that by Raffles in 1818.

From this very brief historical sketch there is evidence that homegardening is a very old tradition which may have evolved over a long time from the practices of the hunters/ gatherers and continued in the ancient civilizations up to modern times.

This chapter is not intended to present a literature review of homegardens, but rather to discuss their salient features, based on the author's experience in Indonesia, as related to their potential and opportunities for future development, and the associated constraints and pitfalls.

The homegarden as an agroforestry system

A common interpretation of homegardens is that it is a system for the production of subsistence crops for the gardener and his family. It may or may not have the additional role of production of cash crops. It can be immediately surrounding the home or slightly further away, but still near the residential area. The Indonesian term *pekarangan* is derived from the .word *karang,* meaning a place of residence (Poerwadarminta, 1976) and, hence, *pekarangan* specifically refers to a garden on the residential site.

The term agroforestry denotes land-use systems consisting of a mixture of perennials and annuals, and often also animals. A major concern in agroforestry research is sustainability (Lundgren and Raintree, 1983). It is determined by the structure of the system, its ecological functions and its continued ability to fulfil the socio-economic needs of the people. Thus, as is implied by the term, homegarden as an agroforestry system should ideally combine the ecological functions of forests with those of providing the socio-economic needs of the people (Soemarwoto and Soemarwoto, 1984). The ecological functions of forests include hydrologic benefits, microclimatic modification and soil-erosion control, and genetic-resource conservation.

The structure and composition of homegardens

Structure is intimately linked to function. Although this relationship is obvious, too often we ignore it. Not surprisingly, manipulations of structure have often led to the loss of valuable functions, or vice versa, which in turn has resulted in conditions contrary to our expectations or even collapse of the system.

A prominent structural characteristic of the homegarden is the great diversity of species with many life forms varying from those creeping on the ground, such as the sweet potato, to tall trees of ten metres and more, e.g., the coconut palm, and vines climbing on bamboo poles and trees. These create the forest-like multistorey canopy structure of many homegardens. Well-known examples are those from Java, but they are also found in other parts of Asia such as Malaysia, the Philippines, Thailand, Sri Lanka and India, as well as in other continents. For example, at the First International Workshop on Tropical Homegardens held at the Institute of Ecology in Bandung in December 1985, such systems were reported from the Pacific, from Africa, and from Latin America by various authors. Such land-use systems are also abundantly reported in the literature (e.g., Anderson, 1954; Kimber, 1973; Fernandes *et al.,* 1984). Fernandes and Nair (1986) gave schematic presentations of the structure of different homegardens from various geographical regions and reported that the canopies of most homegardens consisted of two to five layers. Christanty *et al.* (1980) demonstrated the remarkably close resemblance of the light-interception curve of a homegarden in West Java with that of the Pasoh forest in Malaysia, as measured by Yoda (1974). These measurements provide an objective and precise method for the identification of the canopy layers. The forest-like structure is also derived from the lack of a discernible planting pattern: usually there are no rows, blocks or definite planting distances among components. To the casual observer, homegardens look haphazard.

A popular assumption is that the forest-like structure has been the result of deliberate planning to mimic the forest. This is doubtful, since in Java, where such homegarden structure is very well developed, forests do not have a high cultural value. They are considered dangerous places where wild animals roam and evil spirits dwell. On the other hand *babad alas* (forest clearing), e.g., for the establishment of a settlement, is considered a noble deed. The term *babad alas* still remains in the Javanese vocabulary to indicate pioneering activities of a praiseworthy nature, e.g., the foundation of a university. It is therefore inconceivable that people would deliberately create a "forest" in the surroundings of their home. Indeed a person feels offended when his homegarden is said to resemble a forest. It is more likely that the forest-like structure has been an accidental result of people's efforts to fulfil their many kinds of subsistence needs in the absence of a market economy. Karyono (1981) found associations of plants in rural homegardens in which the plants played complementary roles. For example, Table 1 lists the various economic plants with which pineapple is associated in the homegardens in rural areas of Citarum watershed, West Java, Indonesia.

Table 1   Association of pineapple (*Ananas comosus*) with various other plants in the homegardens in rural areas of Citarum watershed, West Java, Indonesia.

Associated plant	Economic use
Lemon grass (*Cymbopogon citratus*)	Spice
Hibiscus spp.	Ornamental
Ginger (*Zingiber officinarum*)	Medicinal, spice, cash crop
Muntingia calabura	Shade, fuelwood
Crescentia cujete	Medicinal, construction material
Banana (*Musa* spp.)	Fruit, subsidiary staple food, cash crop
Jackfruit (*Artocarpus heterophylla*)	Fruit, fuelwood, timber
Mango (*Mangifera indica*)	Fruit, fuelwood, cash
Custard apple (*Anona squamosa*)	Fruit, fuelwood, cash
Taro (*Colocasia* spp.)	Subsidiary staple food
Turmeric (*Curcuma longa*)	Medicinal, spice, cash
Sweet potato (*Ipomoea batatas*)	Subsidiary staple food, animal feed
Pithecellobium jiringa	Fruit, vegetable, cash, fuelwood
Sandoricum koetjape	Fruit, medicinal, construction material

Source: Karyono, 1981.

With the development of a market economy the complexity of the homegarden diminishes, and its resemblance to a forest tends to disappear, as described in a later section.

Species diversity and plant density vary from place to place, influenced by ecological and socio-economic factors. Kimber (1973) found 31, 30 and 67 species in three gardens in Martinique. The smallest garden was about 1,500 sq m. In Grenada there were 18 vegetable varieties and 13 distinct types of food trees in a sample garden of less than 2,000 sq m (Brierley, 1985). Thaman (1985) reported from random surveys of homegardens in Port Moresby, Papua New Guinea; Suva, Fiji; Nuku'alva, Tonga; South Tarawa, Kiribati;

Nauru Island; and the "Location" contract settlement on Nauru at least 85, 114, 79, 61, 33 and 65 different species and distinct varieties of food plants, respectively, in the homegardens in those areas. In addition, a very wide range of non-food plants was found in mixed homegardens which were of considerable importance for handicraft, fuel, medicine, fibre, dyes, ornamental purposes, perfumes and deodorants, livestock feed, and shade and construction materials. A total of 100 species were found in kitchen gardens and 77 species in hut gardens (a hut being a temporary shelter built in the field during the growing season of the field crop) in the Knon Kaen province in north-east Thailand (Kamtuo *et al.,* 1985).

The dynamics of species succession and plant density and composition of homegardens have been intensively studied by Indonesian researchers. In a West Javanese village of 41 households the average number of plant species per homegarden was 56. The total number of species in the village was 219 in the dry season and 272 in the wet season, i.e., an increase of almost 25 percent in the wet season. The highest increase was in the number of vegetable species, which was almost 75 percent in the wet season. The number of spice plants also increased considerably during the wet season, although the number of species remained constant. Another steep increase was in the number of subsidiary staple food plants. Obviously the villagers were taking as much advantage as possible of the rains.

In an extensive survey from the lowlands to the highlands of West Java, Karyono (1981) recorded that the average size of 351 homegardens sampled was 229.1 m². The size decreased with altitude. The total number of species found in the survey was 501 in the dry season and 560 in the wet season, with a cumulative number of 602 in the two seasons. The average number of species in the dry season was 19.0 per homegarden and 24 in the wet season. Species density was 8/100 m² in the wet season. The highest number of species per homegarden was in the altitude between 500 and 1,000 m, species density increasing with increasing altitudes.

Both the number of species and the number of plants were highest in the lowest canopy layer, being 62 percent and 35 percent, respectively, of their total numbers and gradually decreasing to 1 percent and 6 percent, respectively, in the highest layer. However, the highest canopy coverage was found in the fourth layer from the ground.

Poor people tend to grow more staple crops, vegetables and fruit trees, whereas well-off people grow more ornamentals and high-economic-value cash crops (Ahmad *et al.,* 1980). Generally, in the more remote areas more subsistence crops are grown and nearer cities the number of cash crops increases. When labour is scarce, people grow more perennials and less annuals since perennials require less labour than annuals (Stoler, 1975). Penny and Singarimbun (1973) observed that when there were not many off-farm jobs, people spent more time on their homegardens and crop diversity increased. A similar situation was found in a widow's garden in Nicoya (Wagner, 1958).

Culture and tradition are the other factors which influence homegarden composition. In a village on the border between West and Central Java, where the traditions of the Sundanese and the Javanese meet, more vegetables were found in the Sundanese homegardens than in the Javanese ones, but more medicinal plants in the Javanese than in the Sundanese (Abdoellah, 1980). It is known that the Sundanese have a strong preference for vegetables in their diet, while the Javanese customarily consume *jamu,* i.e., traditional herbal medicine for curative and other purposes. Terra (1954) concluded from his extensive studies that intensive homegardening was found in societies with matrilinear traits, e.g., in Central Java, West Sumatra, Aceh, southern Burma, in the Kasi region of Assam in India, in northern Thailand, Kampuchea and in some parts of the mountain ranges between Vietnam and Kampuchea. In the Muslim districts of southern Ethiopia, tobacco, coffee and drugs were intensively cultivated, but they were absent from the homegardens in Christian northern Ethiopia (Simoons, 1965).

Homegardens commonly have animals as components. Brownrigg's literature review (1985) indicated that animals were found in virtually all types of gardens, e.g., poultry, including doves, and fish in the Near-East homegardens, and poultry and livestock in the Luso-Latin and Caribbean homegardens. Other examples of homegardens with animals are the Chagga gardens in Tanzania (Fernandes *et al.*, 1985), the homegardens in Ghana (Asare *et al.*, 1985), Grenada (Brierley, 1985), Indonesia (Soemarwoto *et al.*, 1975; Soemarwoto and Soemarwoto, 1984), Kerala, India (Nair and Sreedharan, 1986) and Bangladesh (Leuschner and Khalique, 1987).

Culture, religion, and economic and ecological factors influence the animal species kept in the homegardens. In Muslim regions, pigs are an absolute taboo, e.g., in West Java and Aceh, but they are very common in the non-Muslim regions, e.g., in Christian North Sumatra and in Bali, Indonesia, where the dominant religion is Hinduism. In West Java with its high rainfall and long wet season, people almost always make a fishpond in their homegardens.

The function of homegardens

The interest in homegardens has primarily been focused on their function of producing subsistence items and generating additional income. They are known for their stable yields, very varied products, continuous or repeated harvests during the year and low inputs. Although light intensities decline in the successive layers because of the stratified structure of the canopies, Christanty *et al.* (1980) showed that there was no appreciable decrease in photosynthetic rates. Field experiments demonstrated that in these lower layers the yields were still satisfactory (Omta and Fortuin, 1978; Noor, 1981). Obviously the lower-storey plants were selected for their shade tolerance. This shows that the people are capable of effectively utilizing the spatial niches of the homegardens. However, their yields are generally low, although for poor people they can significantly contribute income and/or nutrients to their minimal income and poor diet, particularly during lean periods. For these reasons homegardens are being promoted in many countries, e.g., in Lima, Peru (Ninez, 1985), Ghana (Asare *et al.*, 1985), the Pacific Islands (Falanruw, 1985; Sommers, 1985; Thaman, 1984), Sri Lanka (Jacob and Alles, 1987) and Indonesia.

The Lima gardens, with an average size of 200 sq m, produced an income of US $ 28.33 in five months. Not impressive, but still it added almost 10 percent to the family income during that period (Ninez, 1985). Higher productivity has also been reported. In Indonesia, homegardens can contribute 7–56 percent of the owner's total income (Ochse and Terra, 1937; Ramsay and Wiersum, 1974; Ahmad *et al.*, 1980; Danoesastro, 1980). Homegardens may also produce higher incomes than other land uses. Gonzales-Jacomes (1981) found that in Central Mexico the average income per square metre of homegarden was almost 13 times that of irrigated plots. From our studies it was estimated that in West Java the annual income from homegarden fishponds per unit area was 2–2.5 times that of rice fields, and in tourist areas, where homegardeners sold ornamental plants in pots or plastic bags, the average income could reach almost 20 times that of rice fields. In a study by Stoler (1975) in a village in Central Java, the cash value of production from homegardens was found to be influenced by the size of the homegarden units, the smallest gardens being relatively more productive. She also found that there was an inverse relationship between labour, in terms of hours worked per hectare, and garden size, i.e., there was more labour input in smaller homegardens than in larger ones.

Indonesian data are illustrative of the relationship between productivity and cost. Ochse and Terra (1937) calculated that in Kutowingangun, a village in Central Java, 44 percent of the total calories and 14 percent of the protein consumed came from the homegardens, but only 8 percent of the total costs and 7 percent of the total labour were spent on them. Terra and Satiadiredja (1941) reported that in the same area the gross income as well as net income from homegardens was more than that from dry fields though less than that from rice fields. Similarly, Danoesastro (1980) reported that in three sub-districts in East Java the percentage values of average gross income from homegardens was 39 for rice fields and 72 for dry fields, but the average net income (percentage values) was 84 for the rice fields and 184 for the dry fields. The figures of net income from homegardens varied from 6.6 to 55.7 percent of total income with an average of 21.1, while the cost of production ranged from 4.7 to 28.5 percent of gross income from homegardens (with an average of 15.1). The cost of production from rice fields varied between 47.4 and 66.6 percent, with an average of 55.9 percent of gross income from this form of land use. Thus, the cost of production of homegardens was much lower compared with that of rice fields.

Ahmad et al. (1980) observed that income from homegardens before the rice harvest, the so-called *paceklik* or lean period, was 25.5 percent of total income, and declined during and immediately after the rice harvest to only 6.4 percent of total income. Terra and Satisdiredja (1941) reported that income from homegardens increased sharply before *Idhul Fitri*, the major Muslim festival, when people needed more cash. These fluctuations show the versatility of homegardens for meeting varying needs at different times of the year. Because of this, the homegarden is popularly known in Indonesia as *lumbung hidup* (living store).

It has also been recognized that homegardens are an important source of nutrients. Covich and Nickerson (1966) reported that consumption of 100g each of *Manilkara sapotilla*, *Persea americana*, *Manihot esculenta* and *Guilielma utilis* grown in Choco Indian dwelling clearings in Darien, Panama, was equivalent to approximately 10, 12, 150, 15.5, 31, 25 and 113 percent of the minimum daily requirement of protein, calcium, carotene, thiamine, riboflavin, niacin and ascorbic acid, respectively. Homegardens in villages in Lawang, East Java, produced a daily average of 398.4 calories, 22.8 g protein, 16.4 g fat, 185 g carbohydrate, 818.4 mg calcium, 555 mg phosphorus, 14 mg iron, 8,362 IU vitamin A, 1,181.2 mg vitamin B and 305 mg vitamin C (Haryadi, 1975). Similar results were reported from the Philippines and Nigeria (Fernandes and Nair, 1986). In Indonesia, the author and co-workers have found that rice fields gave higher yields of protein and calories than homegardens, but more calcium, vitamin A and vitamin C were obtained from homegardens than from rice fields. As mentioned earlier, Ochse and Terra (1937) reported that homegardens could supply up to 18 percent of the calories and 14 percent of the protein consumed in the whole village.

In general, a large proportion of the products of homegardens is consumed by the gardeners themselves. This is especially true in remote areas where the market economy has not yet developed. Danoesastro (1980) reported that in his study area in Central and East Java, products consumed varied from 21 to 85 percent, with an average of 44, while in Stoler's (1975) study, an average of 67 percent was consumed. Stoler noted that the cash value of the produce consumed was, on average, 16 percent of total consumption with a breakdown of 12, 17 and 19 percent, respectively, for the smallest, the medium-sized and the largest homegarden groups. Ahmad *et al.* (1980) reported that the percentage of total produce consumed by the household were as follows: fruit 46, coconut 83.7, vegetables 94.7, medicinal plants 95.5, and tubers and roots 97.3. Other important products for home consumption are fuelwood, construction material and materials for handicraft and home industry.

The socio-cultural functions of homegardens have not received much attention so far. In many areas products for religious rituals and ceremonies are very important, e.g., in Bali and Thailand. Thaman (1985) noted the importance of sacred or fragrant plants in homegardens. He gave a list of 35 plant species considered sacred in Tonga. There is an abundance of examples from all parts of the world of plants and animals considered sacred or supposed to possess magical or mystical powers. They play a very important role in the lives of the people. In areas where the market economy has developed, many of these plants and animals also have economic importance because of the demand for them in daily rituals and particularly during major festivals. In cities, the aesthetic role of homegardens is important.

Studies in villages in West Java have shown that homegardens are an important social-status symbol (Ahmad *et al.,* 1980). People who do not have a homegarden and, hence, have to build their house on someone else's homegarden, are considered of low status. In traditional Indonesian villages, people can freely enter homegardens, e.g., to get water from a well, or just to pass through them. Although there may be fences around them, they are seldom completely closed nor are there locked gates. The concept of trespassing does not exist. Those who close off their gardens completely are considered conceited. Fruits and other products are traditionally shared with relatives and neighbours, and products for religious or traditional ceremonies and medicine are given away freely when requested. However, this equitable social situation is now gradually changing (see a later section in this chapter).

A homegarden is also an important place for children to play and for adults to congregate in their free time. For this purpose a small part of the homegarden is kept clean (not planted), e.g., the so-called *baruan* in West Java or *pelataran* in Central Java. It is shaded by some trees planted at the edges. The owner of the homegarden is responsible for the safety of all children who play there. Young people who get married may build their homes on their parents' homegarden. Thus homegardens play an important role in family and community life. They are more than just a production system.

Soil erosion under homegardens

The ecological functions of homegardens have generally been taken for granted and only mentioned in passing. The protective effect of the homegarden on soil erosion has been deduced from its multistorey structure. However, the relationship between canopy structure and soil erosion is complex.

Soil erosion consists of two processes, i.e., splash and surface erosion. Splash erosion is the detachment of soil particles from the soil surface by the kinetic energy of falling water drops, which in homegardens are drops of throughfall, i.e., water dripping from the canopies. The kinetic energy of a falling drop is determined by its mass and velocity (Chapman, 1948). Williamson (1981) showed that the volume of a drop bears a linear relationship to the logarithm of the width of the driptips of leaves — the wider the driptips the larger the volume. Therefore, throughfall from canopies with leaves which on the average have large driptips has high kinetic energy. Due to gravitational force, the velocity of the falling drop is accelerated. However, as the velocity increases the drag force of the air, working in the opposite direction of the movement, also increases until an equilibrium is reached between gravitational force and drag, at which time the velocity becomes constant — the so-called terminal velocity. At a free-fall distance of 7.8 metres, 95 percent of the terminal velocity has been attained (Laws, 1941). Hence, a canopy base that is more than 8 metres high does not have much additional effect on the velocity of the falling drop. However, there are large differences in the drop's velocity with canopy heights lower than 8 metres and, therefore, its kinetic energy. Consequently, when a homegarden has a multilayer structure, with the lowest canopy base being about 3 metres high and the leaves not having narrow driptips, the protective effect on splash erosion will be much less than in a homegarden with tall trees and a very low closed undergrowth with narrow driptips. Litter also provides effective protection against the erosive force of falling drops.

Surface erosion is the removal of soil from the surface by water running over it. The erosive force of running water is also determined by its mass and velocity. Litter reduces the run-off coefficient by increasing infiltration and thus decreases the mass of the overland flow and, hence, also its erosive force.

Measurements of the erosivity of throughfall in multilayered homegardens showed that it was 135 percent of incident rainfall (Ambar, 1986). But splash erosion was only 80 percent of that in an open space. The reduction of splash erosion in the homegardens was due to the low-growing crops and litter. Ambar made similar measurements in monoculture bamboo groves which did not have an undergrowth but with the ground being covered with a mat of litter. The erosivity of throughfall was 127 percent of incident rainfall, but the splash erosion was only 47 percent of that in an open space. Therefore, although the bamboo groves did not have a multilayer structure, the splash erosion was very much reduced by the litter. The lower erosivity of throughfall in the bamboo groves compared to that in the homegardens was the result of the narrower driptips of bamboo leaves compared with the average width of those in homegardens.

In another experiment, splash erosion was measured in a mixed garden, in a bamboo grove and in an open space where the weeds and litter had been removed (Soemarwoto and Soemarwoto, 1984). The erosion in an intact mixed garden and bamboo grove was minimal, while in the open, erosion increased sharply with higher rainfall intensities. When the litter and lower-level crops of the mixed garden and the litter of the bamboo grove were removed, erosion increased significantly in both types of garden. The erosion curve of the mixed garden was a function of rainfall intensity and was greater than that of the bamboo grove. This again was due to the narrow driptips of the bamboo leaves. Clearly the erosive effect of throughfall in agroforestry systems is generally higher than that of incident rainfall and driptip, undergrowth and, especially, litter are important factors in reducing the rate of erosion in such systems (Wiersum, 1984).

Animals in homegardens are important elements in the cycling of matter. In West Javanese villages plants, goats, sheep, horses, chicken and fish, and also man, are components of the recycling of wastes. In non-Muslim regions the pig plays the role of fish. Thus man is an integral part of the trophic system from which he obtains nutrients and income (Soemarwoto et al., 1975). Naturally, there is a health hazard attached to this recycling system. Therefore, although the recycling of wastes does present an excellent opportunity for the efficient use of resources and helps in the maintenance of soil fertility, it should not be accepted uncritically.

Prospects and pitfalls

The fact that homegardens have existed for many centuries, if not millenia, in some parts of the world shows that they are ingrained in the tradition and culture of the peoples in those regions. In some parts at least they have not shown any decline, e.g., in Indonesia. Data from Java show that the total area under homegardens has increased from 1.398 million ha in 1933 to 1.417 million ha in 1937 (Terra, 1953) and 1,554 million ha in 1980 (Government of Indonesia, 1982), representing, respectively, 18.0, 18.1 and 18.5 percent of the total agricultural land. In areas which suffer from heavy soil erosion, such as in the upper Solo River basin in Central Java, there are no visual symptoms of soil erosion under homegardens while the fields outside the village are heavily eroded. In a short survey in the Phu Wai watershed in Khon Kaen, north-eastern Thailand, the author and his team also observed that in the plantation forests and upland fields there was slight-to-severe soil erosion, but there was almost no erosion under the homegardens. Homegardens are also a rich genetic resource. Therefore, it seems reasonable to conclude that homegardens are a sustainable production system. Their socio-cultural functions and the recycling systems also contribute to their sustainability. However, in general, this sustainability prevails under conditions of low yield and input. Therefore, if we wish to use the homegardens as a major tool to raise the standard of living of the people to satisfactory levels, the question arises whether the yield and the income from homegardens can be significantly increased without sacrificing their sustainability.

The earlier-mentioned unpublished studies of Stoler showed that homegardens responded positively to higher inputs of labour. Moreover, introduction of new species as well as high yielding and pest/disease-resistant varieties and/or selection of superior lines from the existing stock, adoption of better planting techniques, and post-harvest technology and marketing systems should be able to increase yield and income from homegardens significantly.

Although many homegardens are traditional, especially in the rural areas, they are not static but are capable of responding to new opportunities or adapting to new conditions. For example, in samples of homegardens of transmigrants in Lampung, Sumatra, who migrated from Tulungagung, East Java, we found a total of 138 species, while in the same number of samples in their village of origin we found only 69 species. Of these 138 species, 96 were acquired locally by the transmigrants, and of these, 36 species had been obtained from the local indigenous people. A majority of the transmigrants believed that local plants grew better than the ones they took from their original village.

The homegardeners in the tourist areas who grow ornamental plants to cater to tourist demand and obtain high economic returns are another example of the ability of traditional people to respond to new opportunities.

There are many examples of introduction of higher inputs and improved technology in homegardens, both spontaneously carried out by the people or stimulated by the government. Improved strains of fish, rabbits, chickens, coconut, citrus, cloves and vegetables have been widely introduced in Indonesian homegardens. The trees are planted in rows and the annuals in blocks. Near to and in cities highly valued plants such as orchids and animals such as poultry are raised. Fish and chickens are not left untended, but are fed with high-quality feed, and the chickens are vaccinated against diseases. Chemical fertilizers and pesticides are increasingly being used. Many homegardeners have benefited greatly from this development to the extent that they obtain their income solely or primarily from their gardens — an income which is sufficient to support a relatively comfortable life-style.

However, these developments also carry risks. These risks are associated with the changes in the structure of homegardens, which in turn bring about changes in their ecological as well as socio-economic functions. Paying attention solely to the tangible economic and nutritional gains of homegardens, and agroforestry in general, runs the risk of sacrificing the intangible ecological and social values. For example, when market demand and price offered for a certain plant product becomes high, the cultivation of that species will spread, often replacing those species and varieties which are of little or no immediate economic value. This causes a reduction in the complexity of the homegarden and degeneration of its forest-like structure. In such processes of commercialization, the highly nutritious, yet commercially less valuable local vegetables are usually the first ones to go. It is not easy to achieve homegarden development with both nutritional and economic advantages.

In homegardens with low species diversity, harvesting becomes more seasonal, instead of continuous. Its multifunctional characteristics decline with a corresponding reduction in its versatility. Its function as a living store from which one can harvest according to the nature and time of need has been more or less lost. In these commercialized homegardens control of species composition and harvest has changed from being internal, i.e., by the gardeners themselves, to external, i.e., by the market forces.

Commercialization causes a decline in the diversity of species and/or varieties, and consequently the process of genetic erosion sets in. For example, in the 1920s, 75 varieties of mango had been reported in the Cirebon area in West Java; but in a recent survey in the same area we found only 48 varieties. In Depok near Jakarta, where homegardeners have specialized in commercial fruit growing, we found only one variety of mango in a sample of 15 homegardens. In the same samples we found four banana varieties, while in the rural interior not far from Jakarta there were 25 varieties in a sample of the same number of homegardens.

The dominance of a certain crop on the farm increases the risk of losses due to its specific pests and diseases. Although sometimes a higher number of plant species can lead to an increase of pest losses, the advantage of a species-rich polyculture is undoubtedly that the risk of losses is spread among many species (Ewel, 1986).

In many homegardens with fruit trees, there are signs of soil erosion, even though these homegardens are not monocultures and to a certain extent they still retain the layered canopy structure but without the low-growing, ground-covering crops. Often there was no litter on the ground which led to severe erosion, for example, where clove trees were dominant and the leaves were collected to be distilled for their oil, and under coffee bushes where clean weeding was practised to facilitate easy collection of the beans which fell on the ground. Erosion was also often observed in homegardens in which vegetables were grown in nutrition campaigns.

The availability of chemical fertilizers has reduced the need for organic manures. Composting is deemed cumbersome and time consuming, and its opportunity cost is considered high, while chemical fertilizers can be bought easily in large or small quantities, as needed. As a result the extent and intensity of the recycling systems are declining. This reduces the efficiency of resource use and in the long run will also affect soil structure and fertility.

The social functions have also been lost to a greater or lesser extent. Fences are now often built for security reasons. Sharing of the harvest and amenities of the garden, such as a well, is no longer a common practice. Households have become more individualized and less community-oriented. The concept of trespass is becoming more widespread. Equitability has declined.

Refusal or resistance to improvement of the homegarden is also common. These failures often happen when the opportunity cost is high, when the growth requirements are not met, or when the introduced species conflict with the interest of the people. For example, many poor people have to use the labour of the whole family, i.e., husband, wife and children, to earn money and cannot spare the time for gardening, even though they know they can get some additional income or nutritious food. Many vegetables, which have been recommended in nutrition campaigns, require full sunlight or intensive cultivation for their growth and, consequently, grow poorly in the shade or half-shade of trees or when they are not well tended. These plants disappear soon after the campaign ends, or more fortunately some of the heliophilic species or varieties end up being planted on the dykes of rice fields.

An example of conflict between an introduced variety and the existing homegarden structure and species association was the introduction of the rapid-growing and high-yielding hybrid coconut. Traditionally the coconut, which requires full sunlight for maximum production, occupied the highest canopy. But the hybrid variety has a low growth habit and, hence, it displaces the fruit trees, while the spatial niche above this low hybrid becomes empty. As a result these plants are only grown at the edges of homegardens, just to please the government officers. (Also see Hoskins, this volume.)

From the above discussion we see that while there are many opportunities and potentials for improving homegardens, there are many risks associated with it if we proceed in a manner which only takes into account the economic and nutrition aspects. A system has properties of productivity, stability, equitability and sustainability (Conway, 1985). These are being changed. High-yielding, but high-input and high-risk, species and varieties increase productivity, but at the expense of stability, equitability and sustainability. Pests and diseases, market-price fluctuations and high seasonality reduce stability. High-yielding, high-input, high-risk species and varieties tend to increase inequitability. Inequitability adversely affects sustainability (Crosson, 1986). Soil erosion and loss of soil fertility and genetic erosion also reduce sustainability. Consequently, sustainability is being jeopardized and, hence, long-term gains become questionable.

Erosion hazards can be dealt with relatively easily since the factors influencing them are well known. Good ground cover and litter, or mulch, are essential. Improved homegardens should be designed and managed in such a way that they have both. Leaves with driptips are beneficial and, as far as possible, efforts should be made to grow plants with such leaves. A multilayer canopy structure is not a *sine qua non,* if the above requirements are fulfilled. Therefore, we do not have to aim specifically at having such a structure. Usually terracing is not necessary, since land for settlements is made flat to facilitate easy construction of the houses. But terracing will be necessary for homegardens which are located on slopes.

The needs of the people, present and future, and taking into account their traditional and religious beliefs and ecological conditions, should be the bases for the planning of species and varietal composition. Whenever possible a combination of woody perennials and annuals should be attempted since they combine the stability of a mature system and the high productivity of an immature one. This would reduce the need for high inputs which would therefore make such improved homegardens more accessible to the poor. Usually the perennials play the role of cash crops and the annuals food. The perennial trees will also help the poor through lean periods (Chambers and Longhurst, 1986).

Since homegardens are a part of the total agro-ecosystem and linkages exist between them and the other parts of the system, i.e., the rice and the dry fields, their development cannot be considered in isolation. Information on the agricultural calendar and the seasonality of homegarden crops would enable us to design a species composition which would improve the role of the homegarden so that it could fill in the troughs of the lean periods and the seasonability of labour supply and demand. Considering the earlier-mentioned study of Stoler (1975), presumably the smaller farmers would be more responsive to providing more labour for homegarden improvement than the larger ones.

In conclusion we can say that homegardens do have a promising future. However, while it is relatively easy to increase yields and income, there are difficult problems in achieving long-term sustainability. These difficulties are both in the biophysical and in the socio-economic realm. It is recommended that ICRAF look into these problems and stimulate research to seek appropriate solutions.

REFERENCES

Abdoellah, O.S. 1980. Structure of homegardens of Javanese and Sundanese people in Bantarkalong. First degree thesis, Department of Biology, Faculty of Science and Mathematics, Padjadjaran University, Bandung, Indonesia (Indonesian).

Ahmad, H., A. Martadihardja and Suharto. 1980. Social and cultural aspect of homegarden. In J.I. Furtado (ed.), *Tropical ecology and development*. Kuala Lumpur: The International Society of Tropical Ecology, Kuala Lumpur.

Ambar, S. 1986. Aspects of vegetation and land use in the erosion process in the Jutiluhur lake catchment, West Java. Doctoral thesis. Padjadjaran University, Bandung, Indonesia.

Anderson, E. 1954. Reflections on certain Honduran gardens. *Landscape* 4: 21–23.

Asare, E.O., S.K. Oppong and K. Twung-Ampofo. 1985. Homegardens (backyard gardens) in the humid tropics of Ghana. In *Proceedings of the First International Workshop on Tropical Homegardens*. Bandung, Indonesia: UN University and Institute of Ecology, Padjadjaran University (in press).

Brierley, J.S. 1985. West Indian kitchen gardens: A historial perspective with current insights from Grenada. *Food and Nutrition Bulletin* 7: 52–60.

Brownrigg, L. 1985. *Homegardening in international development: What the literature shows*. Washington, D.C.: The League for International Food Education (LIFE).

Chambers, R. and R. Longhurst. 1986. Trees, seasons and the poor. *Institute of Development Studies Bulletin* (University of Sussex) 17: 44–50.

Chapman, G. 1948. Size of raindrops and their striking force at the soil surface in a red pine plantation. *Transactions of the American Geophysical Union* 29: 664–670.

Christanty, L., M. Hadyana, Sigit and Priyono. 1980. Light distribution in a Sundanese homegarden. Paper presented at the seminar on the Ecology of Homegardens III. Institute of Ecology, Padjadjaran University, Bandung, Indonesia (Indonesian).

Conway, G.R. 1985. Agroecosystem analysis. *Agricultural Administration* 20: 31–55.

Covich, A. and N.H. Nickerson. 1966. Studies of cultivated plants in Chogo dwelling clearings, Darien, Panama. *Economic Botany* 20: 285–301.

Crosson, P. 1986. Sustainable food production. *Food Policy:* 143–156.

Danoesastro, H. 1980. The role of homegarden as a source of additional daily income. Paper presented at the seminar on the Ecology of Homegardens III. Institute of Ecology, Padjadjaran University, Bandung, Indonesia (Indonesian).

Ewel, J.J. 1986. Designing agricultural ecosystems for the humid tropics. *Ann. Rev. Ecol. Syst.* 17: 245–271.

Falanruw, M.V.C. 1985. The traditional food production system of Yap Islands. In *Proceedings of the First International Workshop on Tropical Homegardens*. Bandung, Indonesia: UN University and Institute of Ecology, Padjadjaran University (in press).

Fernandes, E.C.M. and P.K.R. Nair. 1986. An evaluation of the structure and function of tropical homegardens. *Agricultural Systems* 21: 179–310.

Fernandes, E.C.M., A. O'ktingati and J. Maghembe. 1984. The Chagga homegardens: A multistoreyed agroforestry cropping system on Mt. Kilimanjaro, Northern Tanzania. *Agroforestry Systems* 2: 73–86.

Gonzales-Jacomes, A. 1981. Homegardens in Central Mexico, their relationship with water control and their articulation to major society through productivity, labour, and market. *In Prehistoric Intensive Agriculture in the Tropics.* Canberra: Australian National University.

Government of Indonesia. 1983. Statistics of Indonesia. Central Bureau of Statistics, Jakarta, Indonesia.

Haryadi, M.M.S.S. 1975. Potential contribution of homegardening to nutrition intervention program in Indonesia. Seminar on Food and Nutrition, Gadjah Mada University, Yogyakarta, Indonesia.

Hutterer, K.L. 1984. Ecology and evolution of agriculture in Southeast Asia. In T.A. Rambo and P.E. Sajise (eds.), *An introduction to human ecology research on agricultural systems in Southeast Asia.* Los Banos, Philippines: University of the Philippines.

Jacob, V.J. and W.S. Alles. 1987. Kandyan gardens of Sri Lanka. *Agroforestry Systems* 5: 123–137.

Kamtuo, A., K. Lertrat and A. Wilariat. 1985. Traditional homegardening in Hinland village, Khon Kaen Province, Northeast Thailand. In *Proceedings of the First International Workshop on Tropical Homegardens.* Bandung, Indonesia: UN University and Institute of Ecology, Padjadjaran University (in press).

Karyono, E. 1981. Homegarden structure in rural areas of the Citarum watershed, West Java. Doctoral thesis, Padjadjaran University, Bandung, Indonesia.

Kimber, C.T. 1973. Spatial patterning in the dooryard gardens of Puerto Rico. *Geographical Review* 6: 6–26.

Laws. J.O. 1941. *Measurements of the fall-velocities of water drops and raindrops.* Washington, D.C.: U.S. Soil Conservation Service.

Leuschner, W.A. and K. Khalique. 1987. Homestead agroforestry in Bangladesh. *Agroforestry Systems* 5: 139–151.

Lundgren, B.O. and J.B. Raintree. 1983. Sustained agroforestry. In B. Nestel (ed.), *Agricultural Research for Development: Potentials and Challenges in Asia.* The Hague: ISNAR.

Nair, M.A. and C. Sreedharan. 1986. Agroforestry farming systems in the homesteads of Kerala, southern India. *Agroforestry Systems* 4: 339–363.

Ninez, V. 1985. Working at half-potential: Constructive analysis of home garden programmes in the Lima slums with suggestions for an alternative approach. *Food and Nutrition Bulletin* 7: 6–14.

Noor, Z.M. 1981. The effect of various light intensities on the physiological and morphological characteristics of *taleus padang (Xanthosoma atrovirens).* First degree thesis, Faculty of Science and Mathematics, Padjadjaran University, Bandung, Indonesia.

Omta, S.W.F. and F.T.J.M. Fortuin. 1978. The cultivation of *Solanum nigrum* L. as a leaf and fruit vegetable in the homegardens of West Java: Surveys and experiments. Bandung; Indonesia: Institute of Ecology, Padjadjaran University, and Groningen, Netherlands: Department of Plant Physiology, Groningen State University.

Ochse, J.J. and G.J.A. Terra. 1937. The economic aspect of the "Koetawinangoen report". *Landbouw* 13: 54–67 (Dutch).

Penny, D.H. and M. Singarimbun. 1973. *Population and poverty in rural Java: Some arithmetics from Sriharjo.* Ithaca, New York: Department of Agricultural Economics, New York State University.

Poerwadarminta, W.J.S. 1976. *General dictionary of the Indonesian language.* Jakarta: PN Balai Pustaka (Indonesian).

Ramsay, D.M. and K.F. Wiersum. 1974. Problem of watershed management and development in the Upper Solo River Basin. Paper presented at the Conference on Ecological Guidelines for Forest, Land and Water Resources, Institute of Ecology, Padjadjaran University, Bandung, Indonesia.

Simoons, P.J. 1965. Two Ethiopian gardens. *Landscape* 6: 15–20.

Soemarwoto, O. and I. Soemarwoto. 1984. The Javanese rural ecosystem. In A.T. Rambo and P.E. Sajise (eds.), *An introduction to human ecology research on agricultural systems in Southeast Asia.* Los Banos, Philippines: University of the Philippines.

Soemarwoto O., I. Soemarwoto, E. Karyono, E.M. Soekartadiredja and A. Ramlan. 1975. The Javanese homegarden as an integrated agroecosystem. In *Science for a Better Environment.* Tokyo, Japan: Science Council of Japan.

Sommers, P. 1985. Advancing Pacific Island food gardening systems. Some observations and suggestions. Paper presented at the First International Workshop on Tropical Home-garden. Bandung, Indonesia: UN University and Institute of Ecology, Padjadjaran University.

Stoler, A. 1975. Garden use and household consumption patterns in a Javanese village. Department of Anthropology, Columbia University, New York.

Terra, G.J.A. 1954. Mixed garden horticulture in Java. *Malayan J. Trop. Geogr.* 4: 33–43.

Terra, G.J.A. and S. Satiadiredja. 1941. Results of a third budget-research in the district of Koetowinangoen. *Landbouw* 17: 137–178 (Dutch).

Thaman, R.R. 1984. Urban agriculture and home gardening in Fiji: A direct road to development and independence. *Transactions and Proceedings of the Fiji Society* 14.

——— 1985. Mixed gardening in the Pacific Islands: Present status and future prospects. Paper presented at the First International Workshop on Tropical Homegarden. Bandung, Indonesia: UN University and Institute of Ecology, Padjadjaran University.

Wagner, P.L. 1958. Nicoya: A cultural geography. University of California Publication in Geography, Vol. 12, No. 3.

Wiersum, K.F. 1984. Surface erosion under various tropical agroforestry systems. Paper presented at the Symposium on the Effects of Forest Land Use on Erosion and Slope Stability, East-West Center, Honolulu, Hawaii.

Williamson, G.S. 1981. Driptips and splash erosion. *Biotropica* 13: 228–231.

Yoda, K. 1974. Three dimensional distribution of light intensity in a tropical rainforest in West Malaysia. *Japanese J. Ecol.* 24: 247–254.

SECTION FOUR

Impact measurement and technology transfer

11

Economic considerations in agroforestry

J.E.M. Arnold

Oxford Forestry Institute
South Parks Road, Oxford
United Kingdom
Formerly: Chief, Policy and Planning Service
Forestry Department
Food and Agriculture Organization (FAO) of the United Nations, Rome, Italy

Contents

Introduction

Agroforestry systems have existed since the very beginnings of plant domestication. As Ninez (1984) has pointed out, the description of the mythical Garden of Eden in Genesis II is that of a homegarden, containing "every tree that is pleasant to sight and good for food". The area from which that description was drawn, the early Mediterranean, is but one region where agroforestry has long formed an important component of agriculture.

The established presence of agroforestry practices in many land-use systems today has perhaps been overshadowed by concerns about particular contemporary problems that underly much of recent interest in it. As a result, we may have been more concerned with developing new or improved agroforestry practices to cope with such issues as fuelwood shortages than with increasing our understanding of how existing activities function, and what economic contributions they make. Yet some of the most dynamic changes that have occurred in the recent past have been in existing agroforestry systems.

The purpose of this paper is, therefore, to examine selected existing agroforestry practices, most with roots deep in the past, in order to try and identify the economic considerations that have caused farmers to adopt them. This is approached by defining the advantages that agroforestry practices appear to offer to farmers as an efficient means of using available resources in order to meet their production goals.

The choice of situations examined here is largely determined by the availability of sufficient information to permit such analyses. This is most heavily concentrated in two agroforestry systems — homegardens and farm woodlots — which are reviewed in the first part of the paper, together with the more limited information relating to more extensive forms of agroforestry. In the second part, a number of general conclusions emerging from this review are brought together and discussed in terms of possible future directions for agroforestry research, projects and policies.

Benefits, costs and production objectives

The literature about farming systems which incorporate agroforestry practices describes a number of related benefits or opportunities and costs and constraints (e.g., FAO, 1978, 1985; Budowski, 1981; Arnold, 1983; Wiersum, 1981; Chambers and Longhurst, 1986). The principal positive and negative economic features and the underlying hypotheses about agroforestry are summarized in Table 1.

One hypothesis is that many agroforestry practices are characteristic of simple resource-poor situations and are not easily adapted to more intensive agricultural practices. Another, in a sense contradictory to the first, is that the competition between trees and food crops, and the priority that they must give to meeting their basic food needs, will exclude poor farmers with very little land from tree growing.

In the following section the results of economic studies of well-established homegarden situations in Java (Indonesia), eastern Nigeria, and Kerala (India) are reviewed. In the subsequent section, situations in which trees planted as farmer cash crops in parts of the Philippines, India and Kenya are examined. In most of these, the cultivation of trees as a share of total farm activity has recently been increasing, in most cases at a time of heightening pressures on one or more of the farmer's resources of land, labour or capital. In subsequent sections in the first part of the paper, where less intensive forms of agroforestry are reviewed, the focus is again on changes which farmers are adopting that increase the component of managed tree/crop activities.

Homegardens

Homegardens (also known as homestead and mixed gardens and as compound farms) are usually located, where they exist at all, close to the household as one of the more intensively cultivated parts of the overall farm. They are characterized by a mixture of several or many annual or perennial species grown in association, and commonly exhibiting a layered vertical structure of trees, shrubs and ground-cover plants, which recreates some of the properties of nutrient recycling, soil protection and effective use of space above and below the soil surface to be found in forests (Fernandes and Nair, 1986). Homegardens are widely used to supplement outputs from other parts of the farm through the cultivation of a variety of other subsistence and commercial crops (Ninez, 1984).

In Java, homegardens have long existed as the principal farming system on dry lands, accounting for a substantial proportion of total land use, with irrigated rice cultivation

Table 1 The main benefits and costs of agroforestry

Benefits and opportunities	Costs and constraints
Maintains or increases site productivity through nutrient recycling and soil protection, at low capital and labour costs	Reduces output of staple food crops where trees compete for use of arable land and/or depress crop yields through shade, root competition or allelopathic interactions
Increases the value of output on a given area of land through spatial or inter-temporal intercropping of tree and other species	Incompatibility of trees with agricultural practices such as free grazing, burning, common fields, etc., which make it difficult to protect trees
Diversifies the range of outputs from a given area, in order to (a) increase self-sufficiency, or/and (b) reduce the risk to income from adverse climatic, biological or market impacts on particular crops	Trees can impede cultivation of monocrops and introduction of mechanization, and so (a) increase labour costs in situations where the latter is appropriate and/or (b) inhibit advances in farming practices
Spreads the needs for labour inputs more evenly seasonally so reducing the effects of sharp peaks and troughs in activity characteristic of tropical agriculture	Where the planting season is very restricted, e.g., in arid and semi-arid conditions, demands on available labour for crop establishment may prevent tree planting
Provides productive applications for underutilized land, labour or capital	The relatively long production period of trees delays returns beyond what may be tenable for poor farmers, and increase the risks to them associated with insecurity of tenure
Creates capital stocks available to meet intermittent costs or unforeseen contingencies	

forming the other main component of the farm system. The gardens are traditionally dominated by perennials rather than annuals, and by woody rather than herbaceous growth (see Soemarwoto, this volume).

On farms with sufficient rice land to enable the household to meet its basic food needs, priority in labour and capital allocation is given to this. The garden areas in these larger farms are essentially forest gardens, with capital investment in trees of commercial value (Wiersum, 1981; Hunick and Stoffers, 1984; Michon *et al.,* 1986; Stoler, 1978).

With growing pressure on the land and decreasing area of crop land per head, the proportion of land under homegardens has been increasing to up to 75 percent of cultivated land (Stoler, 1978). Access to rice land has, in the meanwhile, declined, and a large proportion of the farmers now have no rice land, or not enough to produce their basic rice requirements. As this process occurs, the homegarden areas are cultivated more intensively, becoming mixed rather than forest gardens as annuals are progressively intercropped to provide food and income.

Management is intensified by increasing labour inputs. The scope for increasing productivity is such that labour inputs in small gardens are reported to be, on average, three times as high as those for larger gardens. Returns to labour are high, and intensive garden management produces up to 20 percent of household income and 40 percent of calorific intake (Stoler, 1978).

Another approach to intensification of garden area use has been to increase the value added from homegarden produce. Penny and Singarimbun (1973) describe how some of the poorest farmers shift from producing just fruit from their coconut palms to production of coconut sugar, a highly labour-intensive process, which, though producing only low returns to labour, increases returns to coconut-bearing land. Other farm-related income-earning activities include pit-sawing and fuelwood gathering (Hunink and Stoffers, 1984).

As land-holding size continues to decline, income is increasingly sought from off-farm employment. At this stage, cultivation of annuals is reduced in order to release labour, and trees and other perennials requiring only low labour inputs come to form the main component again (Stoler, 1978).

The shift to greater dependence on tree gardens in Java as land-holding size decreases and farmers' resource endowments and production objectives change, thus appears to be because they permit more productive use of the land than alternative land management options.

In a study of farming practices in south-eastern Nigeria, Lagemann (1977) found similar relationships between population pressure on the land and intensity of tree cultivation. Farms comprise a mixture of fallow, outer and inner fields and permanently cultivated compounds around the household. The latter contain a variety of tree species, including oil palm, raffia palm (*Raffia* spp.), coconut, and banana and plantains intercropped with cassava, yams and other arable crops.

Growing population pressure is accompanied by decreasing farm size and declining soil fertility. As pressures on the land increase, the proportion of land under compound systems is increased, as is the density of both tree and arable crop cultivation within the compound areas. Lagemann argues that this shift reflects farmers' perceptions that such land use, combined with increased mulching and manuring, offers the most effective way of using their resources to slow down the process of declining soil fertility and thus maintain production. Though labour input per hectare is no higher than in the fields, yields in monetary terms are five to ten times as much per hectare and returns to labour four to eight times as much. Lagemann identifies phasing of planting and harvesting in the compound areas to reduce peak workloads, and the better physical working conditions which the shade of the vegetation provides, as factors contributing to this higher labour productivity.

With increasing population density, compound areas account for up to 59 percent of crop output and a growing proportion of total farm income. The proportion of the latter generated from tree crops rises to a share nearly equal to that from arable crops. Livestock become an increasingly important part of the compound system, both as a source of income and of manure. However, as population density continues to increase, yields and returns to labour eventually decline to the point at which farmers have to turn increasingly to non-farm sources of income. As labour has to move off-farm, a lower input management of the compound areas is adopted, leaving them under a cover in which trees and other perennials predominate.

The overall picture, therefore, as in Java, is one of farmers responding to decreasing land availability by moving to greater dependence on agroforestry systems. Initially this is because these permit more sustainable intensification of land use and higher returns from available labour inputs than alternative land uses. When pressures on the land increase further to the point where income has to be generated mainly from off-farm employments, these agroforestry systems have the added advantage that they can be maintained, in a modified form, as a low-input low-management form of land use.

A study of tree cropping in the homesteads in Kerala, India, records a similar process in a system in which homegardens (where perennial crops such as coconut, arecanut, rubber and pepper are intercropped with seasonal and annual crops such as pulses, bananas, tubers and vegetables) have long existed as part-systems together with more extensively managed areas (Nair and Krishnankutty, 1984; Nair and Sreedharan, 1986). As rising population pressure on the land leads to decreasing land-holding size, uncultivated land is first brought into use resulting in removal of natural tree cover. This is followed by more active management of the homegardens. This includes reducing the range of cultivated trees to those with multiple uses, with priority being given to those species valued for fruit, fodder and mulch and suitable as supporting structures for cultivation of pepper, betel vine and various climbers. In the process, the density of trees and intensity of their cultivation are increased.

As pressure on the land increases further, land-holding size increasingly falls to the point where farming ceases to be the main source of income. As labour inputs into the farm are reduced in order to move to off-farm employment, the tree component increases even further as the vegetation reverts towards a forest condition.

However, in this area of Kerala, capital has become increasingly available, enabling some farmers to intensify land use further through purchased inputs of fertilizer, herbicides, etc. This reduces the importance of multipurpose trees in soil-nutrient maintenance and weed suppression, and they then tend to be removed as an impediment rather than a complement to agriculture. Removal of trees has been accelerated by rapid rises in the prices of timber and land; the latter leading to the shift of land use to cash crops. Trees are then cultivated only where they are competitive as cash crops, for example, *Ailanthus triphysa* grown to supply wood stock to the match industry.

The Kerala experience, which is summarized in Figure 1, thus broadly parallels that of Java and Nigeria up to the point where farmers were able to inject substantial capital into their systems. The subsequent displacement of agroforestry practices seems to confirm that, in the absence of capital, farmers had been employing trees primarily to provide substitutes for purchased inputs, and as crops requiring lower inputs than agricultural crops.

Farm woodlots

The farm woodlots examined in this section essentially involve the growing of trees as a field cash crop. Farmer decisions could therefore be expected to be governed by the question of whether or not this crop is more profitable than alternative crops or other uses of the land. In all three of the situations discussed below tree growing has recently been spreading rapidly.

One of the most fully documented farm-woodlot experiences has been small-farmer growing of trees to produce pulpwood as a cash crop in the Philippines. In an area of previously low-density extensive agriculture, farmers grow *Albizia falcataria* on a 6–8 year rotation for sale to a nearby pulp company (PICOP). Average size of landholding is 11 ha, of which a part is devoted to cultivation of food crops. At least part of the land on 45 percent of the farms had been previously used for growing food crops, and other non-food crops

A. Intensive homegarden cultivation

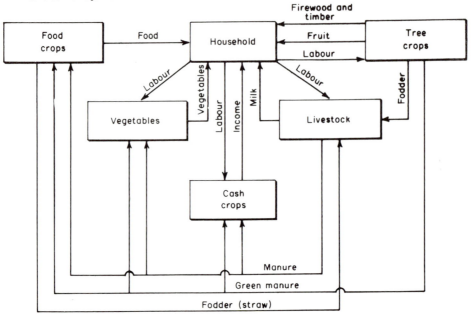

B Changes in intensively managed homegardens

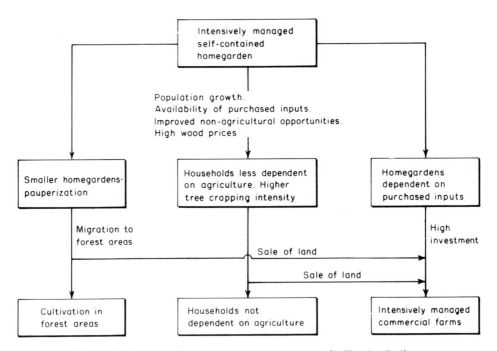

Figure 1   Changes in homegarden management in Kerala, India
Source: Based on Nair and Krishnankutty, 1984.

had been raised on some parts of 31 percent of them. Credit for tree farming was available but was utilized by only 30 percent of the farmers eligible — generally farmers with above-average woodlot size (Hyman, 1983).

Ex-post analysis showed that farmers were able to earn an acceptable return from their woodlots in most prevailing circumstances, with internal rate of return (IRR) of 22 percent to 31 percent for successful tree farmers on costs other than those of land. The data do not exist for comparing these returns with returns from alternative land uses, but farmers cite low labour inputs as the principal reason for preferring tree growing (Hyman, 1983). In an area where average land-holdings considerably exceed the size that can be cultivated under food or other crops with family labour, pulpwood production enables farmers to expand the area they can put to productive use (also see Spears, this volume).

Very few other farm-forestry or agroforestry experiences have been evaluated to the same extent. Even in areas with such extensive farmer tree growing as has been occurring in Gujarat and other parts of India, on-farm information sufficient to permit analysis of actual returns to farmers, and of the relative returns from alternative land uses, is still lacking. In an early study of apparent returns in parts of Gujarat, Gupta (1979) estimated that eucalyptus grown on irrigated land to produce poles for sale would give farmers higher net returns to land and capital than alternative crops. It has been widely reported that one element in farmer decisions in favour of tree growing in India has been the relatively low labour inputs required, reducing farmers' dependence on hired labour at a time of rising wages (Noronha, 1982).

Available information about the extensive cash-crop growing of trees in parts of Kenya has been assembled by Dewees in a study for the World Bank (World Bank, 1986). Prominent among the species being grown are eucalyptus for poles and black wattle for poles, charcoal, fuelwood and sticks for "mud-and-wattle" construction. Markets for these wood products — and locally for pulpwood and saw timber — are growing strongly, with farm-level production accounting for a large part of the supply.

Tree growing tends to be practised by poor farmers who are unable to meet their basic food needs, and for whom it is a principal source of farm income. In Vihiga location in Kakamega District of Kenya, for example, average farm size is about 0.6 ha, of which some 25 percent is under eucalyptus woodlots (Gelder and Kerkhof, 1984).

Gross income per hectare in this area is considerably lower from tree growing than from other agricultural crops. Dewees (World Bank, 1986) suggests that farmer preference for tree crops in these circumstances is conditioned by availability of capital and labour, and by attitudes to risk management. Alternative crops often require investments at levels beyond small farmers' access to capital. Trees, by contrast, require very little expenditure. Tree growing is also attractive to farmers in an area where there is a shortage of labour because of widespread out-migration of male members of the farm households to seek off-farm employment. Where markets for tree products are good, returns to labour from pole production have been estimated to be some 50 percent greater than from maize production (World Bank, 1986). Consequently tree growing is a rational use of resources for poor farmers needing to devote a substantial part of their labour to non-farm employment.

In all three of these situations, therefore, a decision to grow trees has been influenced by two main factors. One is the high cost of labour and capital, and the advantages tree cultivation offers in this respect because of its low input requirements. The other is the prominent part that income generation, as distinct from food production, plays in the farmers' production objectives.

The discussion so far suggests that farmers employ agroforestry practices primarily because they perceive these as being the most efficient ways of meeting their production goals using the resources of land, labour and capital available to them. As their resource endowment changes so their strategies are likely to alter. Equally, their assessment of the most efficient farm strategy to employ is likely to be influenced by the broader framework of needs and opportunities linked to off-farm opportunities. Analysis of what is an efficient use of resources, therefore, has to be carried out in terms of the production objectives being pursued in that situation.

Shifting cultivation

It is now generally recognized that, in the situations in which it has traditionally been practised, shifting cultivation (swidden agriculture) is the most efficient use of farmer resources. The resources at the disposal of the shifting cultivator are predominantly in the form of family labour. Where there is sufficient land to support fallow, no other farming practice will produce a higher return to labour without inputs of capital. The fallow vegetation maintains soil productivity, and the process of clearing and burning provides conditions for crop cultivation requiring minimal inputs for soil preparation and weeding. Though cultivation periods could be extended by increased weeding, it is easier, in terms of requiring less inputs of labour, to clear and burn a new area. Similarly, yields per hectare could be increased by more intensive cultivation, but only at the expense of lower output per unit of labour. As long as they can satisfy their production objectives through less labour-intensive methods, farmers will stick to them (Rambo, 1984; Raintree and Warner, 1986).

A sequential process can be identified whereby, as reduced access to land prevents continuation of shifting cultivation, farmers start to intensify agricultural practices (Olofson, 1983; Raintree and Warner, 1986). This usually takes the form of small incremental changes involving increased inputs of labour, and sometimes of capital in the form of fertilizers, herbicides, etc. This evolutionary process may lead to a move away from agroforestry altogether, but agroforestry practices usually form part of the transition from shifting cultivation.

A widespread practice at an early stage in this process is enriching the fallow by encouraging or planting tree species which either accelerate or enhance the regeneration of soil fertility or produce outputs of subsistence or commercial value, or both. The cultivation in the Sudan, and elsewhere in the semi-arid area of Africa, of *Acacia senegal* as a fallow crop is an example of a species that does both. A leguminous species, it produces gum arabic for sale and fuelwood and fibre and other products for use in the household. Other examples include the planting of rattan as a commercial crop in the swidden cycle in Borneo (Weinstock, 1983), and the multiple-product managed swiddens of the Ifugao in the Philippines (Olofson, 1983). Although quantitative information on inputs and outputs is lacking for these systems, it can be expected that, as they involve only minor increases in labour inputs and minimal changes to the basic swidden system, they will provide fairly high returns to labour (Raintree and Warner, 1986).

At the next step, as pressures on land intensify the transition towards continuous cultivation, various forms of intercropping are encountered. By incorporating soil-enriching species with food crops, these practices introduce the functions of fallow on a continuous basis. Numerous examples of such continuous-fallow strategies are to be found, such as the maintenance of *Acacia albida* in cultivated areas in much of Africa.

An example of planted mixtures of this nature is the intercropping of *Sesbania sesban,* a leguminous tree, with maize in areas of western Kenya (see Figure 1, Sanchez, this volume).

When the maize is shaded out after about three years, the sesbania is left as a fallow crop for one or two years, and then cleared and used for fuelwood. The cycle is then repeated. Practised in an area of labour shortages, this method has been estimated to produce less than half the maize per hectare compared to monocropping of the latter over a ten-year cycle, but it requires less than half as much labour and gives higher maize returns per unit of labour input — in addition to the fuelwood and soil-protection benefits (World Bank, 1986).

Alley cropping

Considerable attention has been directed recently to intensification of managed continuous fallow in the form of alley cropping. This involves the growing of field crops between hedgerows of nutrient-cycling trees or shrubs which are periodically pruned during the cropping season to reduce shading and provide green mulch for the food crops (Kang and Wilson, this volume). Data from economic analysis of the results of alley-cropping research at IITA in Nigeria, involving intercropping of maize with *Leucaena leucocephala* and treatments with nitrogen fertilizer and herbicide, are summarized in Table 2. The intercrop treatment gave the highest economic returns and yields of maize of all the alternatives, but required higher labour inputs than cultivation of maize alone or with fertilizer/herbicide applications. Returns to labour from intercropping, however, were higher than in maize monocropping (Ngambeki, 1985).

Table 2 Evaluation of alley cropping *Leucaena* with maize*

	Increase/decrease over control		Yield/ man-hour	Net returns (US$/ha)	B/C ratio (ha)
	Yield (kg/ha)	Man-hours/ha			
Maize control	-	-	3.3	- 438	-
Nitrogen	752	46	4.1	- 371	0.79
Herbicide	452	- 173	5.9	480	1.18
Herbicide-nitrogen	1,123	- 127	6.8	146	1.10
Leucaena	1,407	259	4.2	496	1.32
Leucaena-nitrogen	1,358	302	3.9	130	1.05
Leucaena-herbicide	1,451	152	4.5	139	1.07
Leucaena-nitrogen-herbicide	1,892	199	4.8	- 58	0.97

* Compared with alternative treatments of maize, from IITA alley-cropping experiment 1981/82; average values per season.

Source: Ngambeki, 1985

The potential for improving the efficiency of such low-input agroforestry practices adopted by farmers in order to progressively intensify their farming therefore appears quite high. However, it is not yet clear how well experimental results reflect on-farm conditions. Further work in testing the research results on-farm is needed to ensure that the labour availability and farmer skills assumed are realizable in practice, that the costs and values are consistent with those actually experienced or perceived by farmers, and that the physical input-output relationships and performance are replicable on-farm (Balasubramanian, 1983; D. Hoekstra, personal communication).

Economic information is also available for some other intermediate agroforestry practices. A study of coconut farmers in an area in Sri Lanka, for example, showed that intercropping increases net returns. Labour-intensive crops such as betel, ginger and turmeric were adopted primarily by small farmers with family labour available, and those such as pepper and coffee, which gave higher returns to labour, were preferred by larger farmers dependent on hired labour (Karunanayake, 1982). Once again, therefore, farmers are seen to be responding to interrelationships between resource availability and production objectives.

Discussion

It would be unwise to base too much on analysis of the limited number of situations covered by the studies discussed in this chapter. Apart from being few in number they are based mainly on two broad forms of agroforestry: the homegarden and the farm woodlot. Nevertheless, the information outlined above, the main elements of which are summarized in Table 3, does suggest some of the main economic factors which encourage farmers to adopt tree/crop/livestock management as a major component of their overall farming system.

In most of the situations, farmers lacked access to capital and consequently were unable to increase their land or labour resources by renting or purchasing. In many instances, farmer decisions were clearly also influenced by considerations of risk management. In such conditions of limited resources and high susceptibility to risk, five overlapping farmer strategies involving adoption of agroforestry practices can be discerned:

1. To maintain productivity of land in situations of scarce capital in which the presence of trees can help substitute for purchased inputs of fertilizer and herbicide and for investments in soil and crop protection;
2. To make productive use of land in situations of scarce capital and labour where trees, as low-input low-management crops, constitute the most effective use of these resources;
3. To increase useable biomass outputs per unit of land area in situations where land and capital are limited, and tree/crop/livestock combinations permit fuller use of available labour than alternative uses of the land;
4. To increase income-earning opportunities from use of farm resources as size of landholding and/or site productivity fall below the level at which the household's basic needs can be met from on-farm production;
5. To strengthen risk management through diversification of outputs, wider seasonal spread of inputs and outputs, and build up of tree stocks which can be sold in order to meet periodic or unforeseen needs for capital.

Such an interpretation of the economic role of agroforestry activities suggests a number of propositions about agroforestry of possible relevance to project and policy analysis and design.

Table 3 Selected situations with agroforestry components in changing farm systems

Agroforestry component	Constraints/opportunities	Farmer response	Contribution of agroforestry
Homegardens Java	1. Declining land-holding size, minimal or no rice paddy, minimal capital	Increase food and income output from homegardens	Highest returns to land from increasing labour inputs, flexibility of ouputs in face of changing needs and opportunities
	2. Further fall in land-holding size below level able to meet basic food needs	Transfer labour to off-farm employment	Most productive and stable use of land with reduced labour inputs
Compound farms Nigeria	Declining land-holding size and site productivity, minimal capital	Concentrate resources in compound area, raise income-producing component and off-farm employment	Improves productivity, highest returns to labour, flexibility
Homegardens Kerala	1. Declining land-holding size, minimal capital	Bring fallow land into use, intensify homegarden management	Multipurpose trees maintain site productivity and contribute to food and income
	2. Capital inputs substantially increased	Transfer land use to high-value cash crops, substitute fertilizer and herbicide for mulch and shade	Trees removed unless high-value cash crop producers
Farm woodlots Kenya Farm woodlots Philippines	Farm size below basic-needs level, minimal capital, growing labour shortage Abundant land, limited labour	Low-input low-management pole cash crops, off-farm employment Put land under pulpwood crop	Lower capital input than alternative crops and higher returns to labour Expands area under cultivation, increases returns to family labour

Resource availability

Land

Agroforestry practices can be seen to be potentially appropriate to a wide range of different land-holding and land-use situations. Given the widespread feeling that tree growing is predominantly an activity for larger farmers, it is worth underlining the extent to which it exists as a viable activity of farmers with little land—often very little land. In certain circumstances, homegardens have proved to be an efficient strategy for capital-poor farmers wishing to intensify land use as land-holding size decreases. Tree crops are equally suitable for those short of capital and labour wishing to put their land under a less intensive form of use. The relationships involved are often complex, but are usually linked partly to the changes in farmer production objectives discussed later.

Labour

The relationship of tree growing to labour availability, and cost of labour, is equally varied. When combined with different land and capital conditions, as well as production objectives, scarce or abundant labour can lead to quite different decisions about agroforestry. However, a number of points seem worth emphasizing.

One is the role that agroforestry may play in helping to even out the peaks and troughs in work which characterize tropical agriculture. Work on tree crops can be scheduled outside the peak periods. The presence of trees can provide the basis for a number of counter-seasonal activities such as fuelwood and charcoal production (Chambers and Longhurst, 1986).

A second is that, as the need to depend on off-farm employment grows among farm families, labour availability has to be assessed not just in terms of other farm activities but also in terms of off-farm opportunities. Attention needs to be paid to the danger that intensifying or rescheduling labour use in on-farm agroforestry activities may prevent farm families from obtaining more remunerative employment off-farm — as may have happened to coconut-sugar farmers in Java (Stoler, 1978).

Thirdly, the growing dependence of rural people on off-farm employment suggests that the potential for linking agroforestry outputs with off-farm employment and income creation probably also deserves more attention. Recent work has shown that small rural processing enterprises based on wood and other raw materials from trees and forests are one of the largest sources of off-farm rural employment (Fisseha, 1987). Their mainly rural markets are themselves largely seasonal, fluctuating with rural income flows so that their peak activities are counter-seasonal with those in agriculture. Most are very small and are operated jointly with agricultural or other enterprise activities. A marked feature of such small-enterprise operations is the high proportion of women involved.

Capital

It is widely argued that the lengthy production period, and the incidence of most of the costs at the time of establishment, create financial problems for farmers in adopting practices involving tree growing. It is this argument that underlies the widespread provision of planting stock, either free or at subsidized prices, in programmes to support tree growing.

However, the evidence that tree systems are favoured by farmers when capital is scarce (because trees require less investment than alternative crops and/or provide substitutes for purchased inputs, e.g., of fertilizer and herbicide), suggests that improved access to capital would not necessarily increase adoption of agroforestry practices. Indeed, as the information from Kerala (India) and Kenya showed, the contrary is likely to be the case; access to capital will frequently enable farmers to adopt alternative higher yielding uses of their land.

Unavailability of capital could, of course, be an impediment to investment in longer rotation timber species grown as cash crops. However, even in this situation the constraint seems to be not the capital cost of establishment but the opportunity cost of locking up land for the lengthy period that elapses before there is any return — and possibly, as in the case of the production of pulpwood in the Philippines, the cost of harvesting (Hyman, 1983).

Production objectives

Much of the discussion of the role of agroforestry in farming systems has been based on the assumption that staple food production is the principal production objective. In most of the situations reviewed in this chapter, however, farmers were working with a resource base too small to enable them to meet their basic household food needs and were seeking to generate income as well as — or even instead of — food. The growth in agroforestry in many situations has been largely a response to this change in production objectives.

Recognition of the importance of income generation as the primary production objective for many poor farmers also changes the frame of reference of the debate about fuelwood self-sufficiency. Observers of the rapid expansion of farm-level tree growing in India (Blair, 1983) and Kenya (World Bank, 1986) have pointed out that although meeting subsistence fuelwood needs may seem to have the highest priority from the perspective of the planner, or of the fuelwood-using women members of the household, the rational choice for the household as a whole is to grow tree crops in order to generate income with which to satisfy its overall needs.

Markets and marketing

The presence of income generation as a primary production objective highlights the importance of correctly understanding the markets for the products of agroforestry. The growth in markets for some agroforestry products has recently been very rapid — notably for fuelwood, charcoal and poles. Though much of the intervention that has been taking place has been intended to increase production for the market, there has been remarkably little systematic study of how such markets are structured and function. Imbalances between supply and demand as a result of poor market information could have severe negative impacts on farmers through falling prices.

Particular interest is attached to supplying urban and other commercial markets with fuelwood — understandably, as this is the wood product used most widely and heavily in most countries. However, even where market-oriented tree growing by farmers has expanded rapidly, as in India and Kenya, it has been to supply higher valued building poles and wood for industry, not fuelwood. Prices of the latter have not yet stimulated significant investment in its production as a farmer cash crop.

In a review of energy-price trends in a number of large cities in south Asia, Leach (1986) has shown that though fuelwood prices varied widely from place to place there had been little increase in real prices in most of the markets over the previous 12–14 years (Figure 2). Interfuel substitution is apparently widespread, with fuelwood usage and prices sensitive to prices of alternative fuels. This would suggest that it is unlikely that there would be substantial increases in fuelwood prices in such markets in the foreseeable future.

However, it is the price at the point of origin (the stumpage price), rather than the

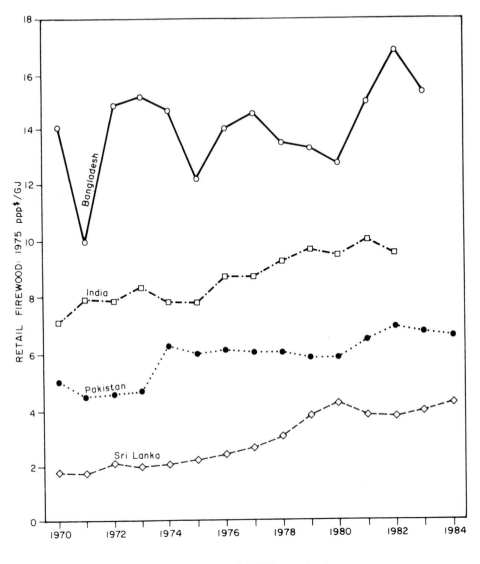

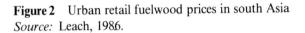
Prices shown in constant currency - US $ 1975, purchasing
power parity exchange rate.

Figure 2 Urban retail fuelwood prices in south Asia
Source: Leach, 1986.

market price which is of relevance to farmer decisions. In the case of fuelwood this is
commonly only a small fraction of the market price. The price of standing wood for sale as
fuelwood in two West African situations, for example, was only 1–1.5 percent of the retail
price, and that of wood cut and stacked at the farm gate was 11–13 percent (Baah-
Dwomoh, 1983).

   Attention has recently been directed to studying the potential for raising stumpage
prices of fuelwood as a proportion of market price by reducing transport and distribution

costs and/or through subsidies or bonuses to producers financed from levies on these intermediate stages. It is too early to say whether these will be effective. However, as long as low-cost sources of fuelwood remain, in the form of existing wood stocks, and users can turn to alternative fuels or to energy-conservation measures as an alternative to purchasing fuelwood, it is bound to be difficult to raise prices of farmer-grown supplies of the latter.

Risk management

Risk and avoidance of risk loom large in decision-making by poor farmers, often modifying or overriding other economic considerations. Living, as they do, at or close to the margins of existence, poor farmers are concerned to avoid any change which, though it might improve their situation if it functions well, could leave them even worse off if it does not. Equally, there is likely to be a preference for choices that reduce existing risk even if these offer less potential for economic improvement than alternatives.

As has been noted earlier, an important factor in the widespread adoption of homegardens has been their contribution to risk reduction, by spreading output across several products and over the different seasons. However, it is evident that with reduction in land-holding size the more fundamental risk of not being able to meet basic food needs becomes more important than the risk of intermittent shortfalls or failures, and management of garden areas is concentrated on species which can produce food and income in the short term. In other words, where a small homegarden is the only or main source of support for the farm family, the latter cannot risk depending on long-term tree crops — particularly where their tenure of the land may be insecure.

On the other hand, where income from the land has become only a minor or supplementary component of overall income, tree crops can again contribute positively to risk reduction. As a stable low-input low-management form of land use, it enables those who depend primarily on off-farm employment to maintain their land in productive use.

The disadvantage that income is available only periodically may be offset by the value the trees provide as a capital asset to be drawn upon for intermittent payments, such as education (Chavangi, 1984), or to deal with contingencies such as floods, famine or sickness. Chambers and Leach (1987) have pointed out that vulnerability to such contingencies is an important dimension of poverty. Trees apparently often do serve as assets upon which the rural poor draw in emergency. Various instances are cited in which people have had to liquidate trees (even fruit- and cash-crop bearing trees) because they have been their only remaining asset. A tree component in a farming system may therefore be there at least in part to provide a measure of insurance.

Agroforestry practices can therefore mitigate, or aggravate, risk in a number of ways. As the nature of the risks that threaten farmers change with circumstances, so the relevance of agroforestry is likely also to alter. Though lack of data limits the practical application to agroforestry of most of the techniques available for analysing risk, quite a lot can be done to improve our understanding of its existence and nature, and to take it into account in defining agroforestry options (Blandon, 1985).

Economic research

Research of the kind that has been reviewed in part in the present paper—analysis of actual inputs and outputs in existing operational agroforestry practices, and of the economic efficiency of the latter — clearly needs to be extended in order to improve the basis for defining the applicability of such systems more accurately. A related priority area is that of

establishing input and output values which correctly reflect costs and benefits to the farmer
— as distinct from costs and benefits to the supporting agency or donor, and from values as
perceived by the national planner (Gregersen and Contreras, 1979; Magrath, 1984; World
Bank, 1987).

Research also needs to continue on exploring usable ways of incorporating some of the
more complex dynamic dimensions of agroforestry into the framework of economic
analysis. Particular problems in this respect arise from the nature of the relationships within
multiple-species cropping systems, and the additional temporal relationships added when
trees form part of the latter. Intercropped species may interact to produce supplementary,
complementary or competitive impacts on their individual or aggregate performance — or
some combination of these effects. The relationships are likely to vary with their spatial
arrangement and management, and with soil-moisture and nutrient content and the status
of other growth factors. They are also likely to change over time because of the changes that
trees create in the micro-environment they inhabit as they grow. Various approaches to the
problems of incorporating economic measures of this complex of linkages, in a manner
consistent with the likely availability of data, are being explored, based mainly on variants
of budgeting and partial budgeting techniques (Raintree, 1983; Etherington and Mathews,
1983; Hoekstra, 1985).

Conclusions

Agroforestry practices contribute to a wide variety of existing farming systems. In a number
of these, farmers have been responding to growing pressures on their land and other
resources by intensifying or increasing various agroforestry activities. These have proved to
be particularly important in enabling them to intensify use of certain types of land as size of
land-holding decreases, and to reduce the intensity of management as labour becomes
scarcer and more costly. The appropriateness of agroforestry as a response to these
changing pressures is often related to the concurrent shift in farmer production objectives
towards income generation as farm size falls below the level at which they can meet their
basic household food needs. These ongoing changes suggest the need to re-examine the
impact of schemes which provide access to land, labour and capital as ways of promoting
agroforestry activities. The growing importance of income generation as a production
objective of farmers also points to the need to give greater attention to the markets for
agroforestry products, and to the intermediate stages between production and the market.
Increased economic research has an important role to play in clarifying and extending
understanding of these and other economic considerations in agroforestry.

REFERENCES

Arnold, J.E.M. 1983. Economic considerations in agroforestry projects. *Agroforestry Systems* 1: 299–311.

Baah-Dwomoh, J. 1983. Estimating stumpage value of wood in the Sahel. World Bank, Washington, D.C. (Mimeo.)

Balasubramanian, V. 1983. Alley cropping: Can it be an alternative to chemical fertilizers in Ghana? Paper prepared for the Third National Maize Workshop, Kumasi, Ghana, 1–3 February 1983.

Blair, H.W. 1983. Equity and public policy in natural resource management: Social forestry in Maharashtra State, India. Paper prepared for the Annual Meeting of the Association for Asian Studies, San Francisco, California, 25–27 March 1983.

Blandon, P. 1985. Agroforestry and portfolio theory. *Agroforestry Systems* 3: 239–245.

Budowski, G. 1981. Applicability of agroforestry systems. In L.H. MacDonald, (ed.), *Agro-forestry in the African humid tropics.* Tokyo: United Nations University.

Chambers, R. and M. Leach. 1987. Trees to meet contingencies: Savings and security for the rural poor. Discussion Paper 228. IDS, University of Sussex.

Chambers, R. and R. Longhurst. 1986. Trees, seasons and the poor. In R. Longhurst (ed.), *Seasonality and poverty. IDS Bulletin* 17(3): 44–50.

Chavangi, N.A. 1984. Cultural aspects of fuelwood procurement in Kakamega District. Working Paper No. 4, Kenya Woodfuel Development Programme. Nairobi: Beijer Institute.

Etherington, D.M. and P.J. Matthews. 1983. Approaches to the economic evaluation of agroforestry farming systems. *Agroforestry Systems* 1: 347–366.

FAO. 1978. *Forestry for local community development.* Forestry Paper No. 7. Rome: FAO.

————. 1985. *Tree growing by rural people.* Forestry Paper No. 64. Rome: FAO.

Fernandes, E.C.M. and P.K.R. Nair. 1986. An evaluation of the structure and function of some tropical homegardens. *Agricultural Systems* 21: 179–310.

Fisseha, Y. 1987. Basic features of rural small scale forest based processing enterprises. In *Small Scale Forest Based Processing Enterprises.* Rome: FAO.

Gregersen, H.M. and A. Contrerars. 1979. *Economic analysis of forestry projects.* Forestry Paper No. 17. Rome: FAO.

Gelder, B. van and P. Kerkhof. 1984. The agroforestry survey in Kakamega District: Final report. Working Paper No. 6, Kenya Woodfuel Development Programme. Nairobi: Beijer Institute.

Gupta, T. 1979. Some financial and natural resource management aspects of commercial cultivation of irrigated eucalyptus in Gujarat, India. *Indian Journal of Forestry* 2(2).

Hoekstra, D.A. 1985. The use of economics in diagnosis and design of agroforestry systems. Working Paper No. 29. ICRAF, Nairobi.

Hunink, R.B.M. and J.W. Stoffers. 1984. Mixed and forest gardens on central Java: An analysis of socio-economic factors influencing the choice between different types of landuse. Department of Developing Countries, University of Utrecht, Utrecht.

Hyman, E.L. 1983. Pulpwood treefarming in the Philippines from the viewpoint of the smallholder: An ex-post evaluation of the PICOP Project. *Agricultural Administration* 14: 23–49.

Karunanayake, K. 1982. An economic assessment of intercropping under coconuts in Sri Lanka. M.Sc. thesis, Australian National University, Canberra.

Lagemann, J. 1977. *Traditional African farming systems in eastern Nigeria: An analysis of reaction to increasing population pressure.* Munich: Africa Studien, Weltforum Verlag.

Leach, G. 1986. *Household energy in South Asia.* London: Institute of International Education and Development (IIED).

Magrath, W. 1984. Microeconomics of agroforestry. In K.H. Shapiro (ed.), *Agroforestry in developing countries.* Ann Arbor: Center for Research on Economic Development, University of Michigan.

Michon, G., F. Mary and J. Bompard. 1986. Multistoried agroforestry garden system in West Sumatra, Indonesia. *Agroforestry Systems* 4: 315–338.

Nair, C.T.S. and C.N. Krishnankutty. 1984. Socio-economic factors influencing farm forestry: A case study of tree cropping in the homesteads in Kerala, India. In *Community forestry: Socio-economic aspects.* Bangkok: FAO/East-West Center.

Nair, M.A. and C. Sreedharan. 1986. Agroforestry farming systems in the homesteads of Kerala, southern India. *Agroforestry Systems* 4: 339–363.

Ngambeki, D.S. 1985. Economic evaluation of alley cropping leucaena with maize-maize and maize-cowpea in southern Nigeria. *Agricultural Systems* 17: 243–258.

Ninez, V.K. 1984. *Household gardens: Theoretical considerations on an old survival strategy.* Potatoes in Food Systems Research Series, Report No. 1, International Potato Center, Lima, Peru.

Noronha, R. 1982. Seeing people for the trees: Social issues in forestry. Paper to the Conference on Forestry and Development in Asia, Bangalore, India, 19–23 April 1982.

Olofson, H. 1983. Indigenous agroforestry systems. *Philippine Quarterly of Culture and Society* 11: 149–174.

Penny, D.H. and M. Singarimbun. 1973. *Population and poverty in rural Java: Some economic arithmetic from Sriharjo.* Cornell International Agricultural Development Monograph 41, Department of Agricultural Economics, Cornell University, Ithaca, New York.

Raintree, J.B. 1983. Bioeconomic considerations in the design of agroforestry cropping systems. In P.A. Huxley (ed.), *Plant research and agroforestry.* Nairobi: ICRAF.

Raintree, J.B. and K. Warner. 1986. Agroforestry pathways for the intensification of shifting cultivation. *Agroforestry Systems* 5: 39–54.

Rambo, A.T. 1984. Why shifting cultivators keep shifting: Understanding farmer decision-making in traditional agroforestry systems. In *Community forestry: Some aspects.* Bangkok: UNDP/East-West Center/FAO.

Stoler, A. 1978. Garden use and household economy in rural Java. *Bulletin of Indonesian Studies* 14: 85–101.

Weinstock, J.A. 1983. Rattan: A complement to swidden agriculture. *Economic Botany* 37: 56–68.

Wiersum, K.F. (ed.). 1981. *Viewpoints on agroforestry.* Department of Forestry, Agricultural University, Wageningen, Netherlands.

Wiersum, K.F. 1982. Tree gardening and taungya on Java: Examples of agroforestry techniques in the humid tropics. *Agroforestry Systems* 1: 53–70.

World Bank. 1986. Economic issues and farm forestry. Working paper prepared for the Kenya Forestry Sector Study, World Bank, Washington, D.C. (Mimeo.)

World Bank. 1987. An interim review of economic analysis techniques used in appraisal of World Bank-financed forestry projects. Draft staff working paper, World Bank, Washington, D.C. (Mimeo.)

12

Agroforestry and the social milieu

Marilyn W. Hoskins

Community Forestry Officer, Policy and Planning Service
Forestry Department
Food and Agriculture Organization (FAO) of the United Nations
Via delle Terme di Caracalla, 00100 Rome, Italy

Contents

Introduction

During the Workshop on Agroforestry held in Freiburg, Germany, in 1982 (Jackson, 1984), it was evident that the participants, selected by the United Nations University as people already actively involved in agroforestry, had a basic division in approach — those who felt agroforestry was a value-neutral science or technique, and those who looked to agroforestry as a promising development tool. It was not that the second group felt there was no need for basic research, but rather that they were hopeful that research designs, training, and other foci in what was then considered a new professional arena would take into consideration the socio-economic and political issues to make agroforestry useful in development efforts.

Since that time, research on physical interrelationships among biological components of agroforestry has continued. But instead of taking research priorities from the more customary technical issues involved, mainly in production-increase potential, much agroforestry research has been based on topics selected through an expanded farming-systems type approach, the diagnosis and design (D & D) methods. D & D uses a multidisciplinary problem-solving focus originating from the perspective of farmers. Establishing research priorities from within the social milieu in this manner promises to put agroforestry research results ahead of much traditional forestry and agricultural research in being of great relevance to farmers. However, the methods of making new innovations available to farmers are not yet clear.

During recent years, various social scientists also have been looking at factors involved in local tree growing and have helped to point out a number of issues which must now be taken into consideration if new techniques in agroforestry are to be adopted by farm families. The combination of social and political factors which are involved in promoting trees in traditional farming and livestock systems are unique, and as such deserve special emphasis. Some specific elements are similar to those found in organizing community development or in water resource management. Agroforestry extension, however, is more commonly compared to agricultural extension, with which a number of the more temptingly obvious similarities prove to be superficial. Current agricultural extension methods are not tailored to include consideration of the special legal status which trees may have compared to other crops, the time horizon for farmers before tree benefits may be available, the different seasonal rhythm of labour and other requirements of perennials compared to annuals, as well as the changing availability of many specific trees and tree products which have formerly been available as a free good.

These considerations must be recognized and highlighted, whether the situation calls for adequately training, retraining and co-operating with agricultural extension and other services and/or for developing special forestry/agroforestry extension units within the forestry departments. The uniqueness of the socio-economic factors involved in tree promotion is perhaps even more important to examine if we are to work with extension and development ministries and agents already in place. Probably agroforestry awareness and training will be needed for various extension agents dealing with farmers. However, if methods commonly used in agricultural promotion programmes are adopted without careful modification for use with agroforestry they may indeed defeat the promotion of farmer adoption, causing such programmes to come to a dead end, if they begin at all.

It is not easy to select and describe the crucial socio-economic variables in a universal way: situations differ depending on the locality, environment and the major traditional production activities; issues overlap and are not easily considered in isolation. Variables will need to be studied case by case. However, examples of common issues, even if incomplete, may serve to highlight some of the questions which should be raised in order to tailor agroforestry promotion policies and the training of promoters in an effective manner. Common issues include: local uses and knowledge, tenure, organization, conservation, landlessness/distance, enterprises and marketing, labour, nutrition, and gender/age.

Local use and knowledge

Neither forestry nor agricultural training focuses on learning from farmers how they use and manage tree resources. Even farming-systems approaches, which ideally put agriculturalists and farmers together to look at the family farming system, generally stress maximizing crop production and omit trees altogether. Foresters ordinarily consider timber or, more recently, polewood and fuelwood production. What the tree means to local residents in terms of desired production techniques and wood qualities is seldom explored. Trees which split into planks using a wedge are required for a type of housing and technology used in the Congo. "Y"-shaped posts are always used for traditional housing and granary supports in Malawi and many other locations throughout Africa. Foresters intent on looking at availability of fast-growing, straight trees with good crowns may miss seeing the collapsed granary when these straight trees are substituted for "Y" posts in traditional constructions.

Non-wood items such as seeds, fruits, fibres, gums, saps, leaves and bark are seldom noted by extensionists unless they relate to a cash crop such as gum arabic or certain fruits. Yet farmers propogate trees for many purposes. For example, palms are grown in places as far apart as Benin and Thailand to provide palm sap for wine or coconuts for family use or palm fibres to weave baskets for a small family enterprise.

Local women, men and youth can identify not only numerous uses they make of trees and shrubs, but in some areas they have a whole system of classifying interplant relationships. In Senegal, for example, some plants are classified as "co-wives" and others as "brothers". In this case co-wives is the classification for plants that inhibit and brothers are plants which foster each other's growth. In the Bhil ethnic group in India, some plants are also called co-wives but these are considered plants which grow well together. In Nepal they are called husband and wife. Where such classifications of local plants are well developed, they could be very important in furthering agroforestry development and adoption.

Other local knowledge relates to such scientific management as biological pest control. In Madhya Pradesh, India, residents have long grown millet in combination with cowpeas, having learned that the insect pests which live on the latter attack the insects living on the millet and thereby help prevent crop loss.

Learning about the scientific knowledge gained from generations of informal research into making a livelihood in a selected environment can help further the knowledge of professionals, who, after all, must realize that they are newly facing each specific environment and its vagaries.

When people promoting agroforestry do not establish two-way communications, they run the risk of dissipating great sums and energy on something irrelevant in the environment in question. An example of this comes from Indonesia. A dwarf coconut palm was developed which was an obvious scientific success. Although it did not live as long as the local variety, it had more coconuts, matured sooner, and was much easier to harvest as its fruits could be reached while standing on the ground. All of these features might be desirable in cash-crop palm orchards, but they were not appreciated by families who wished to grow them in their homegardens. Homegardens are carefully planned to use various levels and space above and below the ground and to manage light. The leaf crown of the tall palm tree was in a space above all the other plants and caused little shade; the shorter variety competed in the space reserved for bananas. The faster production of more fruit was not as highly valued as the more limited number of coconuts, adequate for family use, being produced over a longer number of years before replanting. Finally, the shorter plants were harder to monitor to keep passersby from stealing the fruit (O. Soemarwoto, personal communication).

Clearly, in agroforestry promotion there is no room for one-way communication. Listening skills are essential. The selection of the specific agroforestry innovation to be tried in a certain environment must not be decided from outside. The popular "Training and Visit (T&V)" approach to extension selects the most promising technique for immediate increase in production. This focus may be quite misleading for agroforestry techniques which require long periods of time to test and to obtain results. Pre-selected topics may destroy the potential for creating a system in which the farmers are an integral part of the farm-level research team.

There is a vast literature on indigenous technical knowledge (for example, see FAO, 1985b; Brokensha et al., 1980). There is also a newly developing literature on rapid appraisal which examines farmers' use of trees within the farming systems and their attitudes toward agroforestry, such as that being produced by Khon Kaen University in Thailand. This second category of information can be especially useful in developing appropriate training of agroforestry promoters in those multidisciplinary approaches which help outsiders quickly gain the background to be able to speak with farmers in terms relevant to the local situation.

As Dani and Campbell (1986) point out, in the Himalayas rural people are developing their own effective watershed management techniques. Many of these are agroforestry practices which result in more fuelwood being produced and more trees being planted on the farmland. Sajise (1985) has described customary agroforestry practices developed by isolated ethnic groups in the mountains of the Philippines, which include enriched fallow combined with using tree stems as the basis of terraces. A number of foresters have also contributed to understanding local agroforestry practices. For example, in Africa farmers have various methods of pollarding and protecting trees within the rural landscape (Polsen, 1986; Weber and Hoskins, 1983). In many cases, then, the challenge for agroforestry extension seems to be to identify and understand what positive steps farmers are already taking and to find ways to support and strengthen these. This appears to be true whether the agroforestry intervention relates to fostering trees in agricultural or herding systems or successfully integrating annuals and animals into forested areas.

Tenure

Tenure issues for agroforestry may be similar to those faced in promoting agricultural programmes only when land use is privatized, tenure of trees is synonymous with tenure of land, and the farm operator is the owner. But this is frequently not the case. There is a growing literature pointing to the complexity of tenure issues. Raintree (1985) has described seasonal access to the same piece of land by different groups, such as farmers and herders. Serial tenure of this type is practised in much of arid Africa as well as in the savanna belt of southern Africa. Although policy makers and farmers themselves may consider the farmers as holding the land tenure, little support will be given to claims for cattle damages if the crops are not harvested before the accepted time for transhumant or local herders to bring their herds to the fields.

A favourable balance of land area and its use by farmers and herders allows for more passive management with natural regeneration (see FAO, 1984; FAO, 1985a; Dove, 1983). However, where land pressure calls for the introduction and/or active protection of small perennials, a new sort of year-round protection, either social or physical, is required. This new land management may impinge on the traditional use-rights of others.

Sometimes the land tenure is controlled by a group such as an extended family or clan and use-rights are individualized on a short-term basis. In this case the introduction of hedgerows, windbreaks, or even individual tree planting may be considered anti-social or may limit highly valued flexibility. Fortmann (1985) has documented a number of instances when tree tenure is not the same as land tenure or when rights to tree products are considered separately from rights to tree removal. These cases, more common than previously realized, lead to a variety of incentives for tree protection or destruction.

Planting or, on the contrary, removing trees may also relate to establishing or assuring land-use rights. For example, case studies in Peru (Gutzman, 1987) have documented farmers planting trees for the purpose of securing land rights. This is also a customary practice among certain ethnic groups in Cameroon, whereas in others land use is traditionally or currently given to those who clear the land of trees. The Cameroonian landscape around neighbouring villages with these contrasting tree/land relationships is strikingly different; the first is lined and dotted with trees while the second is almost denuded.

Many complex tenure rights are supported through local custom; others, often in conflict with customary rights, have been codified in modern law (see von Maydell, this volume). Numerous countries have protected valued tree species through special regulations. This has mixed results (FAO, 1986). In Thailand, when young teak trees are found on farmlands they are sometimes removed in favour of less valued trees. Teak is controlled by the forest service and its cutting requires obtaining a permit.

A number of countries, such as Burkina Faso, have recently introduced national controls for cutting trees in an effort to slow desertification. Depending on how such regulations are implemented, it could result in farmers removing volunteer seedlings to favour more space for privately controlled annual crops. This reaction has been noted in countries such as Niger where forest land belongs to the forest service and farmers fear that their tree growing also will result in loss of control over land.

Where the farm operators do not have permanent or long-term use rights, they may be forbidden to plant trees for fear that this be used as a tactic by which the tenant will gain control over the land, or conversely, these farmers may not be interested since the benefits of the investment are not assured. This may be the case with women in a family who are given temporary use of fields which are under the ultimate control of the male head of household.

In the above examples, both traditions and more newly created legislation have treated trees and forests differently from annual crops. Promoters of agroforestry will need to understand these differences as they relate to traditional tree tenure, legal control and current practice of access to and control over trees and tree products. Policy makers will want to examine legislation in the light of policies which support farmers who wish to intensify agroforestry.

Organization

Related to land tenure, farming systems and the accessibility of tree products, is the question of local organization. Chandrasekharan (1985) has identified active and passive participation in tree management. He enlarges the concept of passive support to even include changes in habits necessary to allow others to grow trees. Inherent in the concept that tree planting and/or protection may concretize land use, and may eliminate or vastly reduce free access of land to others, is the idea that to be successful tree planting must be accepted by the community at large, whether the activity itself is carried out by a group or by individuals.

In a case study in Burkina Faso, the farmer was supported by various extension and development groups because of his desire to integrate trees in his land-use system. He reported that neighbours carried their animals and put them over the fences to browse on his trees at night. Neighbours had been angered by having their land use limited by his new activities. The need in this case for community-level land-use planning was evident (FAO, 1987a).

In the dune stabilization projects around villages near Shendi, Sudan, the non-governmental group "SOS Sahel" found it essential to request villages to organize committees with representation from each physical area of the village. This organization was necessary in order to design projects which obtained continuous and universal support.

A case study involving conflicts between various ethnic groups in Thailand which were integrating crops into different levels in forest land on contiguous watersheds, highlighted the need for intra-community negotiations and regional land-use planning (FAO, 1987b). Rangeland enrichment frequently fails for lack of user negotiations. In Nepal, the legislation allows for local control and management of forests by user groups after defining management objectives. In this case groups make agreements to manage forest lands in a way that will sustain adequate tree cover, which may include various agroforestry options. Some communities integrate annual crops into the understory and others manage such plants as daphne (used for making local paper) which frequently grows naturally in the understorey. In Burkina Faso a new FAO/SIDA project is focused on agroforestry through integration of annuals or livestock on a carefully planned design into forest reserves. Effective local management of common or group resources requires having or developing local organization.

Existing local organizational structures will give clues to the type of activities which have the most promise in the given situation. In some cases, such as in the highlands of Peru, activities may be organized on traditional community-wide structures. In some localities, residents traditionally work on some community activities as a group but have family ownership patterns. This situation was successfully handled in an agroforestry project run by the non-governmental organization, CARE, in a Karen community in Thailand by residents building a community nursery together but planting seed and tending and owning seedlings individually which they then planted out on their own farmland.

Promoting trees in farmland, integrating animals and crops into forest land, and protecting perennials in rangeland requires organizational support to be self-sustaining. The ability to identify and strengthen local organizations to bring the innovations to local attention and to support their successful implementation either on a group or individual basis is a great deal more important for agroforestry than for either annual crop production or for block forest management.

Agroforestry promoters need to be trained in organizational and negotiation skills, not like those used in agriculture, but more like those used in both community development and in forming water user groups. For policy makers it may require legalizing group management of resources and support of the land-use and management policies adopted through local negotiations.

Conservation

Overlapping with tenure and organizational issues is that of conservation. Agricultural extension is frequently focused on maximizing production. As was dramatically demonstrated in the peanut basin of Senegal, and is an upcoming issue for mechanized farming in the Sudan, in the large majority of tropical farming systems there can be no continuous farming without trees. Conservation is an integral part of long-term development planning.

Since agroforestry is frequently the proposed answer to improved long-term production prospects, conservation and management of soil and water through integration of trees must become part of the ordinary training of agricultural and other extension agents. Foresters, who may be trained in precisely this approach, as in the case of watershed management, must begin to learn traditional conservation techniques from local people and, with those involved, decide whether these traditions should be supported. The labour, cultural, and organizational issues involved in such technologies as stall feeding, fire control, interplanting or enriching fallow lands must be thoroughly considered. Community or regional planning must be extended to include those who may have to give up traditional practices for the benefit of distant water- or land-users. Negotiation skills may frequently be needed to help communities arrive at equitable but effective long-term resource-management decisions.

Policy issues will include first finding ways to protect forest dwellers and transhumant herders and others who successfully practise sustainable agroforestry on an extensive land-use basis. When land pressure makes this land use no longer sustainable, ways must be found to help compensate groups who must then change their traditional way of life. However, the choice of these options is a delicate policy decision and must be weighed with the overall development goals foremost, as will be discussed further.

Landlessness/distance

Closely related to conservation, organization and other agroforestry issues are those of access to tree resources for the landless and those living at a distance from forests.

Landlessness has seldom been the focus of training for agricultural extension agents who, in all fairness, require land upon which to extend their message. At the same time, tree resources in the past have frequently been available to large numbers of people as a free good. In areas where natural vegetation exists in adequate amounts, free access is common, especially for non-wood products or dead wood, i.e., for non-consumptive uses of trees.

But as situations have changed, new legal regulations exert more control over remaining natural vegetation. At the same time, as value increases and supplies are more limited, customary rights to tree resources found on farmland become privatized. Perhaps more importantly, the rules of "good manners" are fading for the weakest groups in society, as noted by Barona (1985). These might allow gleaning for the poor or land use to widows even when not regularized by either tradition or legislation.

In countries such as the Sudan, mechanized farming is providing the technology for clearing large tracts of forest lands. At the same time it is cutting off access to tree fodder and cattle routes for herders and to tree and land resources for an increasing number of rural people. This development strategy allows large-scale farmers with access to credit for purchasing equipment to plant many times the amount of land previously cultivated. It also has had the effect of absorbing land in fallow and thereby creating an increasing number of landless. In addition, the wholesale clearing of vegetative cover over large areas of semi-arid land is threatening to increase both wind and rain erosion and thereby to destroy the productive land base. Such a system offers a challenge for agroforestry promoters. For example, they might be able to help policy makers and large-scale farmers recognize that windbreaks or other introduced tree-growing strategies on the farmland could contribute to increased crop production. It could be promoted to everyone's advantage, including overall national production goals, even if planted lots or protected strips of natural forest left in the fields were used as controlled cattle routes and various tree products were made available to local residents.

In India the National Wastelands Development Council's programme has been an interesting, though complex, case where a national policy has tried to take into account involvement of the landless. Agroforestry approaches have been promoted for restoring large areas of lands classed as non-productive. In some of the schemes, rural poor and/or landless have been offered financial incentives for reconstituting degraded lands with the potential of long-term use-rights.

Access to fuel for cooking and housing materials and other tree products is essential for the well-being of many families. In urban areas, as well as in highly populated rural areas such as those in Java and the Dominican Republic, mixed tropical homegardens are an agroforestry survival strategy of many families (see Soemarwoto, this volume).

Agroforestry promoters, when supported by adequate national policies, can develop strategies which improve access to tree products to the landless or those who live at a distance from forests.

Enterprise/marketing

One of the essential considerations for rural populations, especially in areas of increased pressure on basic productive rsources, is additional off-farm income provided through processing and/or sale. This is a common consideration in agriculture extension in such activities as milling grains or extracting oils.

However, recent studies of forest-based small-scale enterprises have indicated that in a number of the countries studied these enterprises are among the top three employers of rural people. Such a source of off-farm income is especially essential for minimum-resource or landless rural people. Until recently, these enterprises were hardly noticed. Studies show, however, that access to markets and raw materials, and to organizational and management skills, are among the major constraints to increased viability of these enterprises (FAO, 1987d). Selection of appropriate species can increase availability of essential raw materials. Policies which support appropriate market infrastructure and needed skills-training and promote stable access to local supplies of raw materials would appear to offer opportunities for effective rural development through agroforestry.

Many products from trees which have been a free good until recent times are newly seen in markets. This means that the increasing problem of provisioning local needs for tannins, medicines, oils and other essential products at the same time creates opportunities to earn income from integrating selected trees into local production systems. However, the markets for these products are often not clearly established. Market support, then, offers a slightly different challenge to one usually faced by agricultural extension agents whose products are more frequently, but not always, fed into established market systems.

The issue of harvesting and marketing wood products can offer another challenge, based on weight and bulk, and the resulting difficulty and costs of extraction and transport on the scale of the artisan or small farmer. Where large trees are involved, harvesting may be much more complicated and dangerous and may demand different skills and equipment than the crops with which farmers are more familiar. In many instances, if farmers are not given help with primary processing and marketing, middlemen will absorb the benefits from tree raising. Currently there is no clear understanding by forestry specialists and project designers of the point at which costs to the farmer of legal or illegal cutting of natural vegetation will be equal to or larger than the cost of producing and managing a small or non-industrial wood source either for local use or for sale. A large number of development projects have been oriented around the sale of fuelwood or building poles. Many of these have not been based on realistic market assessments; there has been a ready assumption that a person facing a shortage will necessarily purchase the item. In practice, people often shift to alternative materials such as agricultural by-products and biomass for fuelwood and the results of integrating trees into the production system for this purpose have therefore been economically disappointing.

Intercropping trees with crops requiring fuelwood for processing is one promising approach. Planning such associations can increase the value of both crops. In one case in Sierra Leone, the tobacco-curing factory which found itself in a fuel-shortage agreed to buy tobacco from farmers who could also sell them fuelwood. The neighbouring farmers quickly converted their farmland to agroforestry.

The local farmer's view of the economic benefits of agroforestry providing mixed products from multipurpose trees and from managing the trees with other crops and/or animals must be better understood. The back and forward market linkages must be studied as well as the way rural people may tap into the market system. Agroforestry, therefore, has a special, though not always well-understood, challenge in the area of enterprise and marketing.

Labour

Almost all agroforestry innovations demand change in labour inputs, and the labour requirement is one item in the package of circumstances which rural people weigh before deciding whether or not to adopt a new agroforestry practice.

Farm families have developed labour strategies to use inputs of various family members at various times of the year for different tasks. When these labour cycles of men, women and youths are well understood, agroforestry strategies can be developed which complement them. Obviously, additional labour for persons already fully occupied at peak labour season is considered more costly than when additional demands come during a slack season.

Labour inputs depend upon the management system designed for the agroforestry practice. Innovations such as intercropping require more labour than traditional slash and burn. In Benin, farmers preferred their customary practice of stump-planting small clumps of more slowly growing teak in corners of their compounds or fields than growing strips of trees interspersed with crops when tree strips required weeding at the same time as the cotton crops.

Labour patterns in block planting as done in farm forestry are seen to have some of the same results as mechanized large-scale cash-crop planting. Both greatly reduce labour costs — advantageous to a large-scale producer and an economic blow to the limited-resource farmer or landless worker who depends on labour income. On the other hand, agroforestry may be planned to help the small-scale producer spread out family labour requirements while providing increased overall production.

Before one can predict the adoption rate or the development impact of any agroforestry innovation it will be essential for agroforestry promoters to understand local household-labour patterns. They will need to look with farm families at not only land availability and input requirements but also at how the families and the farm labourers they employ would be affected by any change.

Nutrition

Agroforestry offers the possibility of improving food security by more effectively managing soil and water resources for the sustained production of annuals. It also offers a potential for overcoming many problems of seasonality of food availability by greatly extending the season when green fodder and food supplies are available. Trees and other perennials in the production system can help tide people over drought and pest attacks, etc., when annuals cannot survive (Campbell-Asselbergs, 1987). The bushland, which is frequently fallow land or used for extensive livestock rearing, was long valued for the security it offered through containing less preferred fodder and wild seeds and roots which allowed for survival of both people and animals during what would otherwise be a disaster. Periodic droughts which have proven so catastrophic in recent times have been so severe because of the lack of this fall-back resource. Trees found in homesteads or farmlands are also frequently considered a living savings to be left to grow when not required but to be cut in times of need, thereby offering a more secure livelihood (Chambers and Leach, 1986).

Nutrition is being considered in an increasing number of agricultural extension programmes, but not in relation to the advantages of integrating trees; nutrition is a completely new focus for forestry. If the nutritional role of trees and other perennials is adequately factored into the extension design for agroforestry it can have an extremely positive impact. When the nutritionally vulnerable are the focus for benefits, agroforestry projects can be used to help increase equity.

Gender/age

A great deal has been written about gender considerations, both in farming and forestry (for example, see FAO, 1982; Fortmann and Rocheleau, 1984; Hoskins, 1983; FAO, 1987c). A number of issues previously referred to, such as local use and knowledge, tenure, organization, landlessness/distance, marketing and labour have gender-specific components. Under each of these elements information must be obtained and kept disaggregated by gender.

Certain agroforestry techniques have special promise for women. Homegardens strengthen possible control and access even in cultures such as that in the Sudan where women's mobility is strictly limited. Consideration of the best location for both agricultural and tree crops is needed for women with household responsibilities.

Age-group issues are also frequently relevant. On the one hand, where new activities require more inputs from youth (such as fuel collection or livestock care), young people may be kept home from school. On the other hand, when new resources are made available, the opposite result may be seen.

The promotion of agroforestry, like agriculture, will require gender/age sensitivity. It may also frequently require women on the teams collecting socio-economic information as well as trained female professionals on the teams promoting two-way communication if women are to be included in the development and expansion of agroforestry activities. Policy makers will have to have access to the above information as well as information on gender-related constraints, which policy changes could eliminate, if they are to understand the potential and to support more effective opportunities for women in agroforestry.

What is next?

Agroforestry research has added, and promises to add, much more to the toolbag of community-development, agriculture and forestry extension officers in the overall development effort. It helps focus agriculture on sustainable practices and on ways to make smaller parcels of land produce the range of plant and animal products required for subsistance or for market. It offers hope where land pressure has made traditional agriculture and herding practice unviable. When well designed, it can provide a more diverse production system thereby reducing risks. However, when not designed to respond to the social milieu the benefits can completely miss the poor (Chowdhry, 1985).

There is now a considerable and growing literature describing agroforestry techniques designed to address certain types of problems in various farming conditions. There is also a considerable and growing literature documenting special socio-economic and political issues which are central to effective promotion of development through agroforestry. But the next step will be crucial. That step is to integrate this information in order to develop the process and train the personnel to move new techniques from the research stations to widespread accessibility for farmer adoption. More attention is now being given to designing on-farm research methodology. However, to be most effective and efficient, the design of on-farm research and extension trials will need to ensure that there is a smooth transition in moving scientific results as well as the research/testing methods themselves from research station to the farm, from the researcher to the farmer. Research institutions will have to co-operate with those offering technical support to extension activities to conceptualize how this is to be done most effectively.

It will be a challenge to develop testing methods easily understood and used by farmers in developing realistic but rapid ways to examine plant inter-relationships in the context of their own objectives. This must become a two-way process because only through farmer management and adaptation of these suggested new approaches will the real socio-economic aspects of agroforestry be more fully understood. This final step needs to be designed to complete the information circle, giving data back to the on-station researchers. If attention is given now to planning the full cycle of research and trials and the effective information flow, the speed of providing socially appropriate agroforestry interventions and their adoption will be greatly enhanced. It is only through this testing of methods and information that technical and social scientists can refine their tools and interventions to be, in fact, relevant. Trainers of extensionists are going to have to stay abreast of this dynamic field as more is learned about tree specific issues in different settings. Agroforestry extension agents will need to be trained to approach extension as a service which makes information available and encourages farmers themselves to experiment and to actively participate in the adaptation of research results to fit their needs.

It is the continued integration of the social and technical information of researchers and farmers which will be the key also to project implementation. An example from West Africa illustrates the very common type of self-defeating result which happens when integration is not the case and when implementers do not understand its value.

In a project, including promoting local participation in agroforestry, there were a number of activities scheduled. There was a planned socio-economic study to provide background data, including how farm families use trees and what further tree resources would be useful for the project to make accessible. There was also a training component to train foresters in two-way communication. Finally, there were inputs to develop small local nurseries to give easy access to desired trees. The training was a great success as foresters talked with farm families and learned about farmer concerns. They learned farm families wanted fruit trees but had no interest in raising fuelwood or poles as there was no local market or need for increased availability of these products.

At the same time, the technical decisions were being made in isolation. To the dismay of extension foresters the only tree species planted in the nurseries was a pole-fuelwood species. The socio-economic study was delayed until late in the project implementation. It was seen by the project manager as a way to learn how to get local people to adopt the national strategy of planting woodlots in every village. The project obviously did not meet all of its goals. This common lack of co-ordination in development projects is serious in agroforestry projects given the longer time frame before evaluations clarify the need for modifications in the project.

As newly designed agroforestry techniques move out into farmers' fields, overall development issues will also become increasingly apparent. At this junction, there is need for serious examination of policy issues. A number of specific land- and tree-use and tenure issues have already been noted. But the use of agroforestry in the overall development context needs to be critically assessed. Agroforestry cannot become the development tool of choice only when poor land-use practices by commercial loggers or by poorly designed irrigation or other large-scale agricultural schemes have left denuded hillsides or salt marshes. If the services of the poor are used to reconstitute wasted land, their rights to the rehabilitated resource must be protected. Policies need to be designed to support agroforestry as an integral part of better land-use planning and to strengthen access to these new technologies for the poor not only on wasted lands.

Agroforestry offers solutions to many problems. Its promises are extremely encouraging and attractive. However, as a development tool, agroforestry will be helpful only if it can be put effectively into the hands of men and women farmers, and if the political decision-makers see this as a tool for achieving equity in development.

REFERENCES

Brokensha, D., D.M. Warren and O. Werner (eds.). 1980. *Indigenous knowledge systems and development.*lanham, Maryland: University Press of America.

*Baraona, R. 1985. Agroforestry and swidden cultivators in Latin America. Position paper prepared for Conference on Land Tenure and Agroforestry, Nairobi, Kenya, May 1985.

Campbell-Asselbergs, E. 1986. Nutrition, forests and trees: Linkages, concerns and indicators. Unpublished document, Rome: FAO.

Chambers, R. and M. Leach. 1986. Trees to meet contingencies, savings and insurance for the rural poor. Institute of Development Studies, University of Sussex.

Chandrasekharan, C. 1985. Rural participation in forestry activities. Paper prepared for Ninth World Forestry Congress, Mexico City. Rome: FAO.

Chowdhry, K. 1985. Social forestry: Who benefits? In Community Forestry: Socio-Economic Aspects. FAO/East West Center, Bangkok, Thailand.

Dani, A.A. and J.E. Campbell. 1986. Sustaining upland resources: People's participation in watershed management. ICIMOD, Occasional Paper 3, Kathmandu, Nepal.

Dove, M. 1983. Theories of swidden agriculture and the political economy of ignorance. *Agroforestry Systems* 1:85–99.

FAO. 1982. Follow-up to WCARRD: *The role of women in agriculture production.* Rome: FAO Committee on Agriculture.

————. 1984. *Changes in shifting cultivation in Africa.* Forestry Paper 50. Rome: FAO.

————. 1985a. *Changes in shifting cultivation in Africa: Seven case-studies.* Forestry Paper 50/1. FAO, Rome.

————. 1985b. *Tree growing by rural people.* Forestry Paper 64. Rome: FAO,

————. 1986. *Forest legislation in selected African countries.* Forestry Paper 65. Rome: FAO.

————. 1987a. *La Planification des Projets D'Auto-Assistance en Matiere de Bois de Feu. Session D'Etude,* Burkina Faso, Fevrier 1986.

————. 1987b. Planning self-help fuelwood projects. Workshop Report, Thailand, February 1987.

————. 1987c. *Restoring the balance: Women and forest resources.* Rome: FAO.

————. 1987d. *Small-scale forestry-based processing enterprises.* Forestry Paper (in press), Rome: FAO.

*Fortmann, L. 1985. Tree tenure: An analytical framework for agroforestry projects. Paper prepared for Conference on Land Tenure and Agroforestry. Nairobi, Kenya, May 1985.

Fortmann, L. and D. Rocheleau. 1984. Why agroforestry needs women: Four myths and a case study. *Unasylva* 1984/4.

Gutzman, L. V. 1987. Equidad Y Participacion En Actividades Comunales De Reforestacion En La Region Del Cusco, Peru. Unpublished FAO paper. May 1986.

Hoskins, M.W. 1983. Rural women, forest outputs and forestry projects. Rome: FAO.

————. 1984. Observations in indigenous and modern agroforestry activities in West Africa. In K. Jackson (ed.), *Social, economic and institutional aspects of agroforestry.* Tokyo: United Nations University.

Jackson, J.K. (ed.). 1984. *Social, economic and institutional aspects of agroforestry.* Tokyo: United Nations University.

Polsen, G. 1986. Tree planting: Problems and potentials with special reference to the forester's predicament in the Sudano-Sahelian Zone. Paper presented at SAREC workshop, June 1986.

*Raintree, J. 1985. Agroforestry, tropical land use and tenure. Paper presented to ICRAF workshop on Land, Trees and Tenure. Nairobi, Kenya, May 1985.

*Sajise, P. 1985. Position paper presented to ICRAF workshop on Land, Trees and Tenure. Nairobi, Kenya, May 1985.

Weber, F. and M. Hoskins. 1983. Agroforestry in the Sahel. Virginia Tech. Department of Sociology, Blacksburg, Virginia.

---

* *Editors' note:* These papers will appear in: Raintree, J.B. (ed.), *Land, trees and tenure.* Proceedings of an international workshop on tenure issues in agroforestry. Nairobi: ICRAF, and Madison, Wisconsin: Land Tenure Center (in press).

13

Soil productivity and sustainability in agroforestry systems

Pedro A. Sanchez

Professor of Soil Science
Tropical Soils Program
North Carolina State University
Raleigh, N.C. 27695-7619, USA

Contents

Introduction

Agroforestry systems are generally perceived to be sustainable and to enhance soil properties. Growing trees in conjunction with annual crops or pastures is believed to provide a more thorough plant cover to protect the soil from erosion and a deeper or more prolific root system to enhance nutrient cycling. Shortly after its creation, ICRAF organized a state-of-the-art review of soils research and agroforestry (Mongi and Huxley, 1979). A dearth of available, solid research data about the impact of agroforestry on soils and vice versa was obvious at that time (Sanchez, 1979). Since then, soils research in the tropics has gained considerable momentum, as has soils research in agroforestry (Nair, 1984, 1987; Young, 1986 a,b,c, 1987; Young *et al.,* 1986. The objective of this paper is to critically evaluate the various hypotheses that have been advanced on the soil-productivity and sustainability aspects of agroforestry and suggest future directions.

The soil-agroforestry hypothesis

Although not explicitly stated, the following hypothesis is strongly implied in the agroforestry literature: *Appropriate agroforestry systems improve soil physical properties, maintain soil organic matter, and promote nutrient cycling.* It is also frequently mentioned that agroforestry systems require less purchased inputs and can reclaim degraded lands in the tropics. Many traditional agroforestry systems in the tropics involve the use of trees and shrubs on crop and pasture lands (see von Maydell, this volume; Le Houérou, this volume). Farmers often claim that these trees (many of which are nitrogen-fixing species) improve the yields of associated crops (Figure 1).

Figure 1 Better growth of maize beneath a *Sesbania sesban* tree on farmland in western Kenya (Photo: E.C.M. Fernandes).

As a universal statement, the above hypothesis is probably not correct. Part of the problem is that the best documented successful agroforestry systems are located largely on good soils which are in little need of improvement. Examples of such situations are the stable coffee- or cacao-production systems under shade on volcanic or high base status Alfisols in Latin America (Russo and Budowski, 1986; Budowski, this volume) and the famous homegardens of Asia (Michon *et al.,* 1986) and in Africa (Fernandes *et al.,* 1984) located on similarly excellent soils.

Notwithstanding the importance of such favoured systems, agroforestry is considered specially applicable to marginal soils with severe physical, chemical or drought constraints. Most of the evidence in support of the soils-agroforestry hypothesis, however, is observational, qualitative or extrapolated from other systems. Many of the quantitative data on the effect of trees on soil properties are based on natural systems or pure plantation forestry (Lundgren, 1978; Chijioke, 1980; Singh, 1982; Sanchez *et al.,* 1985; Vitousek and

Sanford, 1986; Andriesse, 1987). Similarly, most of the effects of agronomic practices such as mulching are often taken from crop-production data. Given the critical interaction between trees and crops or pastures, soil-agroforestry relationships should be studied in situations where such interactions take place, and as a function of time. It is the author's proposal that changes in soil properties with time caused by agroforestry systems in well-characterized, long-term data sets be considered as the appropriate test for such hypotheses.

Types of soil-dynamics data sets

Two types of soil-dynamics data sets were recognized in a recent review (Sanchez *et al.,* 1985): Type I, where changes in soil properties are monitored with time on the same site, and Type II where soils of nearby sites of known dates of planting are sampled at the same time. Well-characterized, replicated Type I experiments are preferred, but they are scarce. The problem with Type II experiments is that the initial conditions and soil properties of the sites are unknown. Thus observations at a later stage may be the result of different initial soil properties or the effect of management systems.

The main advantage of a Type II experiment is that it can yield results in a much shorter time as compared to a Type I experiment. Type II experiments can be used to obtain an objective assessment of the effects of an agroforestry system or technology on soil properties, provided comparisons are made on sites with (1) similar soils classified according to soil taxonomy, and (2) little difference in soil texture with depth between sites. In addition, other land factors such as climate, position in landscape and slope aspect should be similar among sites. The greater the variability of a site, the greater will be the need for careful sampling and adequate replication.

Particle-size distribution in the profile is a soil property very difficult to change with management or time, and should therefore be a good indicator of whether the soils were similar when the comparisons started. Even in cases of topsoil erosion, particle-size distribution with depth, coupled with the necessary morphological descriptions for classification, can provide reasonable assurance that initial soil properties were the same. An example from the work of Russell (1983) in Ultisols of the Brazilian Amazon is given in Table 1. Differences in sand and clay content between nearby sites were small and not statistically significant. The effects of agroforestry on soil properties can be adequately estimated from such Type II data sets.

On the other hand, Type II data sets showing major textural differences between sites, or those providing no soil profile data at all, cannot be used to test the soil-agroforestry hypothesis. An example is a thorough study of teak plantations of up to 120 years of age in Kerala, India. Jose and Koshy (1972) reported major changes in soil organic matter and bulk density with age of teak plantations (Figure 2). Their soil-profile data, however, showed considerable differences in clay and sand content among the various sites supporting plantations of different ages. Clay content is usually correlated with organic matter content, while bulk density is positively related to coarse-sand content. It appears likely that the values for organic matter and bulk density were a function of textural differences rather than the effects of teak plantations over time. Jose and Koshy's data, when drawn in Figure 3, do not support the conclusion that age of teak plantation had a major effect on soil organic matter and bulk density.

Table 1 Texture of five Ultisol sites sampled at Jari, Para, Brazil in a Type II experiment comparing the effects of different plantations *vs.* native tropical rain forest.

		Plantations				
Property	Soil depth cm	Native tropical forest	0.5 yr *Pinus caribea*	9.5 yr *Pinus caribea*	8.5 yr *Gmelina arborea*	8 yr *Gmelina* + 1.5 yr pine	Signi- ficance
Sand	1	94	93	93	93	91	ns
	30	84	83	81	83	77	ns
	100	79	80	80	82	76	ns
Clay	1	5	5	6	5	7	ns
	30	15	15	16	13	18	ns
	100	20	16	17	16	20	ns

Source: Russell, 1973.

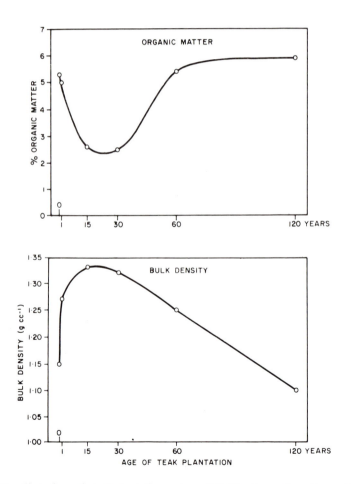

Figure 2   The effect of age of a teak plantation on topsoil (0–30cm) organic matter and bulk density as reported by Jose and Koshy (1972) in a Type II experiment conducted in Kerala, India.

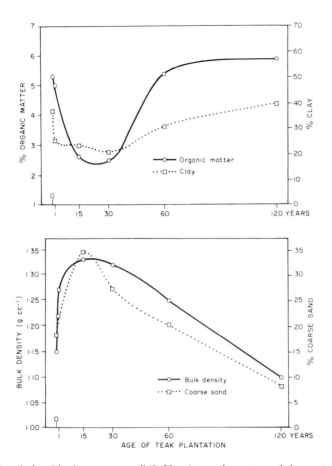

Figure 3   The relationships between topsoil (0–30 cm) organic matter and clay content, bulk density and coarse-sand content in the same Type II experiment described in Figure 2.

Many Type II experiments are not replicated and thus it is difficult to establish statistical significance with the classical analysis of variance. Geostatistical sampling techniques may provide a means of differentiating random soil spatial variability from tree effects under certain circumstances. An example of this approach is the work of Riha *et al.* (1986) comparing the effects of three tree plantations on soil properties in upstate New York. This Type II experiment compared red pine (*Pinus resinosa*) and sugar maple (*Acer saccharum*) growing on one soil, plus a third species, Norway spruce (*Picea excelsior*), growing on an adjacent but different soil, each in unreplicated 0.45 ha plots. The two soils had similar textural profiles and chemical properties, however. The application of geostatistical techniques to compare the two species growing on the same soil appears valid, considering the accuracy of soil maps in that region. The comparison between species growing on adjacent but taxonomically different soils at the great group level, however, may confound the effect of initial soil properties with the effect of trees.

An appreciation and application of this fundamental concept of Type I and II experiments will be a major step towards obtaining objective and valid comparisons of the effects of agroforestry systems on soil properties under a variety of environmental and site conditions.

Soil constraints in the tropics

The effects of soils *per se* on agroforestry systems also deserve more rigorous scrutiny. Major advances have been made on the properties, distribution and constraints of soils in the tropics during the last decade. The widespread use of a quantitative soil taxonomy system (Soil Survey Staff, 1975), the completion of the FAO world soil map, and many local soil surveys provide a vastly improved data base on soil properties and their geographical distribution. Some widely held views of the 1960s are now contradicted by quantitative knowledge. For example, generalizations such as "most tropical soils are low in organic matter", or that such soils "turn into laterite after clearing" are not supported by the data.

    Soil maps or relatively complicated taxonomic names often convey little to agronomists or foresters. Technical classification systems are used to interpret soil taxonomy information in terms relevant to specific users. Foresters frequently use the Site Index, which is a technical land classification system based on the observed rate of tree growth. The Fertility Capability Classification (FCC) system is a technical system that provides interpretations in terms of soil constraints to plant growth (Buol *et al.,* 1975; Sanchez *et al.,* 1982a). An example of the geographical extent of soil constraints by major agro-ecological zones is shown in Table 2, where the relative importance of each FCC constraint varies between the humid tropics, acid savannas, semi-arid tropics, steeplands and wetland regions. Local knowledge of soil taxonomy and FCC constraints, however, is more relevant than such world-wide figures. Local data on soil constraints can be included in the diagnosis, design or improvement of agroforestry systems. Similarly, the interpretation of

Table 2   Geographical extension of main soil constraints in five agroecological zones of the tropics.

Soil constraint	Humid tropics	Acid savannas	Semi-arid tropics	Tropical steeplands	Tropical wetlands
	\% of ecological zone				
Drought stress	-	100	100	77	67
Low nutrient reserves	66	55	17	27	3
Aluminium toxicity	57	50	13	26	4
Acid, but not Al-toxic	18	50	29	16	29
Slopes steeper than 30\%	17	16	15	100	1
High phosphorus fixation	38	32	9	21	-
Shallow soils	7	5	16	54	-
Poor drainage	13	7	10	-	100
Low cation retention	11	4	7	-	-
Gravel	1	3	5	10	2
Shrink-swell	1	-	10	-	-
Salinity	-	-	2	-	7

Source: Sanchez *et al.,* 1982c, based on FAO world soil map data interpreted according to the Fertility Capability Classification System (Sanchez *et al.,* 1982a).

agroforestry data without regard to the soils where the research was conducted severely limits its potential applicability.

The three components of the soil-agroforestry hypothesis are now examined using data sets that meet the Type I or acceptable Type II experiment criteria and are related to soil properties and constraints. Examples from soils of the humid, subhumid and semi-arid tropics are used.

Erosion control and improvement of physical properties

The fundamental reason why agroforestry systems are perceived to improve soil properties is the protection a tree cover gives the soil against surface compaction, runoff and erosion (Young, 1986a). Agroforestry systems can contain one or more such covers (also referred to as canopies): (1) a tree-top canopy, (2) a ground cover provided by annual crops or pasture, and (3) a surface-litter layer produced by the vegetation when any or all of its components are fully established. The onset, duration and thickness of ground cover of each type varies widely in agroforestry systems.

Two ICRAF reviews reported ample data showing that the presence of trees decreases soil erosion but no direct evidence from agroforestry systems (Nair, 1984; Young, 1986c). Most of the information is based on what happens when forests are cut down, but not on what happens when trees are planted in combination with crops or pastures. Although it is generally believed that a plant cover is the best way to protect the soil from erosion, it need not always be so (Wiersum, 1984). The question is, to what extent do individual agroforestry systems produce plant covers that can reduce erosion, how soon, and for how long a period of time?

Most forested land in the humid and subhumid tropics that has been cleared by mechanical means suffers some physical deterioration (Seubert *et al.,* 1977; Lal *et al.,* 1986) ranging from outright topsoil removal to mild surface-soil compaction. The detrimental effects of improper land clearing similarly affect crop, pasture, or tree growth (Sanchez *et al.,* 1985). The land-clearing method used, therefore, is a major starting point for interpreting changes in soil physical properties with time.

Furthermore, improvement in physical properties implies that soils need physical improvement. This is less the case with well-aggregated Oxisols and Andepts and in well-aggregated oxic Ultisols and Alfisols than with many sandy Alfisols and Ultisols that have very unstable structural aggregates.

One fact not often mentioned in the agroforestry literature is that most trees develop a soil-protecting plant canopy more slowly than annual crops or pastures (Broughton, 1979). Most humid tropical tree crops require at least two years to close their canopy, whereas annual crops provide adequate cover within 30 to 45 days and pastures within two to six months. Although there is sufficient evidence that fully developed tree canopies reduce runoff and erosion losses to levels similar to the original forest vegetation, the soil is particularly exposed during the tree-establishment phase unless an annual crop or a legume ground cover is planted shortly after land clearing.

A Type II study of a five-year-old *Acacia auriculiformis* plantation under a lowland humid climate in Java compared the effects of tree-canopy removal, undergrowth and litter on soil erosion. The tree-canopy itself had little effect on soil erosion and the effect of undergrowth was small. Litter cover alone reduced soil erosion by 95 percent as compared with bare soil (Wiersum, 1985). To maintain continuous litter, however, fresh inputs from the tree/shrub canopies were required.

There is ample information on the effects of annual crop canopies on runoff and erosion from a variety of soils and slopes (Roose, 1970; Lal, 1975). Agroforestry systems may provide adequate soil protection at the critical establishment phase, but mainly by crops or legume ground covers not by the trees. This will be an important factor in the design of agroforestry technologies for minimizing soil erosion. Once established, however, an agroforestry technology can provide significant soil protection. This is shown in a successful alley-cropping trial on a pH 6 sandy Entisol at Ibadan, Nigeria which is highly susceptible to soil erosion (Lal, 1987) (Table 3). The existence of trees did not affect maize yields, but did reduce cowpea yields considerably as compared with the best management for annual crop rotations on the site.

Table 3   Alley-cropping effects on runoff and soil erosion under maize-cowpea rotation on a sandy Entisol at IITA, Ibadan, Nigeria (Type I experiment).*

Treatment	Runoff (mm)	Soil erosion $(t\ ha^{-1}\ yr^{-1})$	Maize yield $(t\ ha^{-1})$	Cowpea yield $(t\ ha^{-1})$
Ploughed, annual crops only	232	14.9	4.2	0.5
No-till, annual crops only	6	0.03	4.3	1.1
Leucaena alleys (4m apart)	10	0.2	3.9	0.6
Leucaena alleys (2m apart)	13	0.1	4.0	0.4
Gliricidia alleys (4m apart)	20	1.7	4.0	0.7
Gliricidia alleys (2m apart)	38	3.3	3.8	0.6

* Each value is an average of 4 years' data. Alleys were ploughed before crop plantings to incorporate the mulch into the soil.

Source: Lal, 1987.

Do deliberately planted trees improve soil structure? Silva (1983) measured considerable changes in soil aggregation in Ultisols of humid tropical Bahia, Brazil four years after establishing several agroforestry systems in an acceptable Type II experiment. His results, shown in Table 4, indicate major differences in soil structure due to planted tree species. A young *Pinus caribaea* stand significantly improved soil structure in comparison with the virgin forest, while oil palm and rubber, both with leguminous cover, did not. The effect of slash and burn *vs.* bulldozed land clearing, however, was quite evident four years after planting. Not all results are applicable to tree/crop interactions but they suggest that tree species may play a different role in affecting soil structure.

In the semi-arid tropics, *Acacia albida* has a reputation for being a soil improver, as evidenced primarily by increasing crop yields under its deciduous canopy. A review of the literature by Felker (1978) indicates that the soil's water-holding capacity increases under *Acacia albida* in comparison with nearby sites devoid of such trees. The reports by Charreau and Vidal (1965), Dancette and Poulain (1969) and Dugain (1960) do not supply detailed soil characterization from the different sites, although it may be reasonable to assume similarity. The effect of increased water-holding capacity is believed to be a consequence of higher topsoil organic-matter content under *Acacia albida* on very sandy

Table 4 Effects of four years of plantation establishment and land-clearing methods on percent of soil aggregates greater than 1 mm in an Ultisol of Barrolandia, Bahia, Brazil (in a Type I experiment).

Treatment	Soil depth (cm)		
	0–5	5–15	15–30
	% aggregates greater than 1 mm		
Virgin forest	61	56	80
Slash and burned:			
Pinus caribaea	76	73	83
Oil palm + kudzu	42	57	71
Rubber + kudzu	51	57	87
Bulldozed:			
Cacao	33	39	65
Brachiaria sp. pasture	34	37	40
LSD .05	9	3	9

Source: Recalculated from Silva, 1983.

Alfisols which are very low in organic matter in the natural state. It is interesting to note that Poschen (1986) concluded that on poorly drained sites, enhanced crop production under *Acacia albida* canopies was due to improved drainage.

No Type I experiment on the effect of *Acacia albida* on soil properties was reported in Felker's review. In addition, the lack of adequate descriptions and/or classifications of soils and the absence of acceptable Type II studies severely limits the validity of many of the inferences about the effects of *Acacia albida* on soils. Clearly, the widespread generalization that *Acacia albida* is found mainly in sandy, infertile sites is not true. Miehe (1986) describes *Acacia albida* systems on high base status soils in the Sudan. Geostatistical techniques could be tried on fields with randomly occurring *Acacia albida* trees in the Sahel to separate the observed positive effects of trees on crops growing under them from random soil spatial variability.

Type II studies on *Acacia albida* in the Sahel (Jung, 1966) and *Prosopis cineraria* in Rajasthan, India (Mann and Saxena, 1980) have revealed increased clay content beneath the canopies of these trees. Such observations could be due to the trees acting as a physical barrier and trapping clay particles during the many dust storms that are prevalent in these regions. It is also possible that the tree's presence on a particular site may be due to the higher clay content on the site to begin with. In either case, such observations provide a good basis for well-designed Type I or acceptable Type II experiments to pinpoint the factors operating under various site and environmental conditions.

Alley cropping, growing leguminous shrubs as a source of mulch within rows along crop fields (see Kang and Wilson, this volume), is one of the most publicized agroforestry options for the humid and subhumid tropics. In spite of the attention presently given to alley cropping in the tropics, published information relating the effect of alley cropping to improved soil physical properties is scarce. Kang et al. (1985) reported that the additions of *Leucaena* prunings in a maize alley-cropping system substantially increased moisture

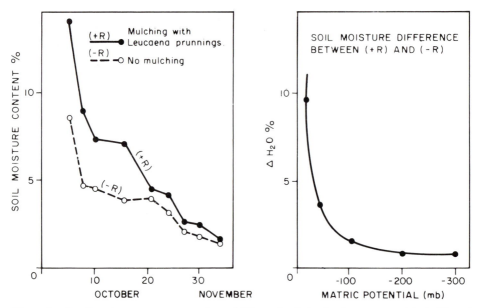

Figure 4   A soil-moisture content sample at 5-15 cm depth in the *Leucaena* mulch (+R) and non-mulch (−R) treatments and difference in soil-moisture content between (+R) and (−R) treatments sampled at the same depth at the different matric potentials for the same period of observation at IITA, Nigeria (*Source:* Kang *et al.,* 1985).

retention of the topsoil (Figure 4). The improvement is quite important because the soil in question is a sandy gravelly Entisol which is very erodible and subject to high temperature fluctuations. Table 3 provides data on the effect of alley cropping on soil erosion.

The effect of mulching is related to many factors, among which reducing soil-moisture losses is one (Lal, 1975; Wade and Sanchez, 1983). Other important mulching effects are soil-temperature amelioration, weed suppression, root-mat development, protection against rainfall impact, and in some soils, increased macro-fauna activity. Negative effects of mulching on plant growth can also occur, the most prevalent being disease enhancement by excessive soil-moisture levels during rainy periods, prevention of seedling emergence when the mulch is too thick, nitrogen immobilization as the mulch decomposes, increased crop damage by termites, and in certain cases allelopathic reactions. A study of agroforestry systems that involve mulching should take into consideration both potentially positive and potentially negative effects.

Furthermore, the protection of the soil surface by mulch layers, whether a layer from free-growing trees or prunings from alley cropping, is a dynamic process, affected by the quantity, timing and rate of decomposition of mulch inputs. Very little quantitative work on these aspects has been done, although their importance is acknowledged in the literature on natural systems (Swift *et al.,* 1979). In agroforestry systems, the three factors (quantity, timing and decomposition rate) may determine whether a large initial addition of mulch will smother weed growth while permitting normal crop development. Rapidly decomposing mulch may not stay on the surface for a sufficiently long period, and thus may provide only temporary cover. The rate of decomposition of organic materials, often called "litter quality", depends on several factors, among which the lignin/nitrogen ratio and total soluble polyphenolic content of leaf blades are believed to be important (Melillo *et al.,* 1982; Spain and Le Feuvre, 1987). No effects of mulch litter quality on agroforestry systems have been found in the available literature in Type I or appropriate Type II experiments.

An assessment of parameters that would predict the time required for a sufficiently thick mulch layer to provide adequate soil cover needs to be undertaken in promising agroforestry systems using properly designed soil-dynamics experiments.

Maintenance of soil organic matter

Agroforestry systems are believed to increase, or at least maintain, the organic-matter levels of the soil (Young, 1986a,b; Young *et al.,* 1986). The range in soil organic matter (SOM) content in the tropics is similar to that found in temperate regions (Sanchez and Buol, 1975; Sanchez *et al.,* 1982b) and the equilibrium content of any soil is a function of organic inputs (addition) and decomposition rates (Greenland and Nye, 1959). Topsoil organic carbon levels usually decrease after clearing tropical forest and reach a new equilibrium after several years of different levels of organic inputs and decomposition rates. Organic-matter dynamics have been reported on annual crops (Sanchez et al., 1983), pastures (Serrao *et al.,* 1979), and tree crops (Sanchez *et al.,* 1985) for various tropical ecosystems.

The known beneficial effects of SOM on crop production were summarized by various workers, including Allison (1973) and Swift and Sanchez (1984). In short, SOM is (1) a source of inorganic nutrients for plants, (2) a substrate for micro-organisms, (3) an ion-exchange material, (4) a factor in soil aggregation and root development, and consequently (5) a factor in soil and water conservation.

From a practical standpoint, it is desirable to have a nutrient source that synchronizes nutrient release with plant-growth demands. This is the basis of our chemical-fertilizer technology. According to Swift (1984, 1985), successful management of organic inputs should be directed at a nutrient-release pattern in synchrony with the crop's nutrient uptake pattern, taking into consideration the soil's ability to regulate related processes such as leaching and denitrification. This "SYNCH" concept is one of the fundamental tenets of the world-wide Tropical Soils Biology and Fertility (TSBF) programme (Swift, 1984; Swift and Sanchez, 1984).

Holland and Coleman (1987) reported slower decomposition of surface litter or mulch in no-till than in tilled systems due to the presence of a higher proportion of fungi in the mulch layer as compared to the decomposer community present in the soil. Increased fungal decomposition may enhance organic-matter retention because fungi retain a higher proportion of metabolized carbon than bacteria (Adu and Oades, 1978). Also, decomposition by fungi can produce more recalcitrant organic fractions than bacteria (Mayauden and Simonart, 1963).

Current work on SOM is focusing on functional pools as opposed to total SOM levels (Parton *et al.,* 1983; Sanchez and Miller, 1987). Consideration of these fractions, defined on the basis of their turnover rates in the soil, may have more relevance for evaluating agroforestry technologies than looking at total soil organic matter content.

Parton *et al.* (1983) define three functional pools: active SOM (turnover rate 0–3 years), slow SOM (turnover up to 25 years) and passive SOM (about 1,000 years). The underlying hypothesis is that maximization of the active and slow SOM pools will result in greater "activity" of SOM with regard to nutrient release, complexing Al, and soil structure. In soil testing for fertilizer use, we no longer analyse "the bones" but "the bloodstream" (for example, not total P, but available P). A similar goal for organic matter is now being set by soil scientists involved in TSBF.

Nutrient cycling in agroforestry systems

One of the advantages commonly attributed to agroforestry technologies is the potential for soil fertility improvement via more efficient cycling of nutrients (Nair, 1984), and it is often recommended to include nitrogen-fixing trees and shrubs in such technologies (Nair *et al.*, 1984; Lundgren and Nair, 1985; Young, 1987).

Good evidence exists from Type II experiments for the nutrient-cycling potential of agroforestry systems on Alfisols and Andepts of moderate to high fertility. Such systems as the *Erythrina poeppigiana* shade trees over *Coffea arabica* in Costa Rica are a good example (Glover and Beer, 1986; Russo and Budowski, 1986; Alpizar *et al.*, 1986). Similarly, Roskoski (1981) reported that *Inga jinicuil*, a leguminous shade tree in coffee plantations, fixed around 40 kg N ha^{-1} yr^{-1}. Increases in nodule biomass and function appeared to be positively correlated with increasing P-fertilizer applications to the coffee, suggesting that phosphorus deficiencies had to be ameliorated prior to adequate functioning of nodules.

Juo and Lal (1977) compared the effects of a leucaena fallow versus a bush fallow on selected soil chemical properties on an Alfisol in western Nigeria in a Type I experiment. After three years, during which leucaena was cut annually and left as mulch, the leucaena fallow resulted in significantly higher effective cation exchange capacity and levels of exchangeable Ca and K as compared to the bush fallow. Similarly, in a Type II experiment, an agroforestry system involving oil palm with leguminous cover crops (*Centrosema pubescens* and *Pueraria phaseoloides*) appeared more efficient at nutrient cycling than an oil-palm plantation with no cover crop. In addition to fixing about 150 kg N ha^{-1} yr^{-1}, the loss of nitrate nitrogen via leaching was significantly lower in the agroforestry system than in the sole crop of oil palm (Agamuthu and Broughton, 1985).

After six years of alley cropping with *Leucaena leucocephala* on a pH 6 Entisol, plots receiving prunings had a higher nutrient status and twice the organic-matter content than plots not receiving prunings (Kang *et al.*, 1984) (Table 5). This is a highly successful agroforestry system.

Attempts to extrapolate the IITA-leucaena alley cropping model to a highly weathered, sandy Ultisol in the Amazon basin of Peru, however, were not as successful (TropSoils, 1987). Table 6 shows a comparison of some of the soil properties between the Nigerian and Peruvian sites. The main differences are (1) the susceptibility of *Leucaena leucocephala* to Al toxicity, and (2) the low nutrient base content of the Yurimaguas subsoil which reduced the recycling pool.

The poor performance of leucaena in acid soils had been anticipated and a number of other woody species were evaluated in alley-cropping trials. These include *Inga edulis, Codariocalyx (Desmodium) gyroides* and *Erythrina* spp. Although yield data and soil properties are still under analysis, *Inga edulis* appears to be a highly promising agroforestry species for acid soils in the tropics (Figure 5).

Although alley cropping works well in moderately fertile soils, current experience suggests that it will be necessary to use inputs such as lime, and possibly P, to allow successful establishment of alley-cropping species and subsequent recycling of nutrients on acid infertile Ultisols and Oxisols (TropSoils, 1986). Much further study is required before this particular agroforestry system can be considered widely applicable to the humid and subhumid tropics.

Some tree and shrub species can selectively accumulate certain nutrients even in soils containing very low amounts of these nutrients. Thus palms and palm litter are rich in

Table 5 Effect of six years of alley cropping maize and cowpea with *Leucaena leucocephala* and nitrogen application on some chemical properties of surface soil of Apomu loamy sand (Psammentic Ustorthent).

Treatments (kg N/ha)	Leucaena prunings	pH (H$_2$O)	Org. C (%)	K	Exchangeable Ca (me 100g^{-1})	Mg	Bray P-1 (ppm)
0	Removed	6.0	0.65	0.19	2.90	0.35	27.0
0	Retained	6.0	1.07	0.28	3.45	0.50	26.2
80	Retained	5.8	1.19	0.26	2.80	0.45	25.6
LSD .05		0.2	0.14	0.05	0.55	0.11	5.3

Source: Kang *et al.*, 1984.

Table 6 Soil differences between sites of Ibadan, Nigeria, and Yurimaguas, Peru, where alley-cropping experiments have been conducted with contrasting results.

Site	Soil taxonomy	Depth (cm)	Clay (%)	Sand (%)	Gravel (%)	pH (H$_2$O)	Org. C (%)	Avail. P (ppm)	Exchangeable (cmol l^{-1})				% Al saturation
									Al	Ca	Mg	K	
Ibadan, Nigeria (Apomu series)	Psammentic Ustorthent loamy sand isohyperthermic	0–9	10	78	2	6.0	0.65	27	0.05	2.90	0.35	0.19	1
		9–36	10	82	3	6.3	0.30	–	0.02	1.05	0.46	0.06	1
		30–87	10	87	4	6.3	0.08	–	0.01	1.67	0.12	0.04	0
Yurimaguas, Peru (Yurimaguas series)	Typic Paleudult fine loamy siliceous isohyperthermic	0–5	5	80	0	4.3	1.00	8	1.75	0.65	0.23	0.04	66
		5–40	16	57	0	4.2	0.50	–	3.02	0.17	0.05	0.04	92
		40–80	28	49	0	4.2	0.20	–	4.97	0.06	0.05	0.03	97

Sources: **Kang** *et al.*, 1985; TropSoils, 1987; data for 9–83 cm depth of IITA soil from Moorman *et al.*, 1974.

Figure 5 Alley cropping with *Inga edulis* at Yurimaguas, Peru (Photo: E.C.M. Fernandes)

potassium (Folster *et al.,* 1976), tree ferns accumulate nitrogen (Mueller-Dombois *et al.,* 1984), *Cecropia* spp. appeared to accumulate Ca and P on acid sites (Odum and Pigeon, 1970) and *Gmelina arborea* accumulates Ca (Sanchez *et al.,* 1985). It is important for researchers looking to incorporate such nutrient-conserving species into agroforestry technologies to appreciate the fact that such plant-nutrient responses will change with location and site/soil characteristics (Golley, 1986).

In addition to translocation of nutrients from soil layers beyond the reach of annual crops or pasture species, enhancement of nutrient status beneath tree canopies has also been attributed to canopy capture of precipitation inputs (Kellman, 1979). Therefore, such factors may enhance nutrient recycling in agroforestry systems but they need to be evaluated in such systems.

Conclusions

Agroforestry as a major approach to sustainable land use in the tropics is now a widely acknowledged concept. Many of its positive attributes relate to the management and conservation of marginal soils of the tropics. While evidence exists for the beneficial effects on soils of certain agroforestry technologies (especially on more fertile soils), there is a tendency for over-generalization and extrapolation of soil productivity and sustainability benefits of agroforestry systems to other more marginal sites. The time has come to bring science into the picture and systematically test the effects of agroforestry systems on different soils, and vice versa.

A soil-dynamics methodology is proposed as a framework for such testing. The above- and below-ground interactions between trees and crops or pastures are likely to provide different results from those obtained in forests, cropped fields or pastures. Recognition of what the major soil constraints are in specific areas would improve the design of agroforestry systems. Science-based soil-agroforestry research will provide a realistic site-specific appraisal of whether agroforestry systems improve soil physical properties, maintain soil organic matter or promote nutrient cycling.

Acknowledgements

Paper No. 11163 of the Journal Series of the North Carolina Agricultural Research Service, Raleigh, N.C., U.S.A.

The author is deeply indebted to Mr Erick C.M. Fernandes, graduate student, for assistance in compiling materials for the paper.

REFERENCES

Adu, J.K. and J.M. Oades. 1978. Utilization of organic materials in soil aggregates by bacteria and fungi. *Soil Biology and Biochemistry* 10:117–122.

Agamuthu, P. and W.J. Broughton. 1985. Nutrient cycling within the developing oil palm-legume ecosystem. *Agriculture, Ecosystems and Environment* 13:111–123.

Allison, F.E. 1973. *Organic matter and its role in crop production.* Amsterdam: Elsevier.

Alpizar, L., H.W. Fassbender, J. Heuveldop, H. Folster and G. Enriquez. 1986. Modelling agroforestry systems of cacao (*Theobroma cacao*) with laurel (*Cordia alliodora*) and poro (*Erythrina poeppigiana*) in Costa Rica. Inventory of organic matter and nutrients. *Agroforestry Systems* 4: 175–189.

Andriesse, J. P. 1987. Monitoring project of nutrient cycling soils used for shifting cultivation under various climatic conditions in Asia. Final Report. Joint KIT/EEC Project No. TSD-A-116-NL. Royal Tropical Institute, Amsterdam, The Netherlands.

Broughton, W.J. 1979. Effects of various covers on soil fertility under *Hevea brasiliensis* and on growth of the tree. *Agro-Ecosystems* 3: 147–170.

Buol, S.W., P.A. Sanchez, R.B. Cate and M.A. Granger. 1975. Soil fertility capability classification. In E. Bornemisza and A. Alvarado (eds.), *Soil management in tropical America.* Raleigh, NC: North Carolina State University.

Charreau, C. and P. Vidal. 1965. Influence de *l'Acacia albida* Del. sur le sol, nutrition minérale et rendements des mils *Pennisetum* au Sénégal. *Agron. Trop.* 6–7: 626–660.

Chijioke, E.O. 1980. Impact on soils of fast-growing species in lowland humid tropics. FAO Rome: FAO Forestry Paper 21.

Dancette, C. and J.F. Poulain. 1969. Influence of *Acacia albida* on pedoclimatic factors and crop yields. *African Soils* 14: 143–184.

Dugain, G. 1960. Rapport de Mission au Niger. Centre de Pédologie de HANN - Dukan. (Mimeo.)

Felker, P. 1978. *State of the art:* Acacia albida *as a complementary permanent intercrop with annual crops.* University of California, Riverside.

Fernandes, E.C.M., A. O'ktingati and J. Maghembe. 1984. The Chagga homegardens: a multistoried agroforestry cropping system on Mt. Kilimanjaro, Tanzania. *Agroforestry Systems* 2: 73–86.

Folster, H., G. de las Salas and P. Khanna. 1976. A tropical evergreen forest site with perched water table. Magdalena Valley, Colombia. Biomass and bioelement inventory of primary and secondary vegetation. *Oecol. Plant.* 11: 297–320.

Glover, N. and J. Beer. 1986. Nutrient cycling in two traditional Central American agroforestry systems. *Agroforestry Systems* 4: 77–87.

Golley, F.B. 1986. Chemical plant-soil relationships in tropical forests. *J. Tropical Ecol.* 2: 219–229.

Greenland, D.J. and P.H. Nye. 1959. Increases in the carbon and nitrogen contents of tropical soils under natural fallows. *Soil Sci.* 10: 284–299.

Holland, E.A. and D.C. Coleman. 1987. Litter placement effects on microbial and organic matter dynamics in an agroecosystem. *Ecology* 68: 425–433.

Jose, A.I. and M.M. Koshy. 1972. A study of the morphological, physical and chemical characteristics of soils as influenced by teak vegetation. *Indian Forester* 98: 338–348.

Jung, G. 1966. Etude de l'influence de l'*Acacia albida* (Del.) sur les processus microbiologiques dans le sol et sur leurs variations saisonnières. Centre ORSTOM-Dakar Senegal. (Mimeo.)

Juo, A.S.R. and R. Lal. 1977. The effect of fallow and continuous cultivation on the chemical and physical properties of an Alfisol in Western Nigeria. *Plant Soil* 47: 567–584.

Kang, B.T., G.F. Wilson and T.L. Lawson. 1984. *Alley cropping: A stable alternative to shifting cultivation.* Ibadan, Nigeria: IITA.

Kang, B.T., H. Grimme and T.L. Lawson. 1985. Alley cropping sequentially cropped maize and cowpea with *Leucaena* on a sandy soil in Southern Nigeria. *Plant Soil* 85: 267–277.

Kellman, M. 1979. Soil enrichment by neotropical savanna trees. *J. Ecol.* 67: 565–577.

Lal, R. 1975. *Role of mulching techniques in soil and water management.* IITA Technical Bulletin No. 1, Ibadan, Nigeria.

————. 1987. Managing the soils of sub-Saharan Africa. *Science* 236: 1069–1076.

Lal, R., P.A. Sanchez and R.W. Cummings, Jr. (eds.). 1986. *Land clearing and development in the tropics.* Boston: A.A. Balkema.

Lundgren, B. 1978. *Soil conditions and nutrient cycling under natural and plantation forests in Tanzanian highlands.* Report in Forest Ecology and Forest Soils No. 31, Swedish University of Agricultural Science, Uppsala.

Lundgren, B. and P.K.R. Nair. 1985. Agroforestry for soil conservation. In S.A. El-Swaify, W.C. Moldenhauer and A. Lo (eds.), *Soil erosion and conservation.* Ankeny, Iowa: Soil Conservation Society of North America.

Mann, H.S. and S.K. Saxena (eds.). 1980. Khejri (Prosopis cineraria) *in the Indian Desert.* CAZRI Monograph No. 11. Jodhpur: Central Arid Zone Research Institute.

Mayauden, J. and P. Simonart. 1963. Humification des microorganismes marqués par ^{14}C dans le sol. *Annales de l'Institut Pasteur* 105: 257–266.

Michon, G., F. Mary and J. Bompand. 1986. Multistoried agroforestry garden system in West Sumatra, Indonesia. *Agroforestry Systems* 4: 315–338.

Miehe, S. 1986. *Acacia albida* and other multipurpose trees on the Fur farmlands in the Jebel Maia highlands, Western Dafur, Sudan. *Agroforestry Systems* 4: 89–119.

Melillo, J.M., J.D. Aber and J.F. Muratore. 1982. Nitrogen and lignin control of hardwood leaf litter decomposition dynamics. *Ecol.* 63: 621–626

Mongi, H.O. and P.A. Huxley (eds.). 1979. *Soils research in agroforestry.* Nairobi: ICRAF.

Moormann, F.R., R. Lal and A.S.R. Juo. 1975. *The soils of IITA.* Technical Bulletin No.3. IITA, Ibadan, Nigeria.

Mueller-Dombois, D., P.M. Vitousek and K.W. Bridger. 1984. Canopy dieback and ecosystem processes in Pacific forests. Hawaii Botanical Science Paper 44, University of Hawaii, Manoa.

Nair, P.K.R. 1984. *Soil productivity aspects of agroforestry.* Nairobi: ICRAF.

————. 1987. Soil productivity under agroforestry. In H. Gholz (ed.). *Agroforestry: realities, possibilities and potentials.* Dordrecht, Netherlands: Martinus Nijhoff (in press).

Nair, P.K.R., E.C.M. Fernandes and P. Wambugu. 1984. Multipurpose trees and shrubs for agroforestry. *Agroforestry Systems* 2: 145–163.

Odum, H.T. and R.F. Pigeon (eds.). 1970. *A tropical rainforest.* Vol. III. Washington, D.C.: Office of Information Services. U.S. Atomic Energy Commission.

Parton, W.J., D.W. Anderson, C.V. Cole and J.W.B. Stewart. 1983. Simulation of soil organic formations and mineralization in semiarid agroecosystems. In R.R. Lowrance, R.L. Todd, L.E. Asmussen and R.A. Leonard (eds.), *Nutrient cycling in agricultural ecosystems.* College of Agriculture Experimental Station, Special Publication No. 23, University of Georgia, Athens.

Poschen, P. 1986. An evaluation of the *Acacia albida*-based agroforestry practices in the Hararghe highlands of Eastern Ethiopia. *Agroforestry Systems* 4: 129–143.

Riha, S.J., B.R. James, G.P. Senesac and E. Pallant. 1986. Spatial variability of soil pH and organic matter in forest plantations. *Soil Sci. Soc. Am. J.* 50: 1347–1352.

Roose, E. 1970. Importance relative de l'erosion, du drainage oblique et vertical dans la pédogénése d'un sol ferralitique de moyenne Cote d'Ivoire. *Cah. ORSTOM Ser. Pédol.* 8: 469–482.

Roskoski, J.P. 1981. Nodulation and nitrogen fixation by *Inga jinicuil,* a woody legume in coffee plantations. I. Measurements of nodule biomass and field acetylene reduction rates. *Plant Soil* 59: 201–206.

Russel, C.E. 1983. Nutrient cycling and productivity in native and plantation forests in Jari Florestal, Para, Brasil. Ph.D. thesis, Institute of Ecology, University of Georgia, Athens.

Russo, R.O. and G. Budowski. 1986. Effect of pollarding frequency on biomass of *Erythrina poeppigiana* as a coffee shade tree. *Agroforestry Systems* 4: 145–162.

Sanchez, P.A. 1979. Soil fertility and conservation considerations for agroforestry systems in the humid tropics of Latin America. In H.O. Mongi and P.A. Huxley (eds.), *Soils research in agroforestry.* Nairobi: ICRAF.

Sanchez, P.A. and S.W. Buol. 1975. Soils of the tropics and the world food crisis. *Science* 188: 598–603.

Sanchez, P.A., W. Couto and S. W. Buol. 1982a. The fertility capability soil classification system: interpretation, applicability and modification. *Geoderma* 27: 283–309.

Sanchez, P.A., M.P. Gichuru and L.B. Gatz. 1982b. Organic matter in major soils of the tropical and temperate regions. *Transactions of the Twelfth International Congress of Soil Science, New Delhi* 1: 99–114.

Sanchez, P.A. and R.H. Miller. 1987. Organic matter and soil fertility management in acid soils of the tropics. *Transactions of the Thirteenth International Congress of Soil Science, Hamburg* (in press).

Sanchez, P.A., J.J. Nicholaides and W. Couto. 1982c. Physical and chemical constraints to food production in the tropics. In G. Bixler, and L.W. Shemilt (eds.), *Chemistry and world food supplies: the new frontiers.* CHEMRAWN II. Perspectives and recommendations. Los Banos, Philippines: IRRI.

Sanchez, P.A., C.A. Palm, C.B. Davey, L.T. Szott and C. E. Russell. 1985. Trees as soil improvers in the humid tropics? In M.G.R. Cannell and J.E. Jackson (eds.), *Attributes of trees as crop plants.* Huntingdon, U.K.: Institute of Terrestrial Ecology.

Sanchez, P.A., J.H. Villachica and D.E. Bandy. 1983. Soil fertility dynamics after clearing a tropical rainforest in Peru. *Soil Sci. Soc. Am. J.* 47: 1171–1178.

Serrao, E.A.S., I.C. Falesi, J.R. Veiga and J. F. Texeira. 1979. Productivity of cultivated pastures in low fertility soils of the Amazon of Brazil. In P.A. Sanchez and L.E. Tergas (eds.), *Pasture production in acid soils of the tropics.* Cali, Colombia: CIAT.

Seubert, C.E., P.A. Sanchez and C. Valverde. 1977. Effects of land clearing methods on soil properties and crop performance in an Ultisol of the Amazon jungle of Peru. *Trop. Agric.* (Trin.) 54: 307–321.

Silva, L.F. 1983. Influencia de cultivos e sistemas de manejo nas modificaoes edaficas dos oxisols de tabuleiro (Haplorthox) do sul da Bahia. Belém, Brazil: CEPLAC, Departamento Especial da Amazonia.

Singh, B. 1982. Nutrient content of standing crop and biological cycling in *Pinus patula* ecosystems. *For. Ecol. Management* 4: 317–322.

Soil Survey Staff. 1975. *Soil taxonomy* (Agricultural Handbook No. 436). Washington, D.C.: United States Department of Agriculture.

Spain, A.V. and R.P. Le Feuvre. 1987. Breakdown of litters of contrasting quality in a tropical Australian rainforest. *Journal of Applied Ecology* 24: 279–288.

Swift, M.J. (ed.). 1984. Soil biological processes and tropical soil fertility: A proposal for a collaborative program of research. *Biology International Special Issue 5.* Paris: International Union of Biological Sciences.

————. 1985. Tropical soil biology and fertility (TSBF): Planning for research. *Biology International Special Issue 9.* Paris: International Union of Biological Sciences.

Swift, M.J., O.W. Heal and J.W. Anderson. 1979. *Decomposition in terrestrial ecosystems.* Oxford: Blackwell Scientific.

Swift, M.J. and P.A. Sanchez. 1984. Biological management of tropical soil fertility for sustained productivity. *Nature and Resources* 20 (4): 1–9.

TropSoils. 1987. Annual report for 1985–86. Soil Science Department, North Carolina University.

Vitousek, P.M. and R.L. Sanford, Jr. 1986. Nutrient cycling in moist tropical forest. *Ann. Rev. Ecol. Syst.* 17: 137–167.

Wade, M.K. and P.A. Sanchez. 1983. Mulching and green manure applications for continuous crop production in the Amazon basin. *Agron. J.* 75: 39–45.

Wiersum, K.F. 1984. Surface erosion under various tropical agroforestry systems. In C.L. O'Loughlin and A.J. Pearce (eds.), *Symposium on effects of forest land use on erosion and slope-stability.* Honolulu, Hawaii: East-West Center.

Wiersum, K.F. 1985. Effects of various vegetation layers in an *Acacia auriculiformis* forest plantation on surface erosion in Java, Indonesia. In S.A. El-Swaify, W.C. Moldenhauer and A. Lo (eds.), *Soil erosion and conservation.* Ankeny, Iowa: Soil Conservation Society North America.

Young, A. 1986a. Effects of trees on soils. In R.T. Prinsley and M.J. Swift (eds.), *Amelioration of soils by trees: A review of current concepts and practices.* London: Commonwealth Science Council.

————. 1986b. The potential of agroforestry as a practical means of sustaining soil fertility. In R.T. Prinsley and M.J. Swift (eds.), *Amelioration of soils by trees: A review of current concepts and practices.* London: Commonwealth Science Council.

————. 1986c. The potential of agroforestry for soil conservation. I. Erosion control (ICRAF Working Paper No. 42) II. Maintenance of fertility (ICRAF Working Paper 43. Nairobi: ICRAF.

Young, A., R.J. Cheatle and P. Muraya. 1986. The potential of agroforestry for soil conservation. Part III. Soil Changes under Agroforestry (SCUAF): A predictive model. ICRAF Working Paper 44. Nairobi: ICRAF.

Young, A. 1987. Soil productivity, soil conservation and land evaluation. *Agroforestry Systems* 5: 277–292.

SECTION FIVE

Research findings and proposals

14

The development of alley cropping as a promising agroforestry technology

B.T. Kang and G.F. Wilson

Soil Scientist and Agronomist, respectively
International Institute of Tropical Agriculture (IITA)
P.M.B. 5320, Ibadan, Nigeria

Contents

Introduction

Shifting cultivation and related slash-and-burn cultivation systems are still the dominant land-use systems in vast areas of the tropics. These farming systems extend over 25 percent (360 x 10⁶ha) of the exploitable tropical lands (Saouma, 1974).

These traditional food-crop production systems are based largely on the restorative properties of woody species. A typical example of the system is shifting cultivation involving partial clearing of the forest or bush fallow in the humid zone, or patches of grass and scattered trees in the subhumid zone, followed by flash burning of the vegetation (for seedbed preparation and partial release of nutrients) and short-term intercropping (Allan, 1965). The cropping period is marked by a random spatial arrangement of crops and "regrowth" of woody perennials. This rotational sequence of temporal agroforestry (Nair, 1985), with long fallow periods that allow regeneration of soil productivity and weed suppression, has sustained agricultural production on uplands in many parts of the tropics for many generations.

A recent survey of traditional agriculture in the humid and subhumid zones of southern Nigeria showed that tree-crop-based systems predominate (Getahun *et al.,* 1982). Bene *et al.* (1977) have pointed out that in most tropical zones food crops and trees do well in combinations. Watson (1983) also stressed the importance of combinations of perennials and food crops in ensuring stable production and satisfactory income for subsistence farmers in the humid tropics. However, due to various socio-economic factors, particularly rapid population growth, these traditional systems have undergone rapid and drastic changes over the past few decades. In tropical Africa, for example, with an annual population growth rate of 3.1 percent (McNamara, 1984), the current population of about 500 million is expected to exceed 900 million by the year 2000. There may not be adequate land to maintain the long fallow that is essential in traditional shifting-cultivation systems. As shown by the examples cited by Prothero (1972), high population pressure has destabilized many traditional production systems: the need for more food has increased deforestation, shortened fallow periods in shifting cultivation cycles, and set in motion a degradative spiral leading to reduced productive capacity of the land and decreased crop yield. In addition, indiscriminate fuelwood gathering, timber harvesting, and grazing have aggravated land degradation in many parts of the tropics (Bene *et al.,* 1977; Poulsen, 1978; Gorse, 1985).

To meet the ever-increasing demand for food in the tropical and subtropical (developing) countries, more land must be brought under cultivation (Dudal, 1980). This is feasible for much of Africa and Latin America where only 18 and 19 percent, respectively, of the potentially-arable lands are under cultivation (IPI, 1986). This will, however, provide only a temporary solution to the food-production problem if it is not followed up by viable and sustainable food-production technologies.

The development of technologies for increasing food production through increase in land productivity thus presents a challenge to scientists. This will involve developing, for the humid and subhumid tropics, highly productive farming techniques that are ecologically sound, economically viable and culturally acceptable.

Soil management problems in the humid and subhumid tropics

Large parts of the humid and subhumid tropics that are currently under shifting cultivation and related traditional farming systems are covered by "fragile" soils. These are predominantly Ultisols, Oxisols and associated soil types in the humid tropics, and Alfisols and associated soils in the subhumid tropics (Table 1). Many of these soils are grouped as low-activity clay (LAC) soils because of their limitations, unique management requirements and other distinctive features that adversely affect their potential for crop production (Juo 1980, 1981; Kang and Juo, 1983).

During the past few decades several institutions in the tropics have been actively engaged in determining the constraints and management problems of these upland soils relative to sustainable food-crop production. The results of these investigations (Charreau, 1974; Lal, 1974; IRRI, 1980; Sanchez and Salinas, 1981; Kang and Juo, 1983; Spain, 1983; El-Swaify *et al.,* 1984) and some of the conclusions are highlighted below. Ultisols and Oxisols have problems associated with acidity and Al toxicity, low nutrient reserves, nutrient imbalance and multiple nutrient deficiencies. Ultisols are also prone to erosion, particularly on exposed sloping land. Alfisols and associated soils have major physical

Table 1 Geographical distribution of soils in the humid and semi-arid tropics (million hectares)

Soil order	Tropical Asia	Tropical Africa	Tropical America	Total	Percentage
*Humid tropics**					
Oxisols	14	179	332	525	35
Ultisols	131	69	213	413	28
Inceptisols	90	75	61	226	15
Entisols	90	91	31	212	14
Alfisols	15	21	18	54	4
Others	39	10	11	60	4
Total	379	445	666	1,490	100
Semi-arid tropics†					
Alfisols	121	446	107	694	33
Aridisols	47	440	33	520	25
Entisols	—	225	17	242	13
Inceptisols	28	38	—	66	3
Ultisols	20	24	8	52	1
Others	103	239	148	572	25
Total	319	1,462	313	2,094	100

* Data from NAP, 1982.
† Data from Kampen and Burford (1980), covering also parts of the subhumid tropics.

limitations. They are extremely susceptible to crusting, compaction and erosion, and their low moisture-retention capacity causes frequent moisture stress to crops. In addition, they acidify rapidly under continuous cropping, particularly when moderate to heavy rates of acidifying fertilizers are used.

Where land is abundant, long fallow periods facilitate restoration of soil productivity, resulting in low productivity but biologically stable production systems. The approach for maintaining the desired soil physical conditions is appropriate management of the surface soil through the use of residue mulch and minimum tillage (Lal, 1974). Loss of nutrients during cropping can be compensated for by judicious chemical inputs (Kang and Juo, 1983; Nicholaides *et al.,* 1984), but, due to the inherent low exchange and buffering capacities of LAC soils, maintenance of adequate levels of soil organic matter and judicious crop-residue management play important roles in sustainable crop production (Ofori, 1973; Sedogo *et al.,* 1979; Lal and Kang, 1982; Kang and Juo, 1983). An integrated soil fertility management system, combining the use of chemical soil amendments and biological and organic nutrient sources, will, therefore, be the most desirable nutrient-management system for these LAC soils.

The role of planted fallow in sustainable crop production

Levels of productivity that can be sustained in cropping systems largely reflect the potential and degree of management of the resource base. High productivity comes only from

systems where management intensities that are necessary for sustainability are attained
without extensive depletion of the resources. Evolutionary trends in tropical cropping
systems show that management intensities capable of sustaining productivity are usually
introduced only after considerable depletion and degradation of resources — especially the
non-renewable soil — have taken place. Conservation methods such as use of planted
fallow and other agroforestry approaches are seldom practised, and, where they are
practised, they have been introduced only after long periods of marginal land management
at low levels of energy input.

The important role of the fallow period for soil-productivity regeneration in traditional
shifting cultivation is well known (Nye and Greenland, 1960). The fertility of the soil which
is depleted during the cropping period is regenerated during the fallow period. The rate and
extent of soil-productivity regeneration depend on the length of the fallow period, the
nature of the fallow vegetation, soil properties and the management intensity. During the
fallow period, plant nutrients are taken up by the fallow vegetation from various soil depths
according to the root ranges. While large portions of the nutrients are held in the
vegetational biomass, some are returned to the soil surface or lost through leaching, erosion
and other processes. In addition, during the fallow period the return of decaying litter and
residues greatly add to the improvement of soil organic matter levels.

From the various descriptions of tropical cropping systems (Ruthenberg, 1979;
MacDonald, 1982; Benneh, 1972), a framework for a logical evolutionary pathway of
traditional crop-production systems in the humid tropics can be developed, as shown in
Figure 1. This pathway highlights the major changes and indicates points at which

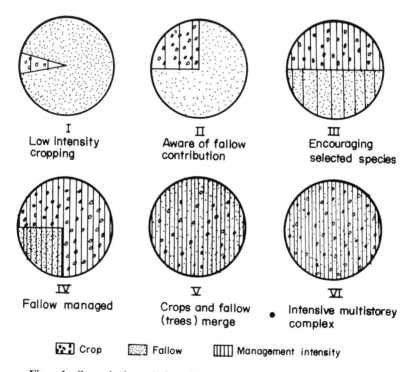

I
Low Intensity
cropping

II
Aware of fallow
contribution

III
Encouraging
selected species

IV
Fallow managed

V
Crops and fallow
(trees) merge

VI
Intensive multistorey
complex

▓ Crop   ▒ Fallow   ▥ Management intensity

Figure 1   Stages in the evolution of managed fallow and multistorey cropping
complex in the humid tropics.

intervention with planted fallows or other agroforestry methods could be introduced, and thus further resource degradation prevented. Raintree and Warner (1986) have also recently described the various agroforestry pathways for the intensification of shifting cultivation.

The pathway depicted in Figure 1 begins with a stage that may be described as a simple rotational sequence of temporal agroforestry. It is characterized by a very short cropping period followed by a very long fallow period. In this fallow period even "inefficient" soil-rejuvenating plant species are able to restore soil productivity. Here the economic return to the input of labour or energy is high; the management input is low and is confined to the cropping period. In the second stage that usually results from population pressure, the cropping period and the area cultivated are expanded. Returns to energy input begin to fall and management intensity increases. At this stage there is an awareness of the contribution of the different species in the fallow system (Benneh, 1972). At the third stage attempts are made to manipulate species in the fallow in order to ensure fertility regeneration in the already shortened fallow period. A good example of this third stage is the retention and use of tree species such as *Acioa barterii, Alchornea cordifolia, Dialium guineense* and *Anthonata macrophyla* as efficient soil-fertility restorers (Obi and Tuley, 1973; Okigbo, 1976; Getahun *et al.,* 1982). Farmers near Ibadan, Nigeria have observed that *Gliricidia sepium,* when used as yam stakes, grew and dominated the fallow and restored the land much quicker than did other species. Consequently, they now maintain *G. sepium* in the fallow even when yam is not included in the cropping cycle. In the fourth stage, mere manipulation of fallow and sole dependence on natural regeneration for the establishment of the desired species are no longer adequate, and a planted fallow of selected species becomes necessary. Though the value and feasibility of planted fallows have been demonstrated experimentally (Webster and Wilson, 1980), the practice has not become widespread. This is the stage at which intervention of techniques such as alley cropping (Kang *et al.,* 1981; Wilson and Kang, 1981) and *in-situ* mulch (Wilson, 1978) can take place.

At each of these successive stages, length of the cropping period extends progressively and that of the fallow diminishes correspondingly. During these extended cropping periods soil degradation continues, and the damage done cannot be repaired by the shortened fallow. Even when the most efficient soil-rejuvenating species dominate the fallow, they can only sustain yield at a level supportable by the existing resource base.

The fifth (merging of cropping and fallow phases) and sixth (intensive multistorey combinations) stages could evolve from the previous stages, but there is no clear evidence for this. In many areas where multistorey cropping, an intensive agroforestry system with trees and crops (Nair, 1979; Michon, 1983), dominates there is no evidence of stages four and five. The most plausible explanation is that, as population pressures grow, and the area available for stage III shrinks, that of stage VI (which is actually the intensively-managed homegardens where fruit trees are always among the major components) expands. As the two stages merge, the more efficient homegarden undergoes modification which results in the development of the multistorey production system.

If one follows the above evolution pattern, sustainability with high productivity can be achieved when conservation and restoration measures are introduced before resources are badly degraded or depleted. In the humid tropics, the multistorey complex which seems to be the climax of cropping-systems evolution, would be the ideal intervention at stages I or II. However, this may not be possible in all cases. Consequently some other types of agroforestry system, such as the planted fallows, are necessary.

Early attempts to use planted fallow in the tropics were dominated by the use of herbaceous legumes for production of green manures (Milsum and Bunting, 1928; Vine, 1953; Webster and Wilson, 1980). Though many researchers reported positive responses, the recommendations were never widely adopted. Later studies indicated that green manuring with herbaceous legumes was not compatible with most tropical climates, especially in areas with long dry periods which precede the main planting season (Wilson *et al.*, 1986): most herbaceous species did not survive the dry season and thus did not have green matter to contribute. However, herbaceous legumes such as *Pueraria phaseoloides, Centrosema pubescens, Calopogonium mucunoides* and *C. caeruleum* are widely used as ground cover in the tree-crop plantations in the humid regions (Pushparajah, 1982).

Following the introduction of herbicides and no-till crop establishment in the tropics, some of the cover crops such as *Mucuna utilis, Pueraria phaseoloides, Centrosema pubescens* and *Psophocarpus palustris* were found capable of producing *in-situ* mulch for minimum tillage production (Lal, 1974; Wilson, 1978). Various reports have also shown that trees and shrubs with their deeper root systems are more effective in taking up and recycling plant nutrients from greater depths than herbaceous or grass fallows (Jaiyebo and Moore, 1964; Nye and Greenland, 1960; Lundgren, 1978; Jordan, 1985).

Milsum and Bunting (1928) were among the earlier researchers to suggest that herbaceous legumes were not suitable sources of green manure in the tropics. They believed that shrub legumes, including some perennials such as *Crotalaria* sp. and *Cajanus cajan*, were more suitable. They even suggested a cut-and-carry method in which leaves cut from special green manure source plots would be used to manure other plots on which crops would be grown. *Cajanus cajan* with its deep roots survives most dry seasons, and at the start of the rains has an abundance of litter and leaves to contribute as green manure. Planted fallow of shrub legumes such as *Cajanus cajan*, already widely used by traditional farmers, was sometimes found to be more efficient than natural regrowth in regenerating fertility and increasing crop yields (Nye, 1958; Webster and Wilson, 1980). With increased use of chemical inputs, serious questions are repeatedly raised as to whether a fallow period is needed and what minimum fallow period will sustain crop production. An objection to the traditional fallow system as illustrated in Figure 1 (phases I and II) is the large land area required for maintaining stable production. On the other hand, modern technologies from the temperate zone introduced to increase food production by continuous cultivation have also not been successful on the LAC soils. Rapid decline in productivity under continuous cultivation continues even with supplementary fertilizer usage (Duthie, 1948; Baldwin, 1957; Allan, 1965; Moormann and Greenland, 1980; FAO, 1985). From the results of a world-wide survey, Young and Wright (1980) concluded that, with available technology, it is still impossible to grow food crops on the soils of tropical regions without either soil degradation or use of inputs at an impracticable or uneconomic level. They further stated, that at all levels of farming with inputs, there may still be a need to fallow, or to put the land temporarily to some other use, depending on soil and climatic conditions. Higgins *et al.* (1982) have given some estimates of rest periods needed for major tropical soils under various climates with different inputs (Table 2). The rest period needed decreased with increasing input levels.

To overcome the management problems of the upland LAC soils, and to incorporate in them the much-needed fallow component, scientists working at the International Institute of Tropical Agriculture (IITA) in Ibadan, Nigeria, in the 1970s opted for an agroforestry approach which had not been tried before then — the use of woody species for managing these soils. This has led to the development of what is now known as the alley-cropping system (Kang *et al.*, 1981; Wilson and Kang, 1981).

Table 2 Cultivation factors* for some major soils in the tropics depending on input levels

Major soils	Inputs					
	Low		Intermediate		High	
	Humid†	Semi-arid**	Humid	Semi-arid	Humid	Semi-arid
Arenosols (Psamments)	10	20	30	45	50	50
Ferralsols (Oxisols)	15	20	35	40	70	75
Acrisols (Ultisols)	15	20	40	60	65	75
Luvisols (Alfisols)	25	35	50	55	70	75
Cambisols (Tropepts)	35	40	65	60	85	80
Nitosols	40	75	55	70	90	90
Vertisols	40	45	70	75	90	90

* Cultivation factor = number of years it is possible to cultivate as a percentage of the total cultivation and non-cultivation cycle.
† Humid — more than 269 days of growing period.
** Semi-arid — less than 120 days of growing period.
Source: Adapted from Higgins *et al.,* 1982.

In both planted fallow and alley cropping, the potential for sustainability comes through more intensive management in which the non-crop-producing component (the fallow or woody species) is managed in such a way that a large portion of the energy flowing through that sector is directed towards crop production, and resource degradation and depletion are prevented. When these practices are introduced early they will maintain the resource base at a high level and thus respond more effectively to intensive management.

The potential of alley cropping as a sustainable farming system in the tropics

The concept of alley cropping

In alley cropping, arable crops are grown between hedgerows of planted shrubs and trees, preferably leguminous species, which are periodically pruned to prevent shading the companion crop(s) (Kang *et al.,* 1981, 1984). Two field examples of the practice are shown in Figures 2 and 3. This production system is classified by Nair (1985) as a zonal agroforestry system. The shrubs and trees grown in the hedgerows retain the same functions of recycling nutrients, suppressing weeds, and controlling erosion on sloping land as those in the bush fallow (Figure 4). Prunings from the trees and shrubs are a source of mulch and green manure. Leguminous woody species also add fixed nitrogen to the system. The alley-cropping technique can, therefore, be regarded as an improved bush-fallow system with the following advantages:

1. Cropping and fallow phases are combined;
2. Longer cropping period and increased land-use intensity;
3. Rapid effective soil fertility regeneration with more efficient plant species;
4. Reduced requirements for external inputs; and
5. The system is scale-neutral, being flexible enough for use by small-scale farmers and for large mechanized production.

Figure 2   Alley cropping: *Leucaena leucocephala* and maize at Machakos, Kenya (Photo by S. Jackson for ICRAF).

Figure 3 Alley cropping: Leucaena and upland rice at IITA, Ibadan, Nigeria

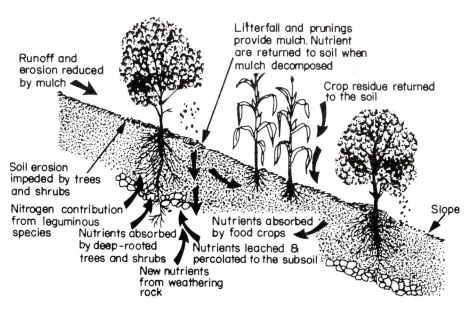

Figure 4 A schematic representation to show the benefits of nutrient cycling and erosion control in an alley-cropping system (Source: Kang *et al.*, 1986).

By integrating small-ruminant production with alley cropping, the International Livestock Centre for Africa (ILCA) project in Ibadan, Nigeria, has developed the alley-farming concept (Sumberg and Okali, 1983) in which prunings from the hedgerows provide high-quality supplementary fodder. So alley farming can be defined as the planting of arable crops between hedgerows of woody species that can be used for producing mulch and green manure to improve soil fertility and produce high-quality fodder.

The alley-cropping concept is currently being evaluated in many parts of the tropics under different names. The International Council for Research in Agroforestry (ICRAF) used the term "hedgerow intercropping" (Torres, 1983), while in Sri Lanka the term "avenue cropping" is used (Wijewardene and Waidyanatha, 1984). Working independently during the late 1960s and the 1970s, the agricultural extension department in Sikka district on the island of Flores in eastern Indonesia also promoted the use of *Leucaena* for controlling erosion and rehabilitating degraded, slightly acid soils on very steep lands in an approach similar to alley cropping under the "Lamtoronisasi" programme (Metzner, 1976; Parera, 1978; Piggin and Parera, 1985). This extension and development work with *Leucaena leucocephala* has been very successful (Parera, 1986). According to him, although *Leucaena* was forced on the farmers in the 1930s for soil rehabilitation, it did not gain early acceptance because it was not accompanied by an appropriate management system. *Leucaena* came into focus in the region only during the seventies when the "Lamtoronisasi" programme was introduced.

The potential for a sustainable farming system

Various field trials were carried out by IITA scientists over the past ten years on strongly acid soils (Ultisols) and slightly acid soils (Alfisols) in the humid and subhumid regions of Nigeria to test the suitability and benefits of alley cropping. Some of the results of these trials have been published (Kang *et al.,* 1981, 1984, 1985, 1986; Kang and Duguma, 1985; Ngambeki, 1985; Wilson *et al.,* 1986). On Alfisols and associated soils *Leucaena leucocephala* and *Gliricidia sepium* were the most promising woody species for alley cropping and alley farming (Atta-Krah *et al.,* 1985). They can be established by direct seeding in association with a growing crop. Once established, the hedgerows can be repeatedly pruned to produce large amounts of biomass that can be used as green manure, mulch or fodder.

Even on degraded land, *L. leucocephala* and *G. sepium* prunings had higher nutrient yields than those of some widely used native fallow species such as *Acioa barterii* or *Alchornea cordifolia* (Table 3). The high nutrient yields are maintained when prunings are added to the soil. However, under a cut-and-carry system where prunings are continuously removed as fodder, the soil can also become impoverished unless nutrients from other sources are added.

The performance of maize, cassava and cowpea in alley cropping with *L. leucocephala* and *G. sepium* has been studied. Higher maize and cassava yields were obtained when alley cropped than in control plots. It is estimated that *L. leucocephala* can contribute about 40kg N ha^{-1} to the companion maize crop (Kang and Duguma, 1985). Ngambeki (1985) also reported large savings in nitrogen fertilizer when maize is alley cropped with *L. leucocephala.* Cowpea yield, however, showed either no increase or reduction in yield when alley cropped with *L. leucocephala.* Upland rice alley cropped with *L. leucocephala* does not respond to added fertilizer nitrogen, but the control plot (not alley cropped) responded to 30 kg of applied nitrogen per hectare.

An important aspect of alley cropping is how it affects yield sustainability. Under long-term observations on a sandy soil, maize yields were significantly higher when alley cropped with *L. leucocephala* than in control plots with or without applied nitrogen

Table 3 Estimated nutrient yield from hedgerow (4-m interrow spacing) prunings (not including woody material) of four fallow species grown in alley cropping on a degraded Alfisol in southern Nigeria

Species	Biomass yield* (t ha^{-1} yr^{-1})	Nutrient yield*				
		N	P	K	Ca	Mg
				(kg ha^{-1} yr^{-1})		
Acioa barterii	3.0	40.5	3.6	20.4	14.7	5.4
Alchornea cordifolia	4.0	84.8	6.4	48.4	41.6	8.0
Gliricidia sepium	5.5	169.1	11.1	148.8	104.3	17.6
Leucaena leucocephala	7.4	246.5	19.9	184.0	98.2	16.2

* Fifth year after establishing of hedgerows; total of five prunings.
Source: B.T. Kang, unpublished data.

(Kang and Duguma, 1985). Similar results were observed in long-term alley cropping trials on degraded Alfisols. With or without applied nitrogen, maize yielded more when alley cropped (Figure 5). This trial also showed that, in addition to nitrogen, improved soil conditions resulting from alley cropping had a positive effect on maize yields.

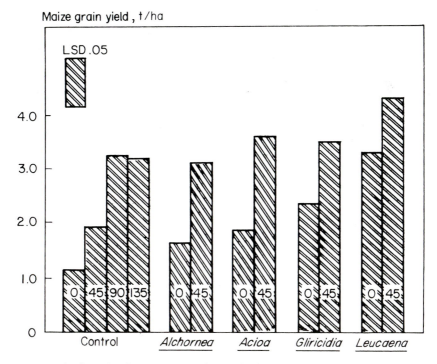

Figure 5 Grain yield of maize on eroded Alfisol (oxic paleustalf) at Ibadan, Nigeria as affected by alley cropping with woody species (*Acioa barterii, Alchoernea cordifolia, Gliricidia sepium* and *Leucaena leucocephala*) and nitrogen rates (kg N/ha) (Source: B.T. Kang, unpublished data).

Results of long-term studies showed significant improvement in soil properties under alley cropping. These soils had higher soil organic matter and nutrient status than in soil receiving no prunings. Prunings added as mulch also substantially increased moisture retention in the topsoil (Kang *et al.*, 1985).

The addition of organic matter and partial shading resulting from alley cropping stimulated increased earthworm activity. Yamoah and Mulongoy (1984) reported higher microbial activity as measured by increased biomass carbon under alley cropping. In addition to improving the soil's chemical, physical and biological condition, hedgerows play an important role in suppressing weeds and reducing runoff and soil erosion. Lundgren and Nair (1985) and Young (1986) have recently illustrated the importance of woody species for soil conservation. Ossewaarde and Wellensiek (1946) had also reported the importance of woody fallow species in soil conservation and weed suppression. Metzner (1982) reported significant results for *L. leucocephala* in controlling soil erosion and improving and maintaining productivity of degraded and sloping lands on the island of Flores in eastern Indonesia.

Kabeerathumma *et al.* (1985) and O'Sullivan (1985) reported remarkable reduction in runoff and soil erosion when *L. leucocephala* was included in the production system. Similarly, observations at IITA showed that, with mechanized alley cropping on sloping land, soil that had been degraded after root-rake clearing and tillage was more stable after *L. leucocephala* hedgerows were introduced than on adjacent land that was shear-blade cleared and maintained under annual no-tillage planting.

Investigations on acidic Ultisols in the humid tropics showed that for these conditions woody species such as *Acioa barterii, Cassia siamea,* and *Flemingia congesta* were suitable for alley cropping. Cassava when alley cropped with *Acioa barterii* or *Cassia siamea* yielded more than in the control (Kang *et al.*, 1986). Although most of the research work on alley farming has been carried out in the humid and subhumid tropics, certain aspects of the concept could be applied in other agro-ecological zones in the tropics. Further evaluation of the technology in semi-arid and highland tropics is needed. Such research should include evaluating the suitability of hedgerow species, and hedgerow and crop husbandry methods for local environments and farming systems. Recent results in the semi-arid tropics of India showed that alley farming has good potential, particularly for providing much-needed fodder (Singh *et al.*, 1986). Similarly, trials at the ICRAF Field Station, Machakos, Kenya (700mm of annual bimodal rainfall, 1500 m altitude) have shown the feasibility of alley cropping during years/seasons of "normal" rainfall (Nair, 1987).

Adoption of alley cropping

Spontaneous spread is the most dependable proof of acceptability of a land-use technology. This has happened in the case of alley cropping over the past few years, and, as stated by Vogel (1986), the incentive to adopt alley cropping will increase as pressures on land increase. The practicability and acceptability of alley-cropping technology can be illustrated by its very successful introduction to "critical" areas of eastern Indonesia (Metzner, 1976; Parera, 1978) and in the southern Philippines (O'Sullivan, 1985). Promising developments have been reported from Sri Lanka (Wijewardene and Waidyanatha, 1984) and parts of Africa.

Examples from Nigeria have shown that the concept is readily accepted in certain parts of the country, but land-tenure systems have been a major constraint to adoption in other areas (Francis, 1986). In the yam growing area of Zakibiam, alley cropping was readily adopted as a source of much-needed staking material. Those farmers also realized that alley cropping with *L. leucocephala* improved soil fertility.

The ILCA project that introduced alley farming at Owa-Ile and Iwo-Ate in Oyo North in southern Nigeria showed high adoption and spontaneous spread of the practice among traditional farmers (Atta-Krah and Francis, 1986). This project will be expanded in the coming years in a joint undertaking by the Nigerian Department of Livestock, the World Bank, and ILCA (L. Reynolds, personal communication).

In introducing the alley cropping/farming technology there are two aspects that have important implications for on-farm research. The first is that alley farming which links several farm enterprises differs from such single-component technologies as improved varieties or fertilizer. The second is that planting and managing the trees implies changes in farmers' behaviour. Since immediate benefits of the system are not directly apparent, introduction and testing of the system in farmers' fields require constant supervision for the first few years. Because of these considerations a group participatory approach appears to be more successful than individual approaches in introducing the technology (Atta-Krah and Francis, 1986; Cashman, 1986). Farmers must be convinced that alley cropping is a long-term investment that will lead to high sustainable productivity.

Outlook

In the traditional system of upland crop production on LAC soils, only a small portion of land is used for food-crop production at any given time. The larger part is under fallow. This extravagant use of land cannot continue, particularly where high population densities prevail. It is also impossible to maintain food production without an adequate fallow period on these LAC soils, unless high inputs are used in combination with short fallow periods. Planted herbaceous fallow, though generally no more efficient than natural regrowth for soil restoration, is useful for reducing adverse effects from cropping. Fallow should be designed to facilitate expansion of production periods. It should arrest degradation, enhance biological recycling, raise labour-use efficiency, and stabilize favourable environmental conditions for crop production. The alley-cropping technology incorporates all the benefits of the fallow period in the food-production period and sustains land productivity for longer periods.

The development of a sustainable production system suitable for large parts of the subhumid ·and humid regions, particularly in Africa, will have the additional benefit of reducing the land area needed for food production. Expanded alley cropping could help to arrest rapid deforestation.

Considering the limited input available to traditional farmers in Africa, low-input regenerative production systems like alley cropping deserve attention and promotion. Even in developed countries, as Wittwer (1983) and Blevins (1986) stated, the new trend is towards production technologies involving greater as well as more efficient use of resources. Wittwer (1983) described this variously as regenerative agriculture, sustainable agriculture, organic farming and gardening. The high-input production systems of these countries are considered wasteful, exploitative of natural resources, and environmentally dangerous because of their excessive use of chemical fertilizers and pesticides.

Further research is needed to select more suitable multipurpose woody species for alley cropping, particularly for acid soils and high elevations. Similarly, testing of the alley-cropping and farming concept for the drier areas needs to be carried out. Alley cropping/farming has good potential for rapid dissemination and adoption in suitable areas.

REFERENCES

Allan, W. 1965. *The African husband man.* Edinburgh: Oliver and Boyd.
Atta-Krah, A.N., J. E. Sumberg and L. Reynolds. 1985. Leguminous fodder trees in the farming system. An overview of research at the humid zone programme of ILCA in southwestern Nigeria. Ibadan, Nigeria: ILCA.
Atta-Krah, A. N. and P. A. Francis. 1986. The role of on-farm trials in the evaluation of alley farming. Paper presented at the Alley Farming Workshop, March 1986, Ibadan, Nigeria.
Baldwin, K.D.S. 1957. *The Niger agricultural project: An experiment in African development.* Oxford: Blackwell.
Bene, J.G., H.W. Beall and A. Côte. 1977. *Trees, food and people: Land management in the tropics.* Ottawa: IDRC.
Benneh, G. 1972. Systems of agriculture in tropical Africa. *Econ. Geog.* 48: 244–257.
Blevins, D.G. 1986. Future development in plant nutrition. *Agron. Abstracts:* 270.
Cashman, K. 1986. A community response to on farm research. A participatory approach for sustainable food production. Ford Foundation consultancy report. Ibadan, Nigeria: IITA.
Charreau, C. 1974. Soils of tropical dry and dry-wet climate areas of West Africa and their use and management. Department of Agronomy, Cornell University, Ithaca. N.Y. (mimeo).
Dudal, R. 1980. Soil-related constraints to agricultural development in the tropics. In *Priorities for alleviating soil-related constraints to food crop production in the tropics.* Manila: IRRI.
Duthie, D.W. 1948. Agricultural development. *E. Afr. Agric. For. J.* 13: 129–130.
El-Swaify, S.A., T.S. Walker amd S.M. Virmani. 1984. Dry land management alternatives and research needs for Alfisols in the semi-arid tropics. Andhra Pradesh, India: ICRISAT.
FAO, 1985. *Changes in shifting cultivation in Africa. Seven case studies.* FAO Forestry paper 50/1. Rome: FAO.
Francis, P.A. 1986. Land tenure systems and the adoption of alley farming. Paper presented at the Alley Farming Workshop, March 1986, Ibadan, Nigeria.
Getahun, A., Wilson, G.F. and Kang, B.T. 1982. The role of trees in farming systems in the humid tropics. In L.H. McDonald (ed.), *Agroforestry in the African humid tropics.* Tokyo: United Nations University.
Gorse, J. 1985. *Fuelwood and domestic energy:- The fuelwood crisis in tropical West Africa.* Washington, D.C.: World Bank.
Higgins, G.M., A.H. Kassam, L. Naiken, G. Fisher and M.M. Shah. 1982. *Potential population supporting capacities of lands in the developing world.* Technical report INT/75/P 13. Land Resources for Populations of the Future. Rome: FAO.
IPI. 1986. *Fertilizer in pictures.* Berne, Switzerland: International Potash Institute.
IRRI. 1980. *Priorities for alleviating soil related constraints to food crop production in the tropics.* Manila: IRRI.
Jaiyebo, E.O. and A.W. Moore. 1964. Soil fertility and nutrient storage in different soil vegetation systems in a tropical rain forest environment. *Trop. Agr.* (Trinidad) 41: 129–130.
Jordan. C.F. 1985. *Nutrient cycling in tropical forest ecosystems.* New York: Wiley.
Juo, A.S.R. 1980. Mineralogical characterisation of Alfisols and Ultisols. In B.K.G. Theng (ed.), *Soils with variable charge.* Lower Hutt, New Zealand: New Zealand Society of Soil Science.
————. 1981. Mineralogical groupings of soils with variable charge in relation to management and classification. Paper presented at International Conference on Soils with Variable Charge, 11–18 February, Palmerston North, New Zealand.
Kabeerathumma, S., S.P. Ghosh and K.R. Lakshmi. 1985. Soil erosion and surface runoff: Multiple systems compared. *Cassava Newsletter* 9: 5.

Kampen, J. and J. Burford. 1980. Production systems, soil related constraints and potentials in the semi-arid tropics with special reference to India. In *Priorities for alleviating soil related constraints to food crop production in the tropics*. Manila: IRRI.

Kang, B.T., G.F. Wilson and L. Sipkens. 1981. Alley cropping maize (*Zea mays*) and leuceana (*Leucaena leucocephala* Lam) in southern Nigeria. *Plant and Soil* 63: 165–179.

Kang, B.T. and A.S.R. Juo. 1983. Management of low activity clay soils in tropical Africa for food crop production. In F.H. Beinroth, H. Neel and H. Eswaran (eds.), *Proceedings of the Fourth International Soil Classification Workshop* (Kigali, Rwanda). Brussels: ABOS-AGCD.

Kang, B.T., G.F. Wilson and T.L. Lawson. 1984. *Alley cropping: a stable alternative to shifting cultivation*. Ibadan, Nigeria: IITA.

Kang, B.T. and B. Duguma. 1985. Nitrogen management in alley cropping systems. In B.T. Kang and J. van der Heide (eds.), *Nitrogen in farming systems in the humid and subhumid tropics*. Haren, Netherlands: Institute of Soil Fertility.

Kang, B.T., H. Grimme and T.L. Lawson. 1985. Alley cropping sequentially cropped maize and cowpea with Leucaena on a sandy soil in southern Nigeria. *Plant and Soil* 85: 267–276.

Kang, B.T. and A.S.R. Juo. 1986. Effect of forest clearing on soil chemical properties and crop performance. In R. Lal, P.A. Sanchez and R.W. Cumming Jr. (eds.), *Land clearing and development in the tropics*. Rotterdam, Netherlands: A.A. Balkema.

Kang, B.T., A.C.B.M. van der Kruijs and D.C. Couper. 1986. Alley cropping for food crop production in humid and subhumid tropics. Paper presented at the Alley Farming Workshop, March 1986, Ibadan, Nigeria.

Lal, R. 1974. *Role of mulching techniques in tropical soil and water management*. Technical Bulletin 1, International Institute of Tropical Agriculture, Ibadan, Nigeria.

Lal, R. and B.T. Kang. 1982. Management of organic matter in soils of the tropics and sub-tropics. In *Non-symbiotic nitrogen fixation and organic matter in the tropics*. Symposia Papers 1, Twelfth ISSS Congress, New Delhi.

Lundgren, B. 1978. *Soil conditions and nutrient recycling under natural and plantation forests in Tanzanian highlands*. Report on Forest Ecology and Forest Soils, No. 31, Department of Forest Soils, Swedish University of Agricultural Sciences, Uppsala, Sweden.

Lundgren, B. and P.K.R. Nair. 1985. Agroforestry for soil conservation. In S.A. El-Swaify, W.C. Moldenhauer and A.Lo (eds.), *Soil erosion and conservation*. Ankeny, Iowa: Soil Conservation Society of America.

MacDonald, L.H. (ed.). 1982. *Agroforestry in the African humid tropics*. Tokyo: United Nations University.

McNamara, R.S. 1984. *The population problem: Time bomb or myth*. Washington, D.C.: World Bank.

Metzner, J.K. 1976. Lamtoronisasi, an experiment in soil conservation. *Bulletin of Indonesian Studies* (Canberra, Australia) 2: 103–109.

————. 1982. *Agriculture and population pressure in Sikka, Isle of Flores. A contribution to the stability of agricultural systems in the wet and dry tropics*. Monograph 28, Australian National University Canberra, Australia.

Michon, G. 1983. Village-forest-gardens in west Java. In P.A. Huxley (ed.), *Plant research and agroforestry*. Nairobi: ICRAF.

Milsum, J.N. and B. Bunting. 1928. Cover crops and manure. *Malayan Agric. J.* 26: 256–283.

Moormann, F.R. and D.J. Greenland. 1980. Major production systems related to soil properties in humid tropical Africa. *In Priorities for alleviating soil related constraints to food production in the tropics*. Manila: IRRI.

Nair, P.K.R. 1979. *Intensive multiple cropping with coconuts in India*. Berlin/Hamburg: Verlag Paul Parey.

————. 1985. Classification of agroforestry systems. *Agroforestry Systems* 3: 97–128.

————. 1987. The ICRAF Field Station, Machakos: A demonstration and training site for agroforestry technologies. *Agroforestry Systems* 5: 383–394.

NAP. 1982. *Ecological aspects of development in the humid tropics*. Washington D.C.: National Academy Press.

Ngambeki, D.S. 1985. Economic evaluation of alley cropping Leucaena with maize-maize and maize-cowpea in southern Nigeria. *Agric. Systems* 17: 243–358.

Nicholaides, J.J., D.E. Bandy, P.A. Sanchez, J.H. Villachica, A.J. Couto and C.S. Valverde. 1984. From migratory to continous agriculture in the Amazon basin. In *Improved production systems as an alternative to shifting cultivation.* Rome: FAO.

Nye, P.H. 1958. The relative importance of fallows and soils in storing plant nutrients in Ghana. *J. W. Afr. Sci. Ass.* 4: 31–41.

Nye, P.H. and D.J. Greenland. 1960. *Soils under shifting cultivation.* Technical Communication 51, Commonwealth Bureau of Soils, Farnham, England.

Obi, J.K. and P. Tuley. 1973. *The bush fallow and ley farming in the oil palm belt of southeastern Nigeria.* Misc. Report 161, Land Resources Division, Ministry of Overseas Development (ODM), U.K.

Ofori, C.S. 1973. Decline in fertility status of a tropical forest Ochrosol under continuing cropping. *Exptl. Agric.* 9: 15–22.

Okigbo, B.N. 1976. Role of legumes in small holdings of the humid tropics. In J. Vincent, A.S. Whitney and J. Bose (eds.), *Exploiting the legume-rhizobium symbiosis in tropical agriculture.* Department of Agronomy and Soil Science, University of Hawaii, Honolulu.

Ossewaarde, J.G. and S.J. Wellensiek. 1946. Capita selecta uit de algemene plantenteelt. (Overview of crop production.) In C.J.J. Van Hall and C. van den Koppel (eds.), *De Landbouw in den Indischen Archipel.* The Hague, Netherlands: W. van Hoeve.

O'Sullivan, T.E. 1985. Farming systems and soil management: the Philippines/ Australian development assistance program experience. In E.T. Craswell, J.V. Remenyi and L.G. Nallana (eds.), *Soil erosion and management.* ACIAR Proceedings, Series 6, Canberra.

Parera, V. 1978. Usaha Kearah memperbaiki pertanian tanah kering di kabupaten Sikka. (Efforts for improvement of dryland farming in Sikka district.) Jakarta, Indonesia: Majalah Pertanian.

————. 1986. The role of *Leucaena leucocephala* in farming systems in Nusa Tenggara Timur, Indonesia. Paper presented at the Alley Farming Workshop, March 1986, Ibadan, Nigeria.

Piggin, C.M. and V. Parera. 1985. The use of *Leucaena* in Nusa Tenggara Timur. In E.T. Craswell and B. Tangendjaja (eds.), *Shrub legumes in Indonesia and Australia.* ACIAR Proceedings, Series 3. Canberra.

Poulsen, G. 1978. *Man and tree in tropical Africa.* Ottawa: IDRC.

Prothero, R.M. 1972. *Population pressure and land use in Africa.* London: Oxford University Press.

Pushparajah, E. 1982. Legume cover crops as a source of nitrogen in plantation crops in the tropics. In *Non-symbiotic nitrogen fixation and organic matter in the tropics.* Symposia Papers 1, Twelfth ISSS Congress, New Delhi.

Raintree, J.B. and K. Warner. 1986. Agroforestry pathways for intensification of shifting cultivation. *Agroforestry Systems* 4: 39–54.

Ruthenberg, H. 1971. *Farming systems in the tropics.* London: Oxford University Press.

Sanchez, P.A and J.E. Salinas. 1981. Low input technology for managing Oxisols and Ultisols in tropical America. *Adv. Agron.* 34: 279–406.

Saouma, E. 1974. In *Shifting cultivation and soil conservation in Africa.* Soils Bulletin 24. Rome: FAO.

Sedogo, M.P., J. Pichot and J.F. Poulain. 1979. Evolution de la fertilité d'un sol ferrugineux tropical sous l'influence de fumures minerales et organiques. Incidences des successions culturals. IRAT, Station de Saria, Haute Volta.

Singh, R.P., R. J. van den Beldt, D. Hocking and G.R. Kowar. 1986. Alley farming in the semi-arid regions of India. Paper presented at the Alley Farming Workshop. March 1986, Ibadan, Nigeria.

Spain, J.M. 1983. Agricultural potential of low activity clay soil of the humid tropics for food crop production. In F.H. Beinroth, H. Neel and H. Eswaran (eds.), *Proceedings of the Fourth International Soil Classification Workshop* (Kigali, Rwanda). Brussels: ABOS, AGCD.

Sumberg, J.E. and Okali, C. 1983. Linking crop and animal production. A pilot development program for small holders in southwest Nigeria. *Rural Development in Nigeria* 1.

Torres, F. 1983. Potential contribution of leucaena hedgerows intercropped with maize to the production of organic nitrogen and fuelwood in the lowland tropics. *Agroforestry Systems* 1: 323–333.

Vine, H. 1953. Experiments on the maintenance of soil fertility at Ibadan, Nigeria. *Empire J. Expt. Agric.* 21: 65–85.

Vogel, W.O. 1986. Socio-economic consideration for alley farming. Paper presented at the Alley Farming Workshop, March 1986, Ibadan, Nigeria.

Watson, G.A. 1983. Development of mixed tree and food crop systems in the humid tropics: a response for population pressure and deforestation. *Exptl. Agric.* 19: 311–332.

Webster, C.C. and P.N. Wilson. 1980. *Agriculture in the tropics.* London: Longman.

Wijewardene, S.R. and P. Waidyanatha. 1984. *Conservation farming for small farmers in the humid tropics.* Colombo, Sri Lanka: Department of Agriculture.

Wilson, G.F. 1978. A new method of mulching vegetables with the in-situ residue of tropical cover crops. *Proceedings of the Twentieth Horticultural Congress.* Sydney, Australia.

Wilson, G.F. and B.T. Kang. 1981. Developing stable and productive cropping systems for the humid tropics. In B. Stonehouse (ed.), *Biological husbandry: A scientific approach to organic farming.* London: Butterworth.

Wilson, G.F., B.T. Kang and K. Mulongoy. 1986. Alley cropping: Trees as sources of green-manure and mulch in the tropics. *Biol. Agric. Hort.* 3: 251–267.

Wittwer, S.H. 1983. Epilogue: The new agriculture: A view of the twenty-first century. In J.W. Rosemblum (ed.), *Agriculture in the twenty-first century.* New York: Wiley.

Yamoah, C.F. and K. Mulongoy. 1984. In *IITA Annual Report,* 1983. Ibadan, Nigeria: IITA.

Young, A. 1986. The potential of agroforestry for soil conservation. Part 1. Erosion control. Working Paper No. 42. Nairobi: ICRAF.

Young, A. and A.C.S. Wright. 1980. Rest period requirements of tropical and subtropical soils under annual crops. In *Report on the second FAO/UNFPA expert consultation on land resources for populations for the future.* Rome: FAO.

15

The role of biological nitrogen fixation in agroforestry

Y. R. Dommergues

Director of Research
Laboratoire de Biotechnologie des Systèmes Symbiotiques
Forestiers Tropicaux
(CTFT/ORSTOM/CNRS)
45bis Avenue de la Belle Gabrielle
94736 Nogent-sur-Marne Cedex, France

Contents

Introduction
Nitrogen-fixing characteristics of the main tree species used in agroforestry
Technology for exploiting nitrogen-fixing trees in agroforestry
Future trends in agroforestry research
Conclusions
References

Introduction

Agroforestry represents an approach to integrated land use involving the more or less intimate association of different plant species, always including trees or woody perennials on the same unit of land (Huxley, 1983). This definition encompasses the association of trees with perennial crops, such as coffee or cacao, pasture species, annual or biennial crops, trees being planted as windbreaks or in alleys and frequently trimmed during cropping, various forms of multispecies combinations such as those found in homegardens, and also mixed forests (mixed perennial cropping).

The most decisive factor for the success of agroforestry is the choice of suitable, useable tree species (Nair *et al.*, 1984). This choice should be based on economic and agronomic criteria. In order to fulfil the second criterion, the highest priority should be given to selecting trees that can improve the soil, and identifying species or clones that will fix or absorb large amounts of N (and also other elements, especially P) and then return them to the soil (Huxley, 1983). Nitrogen-fixing tree species are probably the best choice, if they can actively fix nitrogen and thus significantly contribute to the improvement of the nitrogen status of the soil. Nitrogen-fixing trees can increase the yield of associated crops (annual or perennial) through different mechanisms:

1. The first and main one, which is specific to nitrogen-fixing trees, consists of fixing atmospheric nitrogen and contributing nitrogen to the soil via leaf and fallen fruit litter (litter *sensu stricto*) or the release of root debris and nodules (root litter);
2. The second one consists of the concentration of soil nutrients extracted from the deeper soil horizons, and eventually from the water table, and their return to the soil surface with the litter.

In addition, a series of indirect mechanisms (i) affect soil physical, chemical and biological properties, such as structure (Sprent, 1983), or nitrification, whose reduced intensity alleviates the inhibiting effect of nitrates on nitrogen fixation; (ii) control diseases or pests, such as root nematodes (Mulongoy, personal communication); or (iii) stimulate the activity of beneficial organisms such as earthworms (Wilson *et al.,* 1986). In this chapter the discussion is restricted to the nitrogen-fixation process, starting with the most promising nitrogen-fixing species for the tropics, and continuing with current management techniques, and trends and directions for future agroforestry research.

Nitrogen-fixing characteristics of the main tree species used in agroforestry

Legumes

Only nodulating, that is potentially nitrogen-fixing, legume trees are dealt with in this section. No attention is given to species such as *Parkia biglobosa* or *Cassia siamea* which do not nodulate, and consequently do not fix nitrogen, but are nevertheless sometimes integrated in agroforestry systems.

Tropical legumes may nodulate with two types of *Rhizobium:* fast-growing strains which belong to the genus *Rhizobium (sensu stricto)*, and slow-growing strains which form the cowpea miscellany and are now designated as *Bradyrhizobium* (Elkan, 1984). One category of trees nodulates only with *Rhizobium (sensu stricto)*, e.g., *Leucaena leucocephala* (with a few exceptions) or *Sesbania grandiflora*. Another category nodulates only with *Bradyrhizobium,* e.g., *Acacia mearnsii* or *A. albida*. A third category is more promiscuous since it nodulates both with *Rhizobium* and *Bradyrhizobium,* e.g., *Acacia seyal.*

Acacia
Acacia albida (syn. *Faidherbia albida*)   This tree, native to Africa, is usually considered as a highly valuable component of agroforestry systems, not only as a soil improver, but also as a source of fuelwood and forage in semi-arid zones, especially in Africa (Giffard, 1964, 1971; Felker, 1978; Charreau, 1985; Le Houérou, 1985; Poschen, 1986; CTFT, 1986).

The yield of crops such as sorghum, peanut, or maize grown under a tree canopy of *A. albida* is, on the average, substantially higher than when grown alone, which indicates that the tree improves the soil fertility. However, it is not known whether the beneficial effect is due to nitrogen fixation or to other mechanisms enumerated in the introduction. *Acacia albida* nodulates with strains of *Bradyrhizobium* (Dreyfus and Dommergues, 1981). Since these strains are already present in most soils, *A. albida* can be expected to respond poorly to inoculation except in sterilized nursery soils. Nodulation is normally observed on young seedlings, but nodules are seldom found in the field, which suggests that the nitrogen-fixing potential of this *Acacia* is rather low. However, this conclusion requires confirmation through precise measurements. The nitrogen-fixing potential of *A. albida* could possibly be improved by capitalizing on its great genetic variability.

Acacia senegal This is another African *Acacia* species which is intercropped with food crops in the semi-arid zone (El Houri Ahmed, 1979). *A. senegal* nodulates only with fast-growing strains of *Rhizobium*, i.e., *Rhizobium* (*sensu stricto*) (Dreyfus and Dommergues, 1981). Since these strains are less ubiquitous than strains of *Bradyrhizobium*, one can predict that *Acacia senegal* will require inoculation more often than *A. albida*. The nitrogen-fixing potential of *A. senegal* has not yet been estimated.

Acacia nilotica, A. raddiana (syn. *A. tortilis*) and *A. seyal* Like *A. albida* and *A. senegal*, these acacias are often integrated into different types of agroforestry system in the semi-arid zones. *A. nilotica* and *A. raddiana* nodulate with fast-growing *Rhizobium* whereas *A. seyal* nodulates with both fast- (*Rhizobium, sensu stricto*) and slow-growing (*Bradyrhizobium*) strains (Dreyfus and Dommergues, 1981). The nitrogen-fixing potential of the three species is still unknown.

Acacia auriculiformis and *A. mangium* *A. auriculiformis* has been reported to be used in agroforestry fuelwood production systems in Papua New Guinea (Nair *et al.*, 1984). It produces profuse bundles of nodules, which suggests a good nitrogen-fixing potential (Domingo, 1983).

Acacia mangium hybridizes naturally with its close relative *A. auriculiformis* (National Research Council, 1983). *A. mangium* is assumed to be a good nitrogen fixer and, hence, a promising component in fuelwood production systems when grown in appropriate climatic zones.

Acacia mearnsii (syn. *A. mollissima* or *A. decurrens* var. *mollissima*) A highland tree native to Australia, *A. mearnsii* has been introduced in many countries for its bark which is very rich in tannins. It is also frequently found as a multipurpose tree on farmlands in Java and eastern Africa (Nair *et al.*, 1984). *A. mearnsii* nodulates profusely with strains of *Bradyrhizobium* (Halliday and Somasegaran, 1982), even in very poor soils, provided the pH is not lower than 4.5. Nitrogen fixation is high: it was estimated to be approximately 200 kg N_2 ha^{-1} yr^{-1} by Orchard and Darby (1956). A similar figure was given recently by Wiersum (1985).

Albizia lebbeck and *A. falcataria*
There are about 100 species of *Albizia* distributed throughout Africa, Asia and tropical America. Two species, *Albizia lebbeck* and *A. falcataria*, are ideally adapted to agroforestry systems; they are renowned as soil improvers because of their profuse nodulation.

Native to India, Bangladesh, Burma and Pakistan, *Albizia lebbeck* is widely cultivated in tropical and subtropical regions with an annual rainfall as divergent as 500 and 2,000 mm (National Research Council, 1979). Perennial nodules of large size have been observed on adult trees in Senegal.

Native to the eastern islands of the Indonesian archipelago and the west of Irian, *Albizia falcataria* (syn. *Paraserianthes falcataria*) has been spread throughout South-East Asia. It is one of the fastest growing trees in the world provided that it is grown on sites where annual rainfall is high enough (2,000–2,500 mm) (National Research Council, 1979). It nodulates abundantly, which suggests a good nitrogen-fixing capacity. However, because it is exacting in its soil requirements, *A. falcataria* is probably a poor nitrogen-fixer when it is introduced in relatively infertile soils.

Calliandra calothyrsus
Native to Central America, this legume was introduced in Indonesia in 1936 with seeds from Guatemala, and has been shown to nodulate with a fast-growing strain of *Rhizobium* (Halliday and Somasegaran, 1982). Although prized as a first-class soil improver in rotation schedules and in intercropped systems (Domingo, 1983; Nair *et al.,* 1984), considerably more research is needed to accurately assess its nitrogen-fixing potential and its use as forage and organic fertilizer (Baggio and Heuveldop, 1984).

Erythrina spp.
More than 100 species are planted as shade trees, windbreaks, living fences, plant supports, for food and even medicinal purposes, or as components of alley-cropping systems (NFTA, 1986). *E. poeppigiana,* a fast-growing species, is widely used for shade in coffee and cacao plantations (Budowski, this volume). It has been shown to nodulate with a strain of *Bradyrhizobium* (Halliday and Somasegaran, 1982). Nodules of *E. poeppigiana* tend to be large, spherical, and clustered on the central root system (Allen and Allen, 1981). The biomass of the root nodules varied from 80 to 205 mg (dry weight) dm^{-3} soil, being highest close to the stem of the tree (Lindblad and Russo, 1986). A conservative estimate made in Venezuela and based on the decomposition of nodules during the dry season, indicated that the rate of nitrogen fixation was approximately 60 kg N_2 ha^{-1} yr^{-1} (Escalante *et al.,* 1984).

Gliricidia sepium
Native to Central America, this species is widely used for shading cacao (Mexico) or coffee (Sri Lanka) plantations. It has been shown to nodulate with fast-growing *Rhizobium* (Halliday and Somasegaran, 1982). Further experiments are needed to confirm this observation. Estimates of nitrogen fixation based on nodule biomass and rates of nitrogenase activity are approximately 13 kg N_2 ha^{-1} yr^{-1} in the conditions prevailing in Mexico (Roskoski *et al.,* 1982). *G. sepium* has been introduced in western Africa in alley-cropping systems (Wilson *et al.,* 1986; Kang and Wilson, this volume) (Figure 1), but its nitrogen-fixing activity may be impeded by attacks of root nematodes. Sumberg (1985) reports that different accessions of *G. sepium* exhibit considerable variation. This large genetic variability should be exploited to improve nitrogen fixation.

Inga jinicuil
This species, often found in the same sites as *Inga vera,* is a popular shade tree in coffee plantations in Mexico. In a plantation in Xalapa, annual nitrogen-fixation rates, based on the acetylene reduction method, were 35–40 kg N_2 ha^{-1} yr^{-1}, which, when compared to nitrogen from fertilizers, represents an important nitrogen input. The corresponding nodule biomass was 71 ± 14 kg (dry matter) ha^{-1}. Given a density of 205 trees ha^{-1}, nodule biomass per tree was 346 g (dry weight), a figure similar to that reported by Akkermans and Houvers (1983) for *Alnus* (Roskoski, 1981 and 1982) (see Figure 2).

Figure 1 Alley cropping based on the use of *Gliricidia sepium* at the IITA station, Ibadan, Nigeria (photo K. Mulongoy).

Figure 2 Nodule of *Inga jinicuil* (a leguminous tree). The nodule results from the infection of the roots by the nitrogen-fixing bacterium, *Rhizobium* (photo J.R. Roskoski).

Mimosa scabrella

Native to the Parana region of south-east Brazil, *Mimosa scabrella* is used in humid and subhumid tropical highlands as a multipurpose tree on farmlands and as a fuelwood producer in agroforestry operations (Nair *et al.,* 1984). It was shown to nodulate with a fast-growing strain of *Rhizobium* (Halliday and Samosegaran, 1982). *Mimosa scabrella* responds positively to inoculation (Döbereiner, 1984), but its exact nitrogen-fixing potential has not yet been evaluated.

Leucaena leucocephala

This tree has been the focus of a great deal of research in the past few decades (National Research Council, 1977; IDRC, 1983; Brewbaker, this volume). Native to Central America, it has been planted in many tropical countries, including south-east Asia (Domingo, 1983), Africa (Okigbo, 1984; Sanginga *et al.,* 1986) and South America (Döbereiner, 1984), as a shade tree for commercial crops, alley cropping or agroforestry wood production (Figure 3). The acetylene reduction method (Hogberg and Kvarnstrom, 1982) and the difference method (Sanginga *et al.,* 1985, 1986), which have been used to evaluate nitrogen fixation by *Leucaena leucocephala,* give figures in the range of 100–500 kg N_2 ha^{-1} yr^{-1}. These figures have been confirmed recently by Sanginga, Mulungoy and Ayanaba (personal communication), who used the ^{15}N dilution method to make a precision evaluation of the nitrogen-fixation rate of *Leucaena leucocephala* grown in an Alfisol, pH 6.1, at the International Institute of Tropical Agriculture (IITA) in Ibadan, Nigeria. They showed that *Leucaena leucocephala* fixed 98–134 kg N_2 ha^{-1} in 6 months. The high nitrogen-fixing potential of this tree is related to its abundant nodulation under specific soil conditions, in which the nodule dry weight was reported to reach approximately 51 kg ha^{-1} in a stand of 830 trees ha^{-1} (Hogberg and Kvarnstrom, 1982), and approximately 63 kg ha^{-1} in a stand of 2,500 trees ha^{-1} (Lulandala and Hall, 1986).

Figure 3   Alley cropping based on the use of *Leucaena leucocephala* at the IITA station, Ibadan, Nigeria (photo K. Mulongoy).

Leucaena leucocephala generally nodulates with *Rhizobium* (*sensu stricto*) (Halliday and Somasegaran, 1982), and occasionally nodulates with *Bradyrhizobium* (Dreyfus and Dommergues, 1981). The *Rhizobium* strain specific to *Leucaena leucocephala* is not generally found in soils. This explains the positive response to inoculation obtained in most soils where the level of nutrients (other than nitrogen) is high enough to satisfy the tree's requirements.

Leucaena leucocephala is not a miracle tree. Its sensitivity to soil acidity and its high nutrient demand are reflected in its poor performance in infertile soils, e.g., the sandy soils of the Pointe Noire region of the Congo or of Hainan Island, China, even when properly inoculated and grown in a suitable climate.

Sesbania grandiflora

Native to Asia, this legume tree is popular throughout South and South-East Asia, where it is used as a shade tree, a source of fodder and green manure, and for erosion control. Like other *Sesbania, Sesbania grandiflora* nodules with fast-growing strains of *Rhizobium* (Dreyfus, personal communication). It nodulates profusely and is probably a good nitrogen fixer (Domingo, 1983), but we have observed that in some soils (e.g., Loudima, Congo) its root system reacted badly to nematode attacks.

Actinorhizal plants

About 200 non-leguminous plant species belonging to 19 genera and 8 families nodulate with N$_2$-fixing micro-organisms known as *Frankia* (Figure 4). Since *Frankia* are actinomycetes, these N$_2$-fixing plants became known as "actinorhizal plants" (Torrey and Tjepkema, 1979), a name now used world wide. In tropical agroforestry the main species of actinorhizal plants belong to the genera *Alnus, Casuarina* and *Allocasuarina* and, secondarily, *Coriaria* (Akkermans and Houvers, 1983; Gauthier *et al.*, 1984; Bond, 1983).

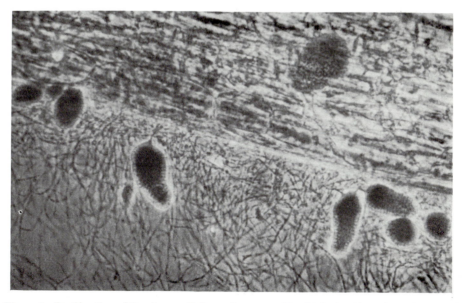

Figure 4 Proliferation of the nitrogen-fixing actinomycete *Frankia* in the rhizosphere of *Casuarina equisetifolia*. Two types of typical actinomycetal structures can be seen: hyphae and sporangia (photo N.G. Diem).

Alnus jorullensis (syn. *A. acuminata*)

In Costa Rica, an agroforestry system composed of this actinorhizal species and pasture grasses has become very popular, probably because of the high nitrogen-fixing potential of *Alnus* (Budowski, 1983). The actual amount of nitrogen fixed and transferred to the pasture is unknown. Introduction of *A. jorullensis* into coffee plantations has been advocated by Fournier (1979). The great genetic variability of *Alnus jorullensis* can be exploited to enhance the nitrogen-fixing potential of this species.

Casuarinaceae

The family Casuarinaceae consists of a group of 82 species mostly from Australia, but also native to South-East Asia and the Pacific islands. Johnson (1982) recognizes four genera: *Casuarina* (*sensu stricto*) (e.g., *C. cunninghamiana, C. equisetifolia, C. junghuhniana* syn. *C. montana, C. glauca, C. obesa, C. oligodon*); *Allocasuarina* (e.g., *A. decaisneana, A. fraseriana, A. littoralis, A. torulosa, A. stricta* syn. *Casuarina verticillata*); *Gymnostoma* (e.g., *G. deplancheana, G. papuana, G. rumphiana*), and a fourth genus not yet described.

Figure 5   Windbreak of *Casuarina glauca* in northern Tunisia (photo Y. Dommergues)

In some parts of the world, *Casuarina* spp. are perfectly integrated in agroforestry systems. One classical example is that of Papua New Guinea, where *C. oligodon* and *G. papuana* (in the highlands) and *C. equisetifolia* (in the lowlands) are intercropped with food crops, and are used as shade trees or in rotation with crops (Thiagalingam, 1983; Bourke, 1985). In India, *Casuarina equisetifolia* plantations are associated with crops such as peanuts, sesame and various grain legumes (pulses) (Kondas, 1981). In many places casuarinas are planted as windbreaks: *C. glauca* in Tunisia (Figure 5) and *E. equisetifolia* in Senegal (protection of market-gardens) and Corsica (protection of cash crops such as *Actinidia sinensis*).

According to Bowen and Reddel (1986), *Casuarina* (*sensu stricto*) are usually well nodulated, whereas nodulation of *Allocasuarina* is variable or often non-existent. There is little cross-inoculation between *Casuarina* and *Allocasuarina,* which means that strains of *Frankia* isolated from *Casuarina* do not usually infect *Allocasuarina,* and vice versa.

The dry weight of a *Casuarina equisetifolia* nodule in the field is comparable or even greater than that of *Alnus* (up to 500 g for a 13-year old *Casuarina equisetifolia,* according to Cao Yue Hua, personal communication) (Figure 6). This profuse nodulation explains its high nitrogen-fixing potential. In an experiment conducted in irrigated microplots at the ORSTOM Station in Dakar, Senegal, a selected clone of *Casuarina equisetifolia* was reported to fix 42.5 g per tree during the first nine months following plantation. Extrapolating this value to 1 ha with 2,000 trees, and assuming a constant fixation rate, would give a figure of 113 kg N_2 fixed per hectare during the first year (Sougoufara, personal communication). The nitrogen-fixation estimate for a 13-year rainfed *Casuarina equisetifolia* plantation located on sand dunes along the Senegalese coast was only 58 kg N_2 ha^{-1} yr^{-1} (Dommergues, 1963).

Figure 6 Nodules of *Casuarina equisetifolia* (an actinorhizal tree). The nodules result from the infection of the roots by *Frankia* (photo H.G. Diem).

A study on nitrogen fixation by *Allocasuarina littoralis,* based on comparisons of litter nitrogen content, suggests that this species could fix 218 kg N_2 ha^{-1} yr^{-1} (Silvester, 1977), but this figure is probably exaggerated.

Some *Casuarina* species are good candidates for the reclamation of salt-affected soils. Recently, Bowen's group in Australia showed that *C. obesa* inoculated with a salt-tolerant strain of *Frankia* grew despite 15,000 ppm chloride (approximately the equivalent of sea water), thus tolerating ten times more chloride than when it grew on nitrate or ammonium nitrogen. This is possibly because the nitrogen-fixing system is considerably less salt-sensitive than the mechanism for uptake of nitrate or ammonium, and/or assimilation enzymes (Bowen and Rosbrook, 1986).

Coriaria

All the 15 species of *Coriaria* are recorded as bearing nodules, which indicates that nodulation is a generic character of *Coriaria,* as it is in *Alnus.* Two species are known to be valuable components in agroforestry systems. One is *C. sinica,* a deciduous bush, widely grown in Hunan, China, as a source of green manure and of feed for silkworms. It grows so fast that the stems can be harvested 4–5 times a year giving 10t of fresh biomass ha^{-1} (Watanabe, personal communication). The other is *Coriaria arborea,* which, grown as an understorey species in many artificial plantations of *Pinus radiata* in New Zealand, fixes up to 192 kg N_2 ha^{-1} yr^{-1}. Its effect on the growth of *Pinus radiata* has not yet been investigated (Silvester, 1977, 1983).

Technology for exploiting nitrogen-fixing trees in agroforestry

This section presents briefly the principles for choosing the species and provenances of nitrogen-fixing trees and describes the practices that are currently recommended to alleviate some of the major environmental stresses inhibiting the nitrogen-fixation process.

Choosing species and provenances of nitrogen-fixing trees

The species or provenances chosen for introduction in any agroforestry system should have the highest nitrogen-fixing potential, be tolerant of environmental constraints, especially low levels of nutrients, and resistant to pests. Results obtained from the sparse data available on the nitrogen-fixing potential of trees suggest that tree species can be put into two broad categories:

1. Species with a high nitrogen-fixing potential (in the range of 100–300 kg N_2 ha^{-1} yr^{-1} and more), e.g., *Acacia mangium, Casuarina equisetifolia,* and *Leucaena leucocephala;*
2. Species with a low nitrogen-fixing potential (less than 20 kg N_2 ha^{-1} yr^{-1}), e.g., *Acacia albida, A. raddiana,* and *A. senegal.*

Species in the first category can be divided into two subgroups:

1. Exacting species, e.g., *Leucaena leucocephala* and *Calliandra calothyrsus,* which require large amounts of nutrients, especially P, K and/or Ca;
2. Non-exacting species, e.g., *Acacia mangium,* which flourishes in marginal acid soils low in nutrients.

Species belonging to the latter subgroup are obviously the most promising in agroforestry, if they also exhibit the specific qualities required for inclusion in agroforestry systems.

The choice of suitable nitrogen-fixing trees for a given agroforestry programme should always be based on well-planned provenance* trials. The objective of these trials should be to screen as many provenances within species as possible in order to identify provenances best suited to the range of soils and climate existing in a region. Although provenance trials for nitrogen-fixing trees are just beginning, studies are being made to explore provenance variations of species such as *Casuarina equisetifolia* in China, the Philippines (Halos, 1983), Senegal (Sougoufara and Corbasson, personal communication), or *Acacia* in Indonesia, Papua New Guinea and Australia (Turnbull *et al.*, 1984), and Congo (D. Diangana and J.C. Delwaulle, personal communications).

The choice of the tree species should also be made taking into account its susceptibility to pathogens and pests such as termites (Mitchell *et al.*, 1986) and, especially, nematodes. Introducing tree species whose root systems serve as reservoirs for pathogenic nematodes (Taylor, 1976) should be avoided because of the risk of contamination of associated crops.

Sterilization of nursery soils

There is a trend towards the generalized use of containers, especially polybags, in many forest nurseries for most indigenous as well as exotic species (Khan, 1985). Taking into account the ease of handling these containers, the limited amount of soil (or any other type of substratum) that is needed for each container, and the relatively low number (2,000–10,000) of plants required per hectare in forest plantations (compared to up to 100,000 ha^{-1} for annual crops), it is economically feasible to sterilize the soil used to fill the containers. A corollary to sterilization is the necessity to inoculate the containerized seedlings with the appropriate strains of *Rhizobium, Frankia* or vesicular-arbuscular mycorrhizal fungi.

In many tropical soils, pathogens can only be eliminated if the soil in the containers is sterilized. The following example illustrates the spectacular effect of sterilization. *Acacia holosericea* seedlings were grown in polybags filled up with non-sterilized soil (treatment A), or with sterilized soil. One part of the plants grown in sterilized soil remained uninoculated (treatment B); the other part grown in sterilized soil was inoculated with *Rhizobium* and a vesicular-arbuscular mycorrhizal fungus, *Glomus mosseae* (treatment C). When transplanted to the field, trees which had been grown in non-sterilized soil (treatment A) exhibited a survival rate which was less than half that of the plants grown in sterilized soils (treatments B and C). The effect of inoculation with *Rhizobium* and *Glomus mosseae* on height was much less spectacular, but still significant (Table 1).

Sterilization of nursery soils is a common practice in temperate countries, but is rarely done in the tropics, probably because the prescribed methods are too expensive (vapour), or dangerous (methylbromide), or because the chemical product is difficult to obtain. Dozomet,† a granulated biocide, which decomposes into gaseous compounds (methylisothiocyanate and formaldehyde) in the soil, leaving no lingering toxic residual effect, can be recommended because it is easy to handle and not too expensive. Its efficiency as a sterilizing agent is satisfactory provided that its rate and duration of application are calculated for the prevailing climatic and edaphic conditions.

---

* In the field of forestry the term provenance refers to natural populations of trees originating in a specific geographic location (also see Burley, this volume).
† Basamid, BASF, 140 rue Jules Guesde, BP 87, 92203 Levallois-Perret Cedex, France.

Table 1   Influence of different treatments in the nursery on survival and growth of *Acacia holosericea* after transplantation* to the field at Sangalkam, Senegal.

Treatment in the nursery†	11 months after transplantation		17 months after transplantation	
	Survival	Height (cm)	Survival	Height (cm)
A (no ster.; no inoc.)	38.2 a	29.4 a	36.8 a	105.3 a
B (ster.; no inoc.)	89.6 b	38.6 b	89.6 b	119.9 b
C (ster.; inoc.)	84.7 b	45.6 c	84.7 b	134.2 c

* Seedlings transplanted at two months.
† no ster. = no sterilization; ster. = sterilization with methylbromide; no inoc. = no inoculation; inoc. = inoculation with *Rhizobium* and *Glomus mosseae*.
Values in columns followed by the same letter are not significantly different, P = .05.
Source: D. Jacques, unpublished data.

Inoculation with *Rhizobium* or *Frankia*

Inoculating the host plant with soil or crushed nodules is a technique that is still recommended but should be discontinued because of the high risk of contaminating seedlings or cuttings with root pathogenic agents, such as *Rhizoctonia solani* or *Pseudomonas solanacearum* in the case of *Casuarina equisetifolia* (Liang Zichao, 1986), or nematodes in the case of Australian *Acacia* introduced in western Africa.

In the past, pure cultures of *Frankia* have not often been used on actinorhizal plants because of the difficulty of isolating and cultivating the strains, especially those of *Casuarina* (Diem *et al.,* 1982, 1983), and consequently of obtaining the inoculants. However, thanks to recent progress in knowledge of *Frankia* physiology, there is reason to hope that actinorhizal plants will be inoculated with pure cultures of *Frankia* in the near future.

Von Carlowitz (1986) has drawn up a list of suppliers of strains or inoculants, most of which still have to be evaluated for nodulation or nitrogen fixation (Somasegaran and Hoben, 1985).

In the case of trees raised in containers, inoculation with *Rhizobium* is best achieved by spraying or drilling the inoculum directly into the container at the time of seeding or planting. When dealing with *Frankia,* it is advisable to mix the soil or substratum of the container with the inoculum because *Frankia,* like vesicular-arbuscular mycorrhizal fungi, is not mobile in the soil (Figure 7). After the containerized plants have been transplanted to the field, the effect of inoculation observed in the sterile nursery soil persists only if the soil does not contain specific native strains.

Pre-inoculated seeds of *Leucaena leucocephala* were sown directly in the field by Sanginga *et al.* (1986) who tested IRc 1045 and IRc 1050 *Rhizobium* strains in agroforestry experiments set up at two locations in Nigeria. At both places, inoculated trees produced more nitrogen and dry matter than the controls. This effect was statistically equivalent to the application of 150 kg ha^{-1} of urea. Further, the strains survived and competed well in the field, as was shown in observations made one year after their establishment. The beneficial effect of *Leucaena leucocephala* on a subsequent crop of maize is discussed later in this chapter.

Figure 7 Nursery of *Casuarina equisetifolia* seedlings inoculated with *Frankia* established along the Senegalese coast (photo Y. Dommergues).

Inoculation with mycorrhizal fungi

Mycorrhizal infections are known to increase the absorption of phosphate and other poorly mobile ions in soil such as Zn, Cu, Mo, and K. Mycorrhizal fungi are most often associated with the roots of nitrogen-fixing trees, endomycorrhizae being more frequent than ectomycorrhizae, the other major type of mycorrhizae.

Nodulation and nitrogen fixation require a high P status in the host plant, which can be facilitated by the mycorrhizal symbiont. The beneficial effect of mycorrhizal infection on nitrogen fixation is similar to that of added P in P-deficient soils. Mycorrhizae can enhance the effects of even a small amount of P fertilizer that is added to soils with a serious P-deficiency (Ganry *et al.,* 1985). In addition to improving nutrient absorption, mycorrhizal fungi also affect the physiology of the host plant, enabling it to increase its water uptake, improve its hormone balance and overcome the first-year dormancy of cuttings (Hayman, 1986).

The technology of inoculation with ectomycorrhizae is now fully applicable (Schenck, 1982). For endomycorrhizal fungi promising results have already been reported in forest nurseries (Cornet *et al.,* 1982), but the technology is not yet ready for extension to the small farmer.

Fertilizer

There is a tendency to neglect the mineral nutrition of nitrogen-fixing trees. This is most irresponsible when dealing with exacting species such as *Leucaena leucocephala* whose exceptional capacity to produce biomass and protein depends on the availability of adequate nutrients (Waring, 1985). Hu and Kiang (1983) estimated the nutrient uptake of a three-year-old plantation of *Leucaena leucocephala* as being P 11–27, K 174–331, Ca

138–305 and Mg 31–62 kg ha^{-1}. These figures are indeed high. *Casuarina equisetifolia* is assumed to have high Ca requirements (Waring, 1985). P is also an important nutrient, not only for the plant itself but also to ensure good nodulation. However, whether a low P supply blocks nodulation by limiting plant growth and hence nitrogen demand, or directly affects *Frankia* in the rhizosphere and in the early stages of nodule initiation is not known (Reddell *et al.*, 1986).

The nutrient requirements of species such as *Acacia mangium* that are less constrained by element deficiency are probably lower but not low enough to be negligible.

It is surprising that some authors still recommend the application of nitrogen fertilizers (together with P and K) on nitrogen-fixing trees (Yadav, 1983). This practice should be prohibited, since it is a well-established fact that mineral nitrogen, especially when applied at high levels, inhibits nodulation and nitrogen fixation.

Obviously, much more research is necessary to quantify the exact fertilizer needs of nitrogen-fixing trees. As suggested by Sanchez and Salinas (1981), research along these lines should keep "inputs to a minimum, aiming only to optimize production under existing constraints rather than to maximize the production per se".

Control of acidity

Soil acidity and related factors (Al and/or Mn toxicity and Ca and Mo deficiencies), which affect many tropical soils (Franco, 1984), influence nitrogen fixation by the direct or indirect effects they have on the host plant and the symbiotic micro-organisms. A typical example is *Acacia mearnsii,* which does not nodulate in the highlands of Burundi where soils have a low pH and a high content of exchangeable Al. The detrimental effects of soil acidity can be overcome by selecting acid-tolerant host plants and symbiotic micro-organisms, an approach that has been adopted with *Leucaena leucocephala* (Hutton, 1984; Brewbaker, this volume) and its competent *Rhizobium* (Halliday and Somasegaran, 1982; Franco, 1984). It is also possible to control the effects of soil acidity by directly applying proper amendments to the soil or by pelleting the seeds in the case of direct sowing in the field.

Different types of amendments such as lime or organic materials can be used. The acidity generated by nitrogen-fixing plants in the long run may lower the pH of weakly buffered soils, and periodic liming may be necessary to maintain high productivity (Franco, 1984). The higher organic-matter content of soil under nitrogen-fixing trees, however, may lead to satisfactory yields even when the pH is lower than usually recommended in conventional cropping systems.

The symbiotic micro-organism can be protected against acidity by pelleting the seeds to be inoculated with calcium carbonate or rock phosphate. This technique, developed in Australia and now used throughout the world, has indeed proved to be a high-value alternative for liming during the introduction and establishment of forage legumes in pastures (Williams, 1984). It could also be used successfully in agroforestry. However, in very acid soils with Al or Mn toxicity, pelleting the seeds alone cannot overcome the effects of acidity regardless of cropping system.

Future trends in agroforestry research

In the last decade a tremendous amount of investigation has been devoted to the genetics and physiology of *Rhizobium,* but until recently forestry and agroforestry had not benefited substantially from the remarkable progress that has been achieved in this field. Hopefully some of the encouraging results that have already been obtained will be transferred to the field. In addition, new concepts and techniques are presently emerging that will, sooner or later, be ready for practical application. The most promising areas of investigation are probably those concerned with improvement of the host plant.

Improvement of the symbiotic micro-organism

Near-term investigations

To date only relatively few effective strains of *Rhizobium* that nodulate nitrogen-fixing trees have been isolated: some of the best known are strains for *Leucaena leucocephala*, e.g., strain TAL 1145 from NifTAL* (Roskoski, 1986) and strains IRc 1045 and 1050 from IITA (Sanginga *et al.*, 1986). There is still much work ahead to collect *Rhizobium* strains for leguminous nitrogen-fixing trees and then screen them for genetic compatibility, nitrogen-fixation effectiveness, and tolerance to environmental stresses, especially soil acidity (Da Silva and Franco, 1984). In addition to preliminary screenings, field trials must be performed to test the response to inoculation since "accumulating data indicate that site variation in performance of selected strains is common" (Halliday, 1984a).

Frankia strains associated with Casuarinaceae exhibit large differences in genetic compatibility (Zhang and Torrey, 1985; Puppo *et al.*, 1985) and effectiveness (Zhang *et al.*, 1984; Bowen and Rosbrook, 1986). Not all *Frankia* strains nodulate all species of Casuarinaceae. There are very large differences in the effectiveness of nitrogen fixation between *Frankia* strains associated with a single species of Casuarinaceae. Furthermore, a *Frankia* strain effective on one species of Casuarinaceae can be very ineffective on another species (Reddell, 1986). Collections of *Frankia* strains already exist (Lechevallier, 1985–6), and some laboratories have already screened and selected strains for use in nursery and field trials.

Mid- and long-term investigations

Using molecular techniques (molecular cloning and recombination), new strains of *Rhizobium* and *Frankia* will probably be engineered to contain multiple copies of the major genes involved in the symbiosis: genes of nitrogen fixation and nodulation, and genes involved in interstrain competition.

Since *Rhizobium* and *Frankia* are exposed to environmental constraints in the field, the new strains should also be stress tolerant. Instead of developing strains with the superior traits indicated above and then studying their behaviour in the field, "bacteria whose genetic libraries already contain adaptation traits to prevailing environmental stresses will be engineered for better nodulation and nitrogen fixation" (Roskoski, 1986) and competitive ability.

Improvement of the host plant

The amount of nitrogen fixed by any nitrogen-fixing tree is related to its nitrogen-fixing potential (NFP), i.e., its ability to-fix nitrogen in the absence of any limiting factor (Halliday, 1984b). The nitrogen-fixing potential is directly conditioned by the genotypes of

---

* NifTAL — Nitrogen Fixation by Tropical Agricultural Legumes, University of Hawaii, 1000 Holomua Road, Paia, Maui, HI 96779, USA.

both the host plant and the associated symbiont. Consequently, to get the maximum nitrogen input into an agroforestry system, the first essential characteristic is to use a nitrogen-fixing tree with a high nitrogen-fixing potential.

The second essential characteristic is that a nitrogen-fixing tree should be maximally tolerant of environmental stresses, be they physical (e.g., excessive temperature, drought), chemical (e.g., excess of combined nitrogen), or biological. Because of these stresses, however, even the most tolerant nitrogen-fixing tree cannot attain its full potential in the field. The amount of nitrogen that is fixed under field conditions is called the actual nitrogen fixation (ANF). The ANF of stress-sensitive nitrogen-fixing trees is expected to be much lower than their NFP; conversely the ANF of stress-tolerant species is expected to be much closer to their NFP. Special mention must be made of the inhibitory effect of high levels of combined (mineral) soil nitrogen, especially nitrate, on nitrogen fixation (as mentioned earlier). This implies that nitrogen-fixing trees in agroforestry systems should be engineered to continue fixing significant amounts of nitrogen even when the intercrop receives nitrogen fertilizers.

A third group of essential characteristics, common to all trees to be introduced in agroforestry systems, has been presented by Huxley (1983). One trait worth mention is the absence of strong plant competitive attributes such as a tendency to overshade understorey plants and to dominate the water economy of the microsite.

Short-term investigations

When the best provenances have been identified, it is mandatory to study differences that occur between the individual trees, especially differences in ANF. Exploiting such spontaneous variations requires two steps. First, the whole population in the provenance must be screened using an adapted procedure to identify the more actively nitrogen-fixing individuals. This procedure might be based on study of the nodulation combined with the measurement of the acetylene-reducing activity of the different individuals. Secondly, the superior phenotypes must be vegetatively propagated, using available techniques (Datta and Datta, 1984; Leaky, 1986; Duhoux *et al.,* 1986). Sougoufara *et al.* (1987) recently used this approach successfully. The result was a spectacular increase of the nitrogen-fixing potential of *Casuarina equisetifolia* (Figure 8). These authors identified a clone of *Casuarina equisetifolia* (called clone B) with a much higher nitrogen-fixing potential than that of a reference clone (clone A), i.e., a clone with a potential similar to that of the seedlings usually grown. Clones A and B were grown in a sterile nitrogen-deficient soil. One set of clones A and B was inoculated with the same *Frankia* strain, the other one remained uninoculated. After seven months, the uninoculated clones displayed poor growth, while the inoculated clones had grown satisfactorily, but their response to inoculation differed markedly. Inoculated clone B produced 2.6 times more biomass (expressed in terms of dry weight and total nitrogen) than inoculated clone A. Concomitantly, nodule weight and nitrogen fixation (expressed as acetylene-reducing activity per plant) of clone B were significantly higher (1.6 times) than those of clone A. The difference in the nitrogen-fixing potential of the clones appeared to be related to their nodule weight (Table 2). Field trials under way in Senegal will, hopefully, confirm the superior nitrogen-fixing potential of clone B.

Mid- or long-term investigations for improving nitrogen-fixing trees

The methods mentioned below can be used either for improving the nitrogen-fixing potential or the stress tolerance, or both traits, of nitrogen-fixing trees.

Figure 8   Colony of elongated buds of *Casuarina equisetifolia* obtained *in vitro*. The buds are ready for excision and rooting (photo E. Duhoux).

Table 2   Comparison of two clones of *Casuarina equisetifolia**

Treatment	Shoot			Nodule	ARA† per	
	d.wt. (mg/plant)	N%	N total (mg/plant)	d.wt. (mg/plant)	plant	g nodule
Uninoculated						
Clone A	130 a	0.73 a	0.95 a	0 a	0 a	0 a
Clone B	90 a	1.02 a	0.92 a	0 a	0 a	0 a
Inoculated with Frankia‡						
Clone A	660 b	1.71 b	11.29 b	54 b	2.88 b	54 b
Clone B	1,730 c	2.02 c	34.93 c	88 c	5.58	c 56 b

* Seven-month old cuttings, nine replicates.
† Acetylene-reducing activity expressed as $u$mol C_2H_4 per plant or per g nodule, dry weight (d.wt.).
‡ Each cutting was inoculated with 2 ml (20ug of proteins) of a 4-week old culture of *Frankia* strain ORSO21001.

Values in columns followed by the same letter are not significantly different, P = .05 (nodule dry weight, ARA), P = .01 (shoot dry weight, total N, and N%).

Source: Sougoufara *et al.,* 1987.

(i) Hybridization of nitrogen-fixing and non-nitrogen-fixing plants Gene transfer by wide hybridization to obtain new nitrogen-fixing systems has not yet been exploited. Sophisticated methods, like somatic hybridization by protoplast fusion or embryo rescue (National Research Council, 1982), are now available, and could be most helpful in this unexplored field of research.

(ii) Micrografting The ability to nodulate (and fix nitrogen) can possibly be introduced into non-nodulating species of the same genus or family by grafting the former plant on to nodulated (and nitrogen-fixing) rootstocks obtained *in vitro*. Preliminary results obtained by Kyle and Righetti (1985) with actinorhizal Rosaceae are most promising.

(iii) Mutagenesis breeding Improved lines of soybean that can fix nitrogen even in the presence of high levels of nitrate in the soil have already been obtained as a result of mutagenesis breeding (Carroll *et al.*, 1985). Similar procedures could be used to generate improved lines of nitrogen-fixing trees.

(iv) Indirect transfer of nitrogen-fixing capability through mycorrhizal fungi The transfer of the whole set of genes required to fix nitrogen, from a bacterium to a mycorrhizal fungus, could also be considered. However, many difficulties must be overcome first, one prerequisite to genetic engineering of nitrogen-fixing endomycorrhizal fungi being the availability of a reliable method to grow this fungus *in vitro*.

(v) Transfer of the stem-nodulating ability Stem nodulated legumes, e.g., *Sesbania rostrata* and *Aeschynomene afraspera,* have the unique capacity to absorb combined nitrogen through their stem nodules, even in the presence of high soil mineral nitrogen (Dreyfus *et al.*, 1984). Transferring the stem nodulation character from these plants to non-stem-nodulated legume trees would probably be one way to develop uninhibited nitrogen-fixing trees.

(vi) Genetic transfer of nitrogen fixation ability to plant cells The transfer of cloned genes from nitrogen-fixing micro-organisms to plants should be considered as a long-term research project since it has not yet been determined whether, after being introduced into a plant, all the genes required for nitrogen fixation are permanently incorporated, expressed, and inherited by the whole plant. In spite of its highly speculative character, this strategy should not be neglected.

Field studies on nitrogen fixation

Estimation of nitrogen fixation

With a few exceptions (*Leucaena leucocephala* and *Casuarina equisetifolia*), the NFP or ANF of most nitrogen-fixing trees is not precisely known. This dearth of information is prejudicial to the development of proper management practices. The principles of the current methods have been discussed in many reviews (e.g., LaRue and Patterson, 1981; Herridge, 1982; Silvester, 1983) and described in Bergersen's treatise published in 1980.

(i) Nitrogen difference The amount of nitrogen fixed is considered to be the difference between the total nitrogen yield of the nodulated (nitrogen-fixing) plant and the total nitrogen yield of a non-nodulated (non-nitrogen-fixing) companion plant, preferably of the same species, serving as a control. Estimates are accurate only when the structure and function of the root systems of both plants are similar. Despite its shortcomings, this method — together with related methods based on nitrogen balance studies — often provides useable evaluations of nitrogen fixation.

(ii) Reduction of acetylene The nitrogen-fixing system is placed within an atmosphere enriched with 1–10 percent C_2H_2. After a short incubation time (1–2 h) a sample of the atmosphere is removed and the C_2H_4 resulting from the reduction of C_2H_2 by the nitrogenase is analysed. Acetylene-reduction assays are converted to estimates of nitrogen fixation using a conversion ratio ($C_2H_2:N_2$) that was originally assessed to be 3:1, i.e., one mole of C_2H_4 being equivalent to $1/3$ mole of N_2 reduced (fixed). It is now realized that this ratio is very variable and therefore must be checked for each system. Various techniques for the use of the acetylene-reduction method have been described in excellent detail (Bergersen, 1980) and will not be dealt with here. The method has already been applied to nitrogen-fixing trees such as *Inga jinicuil* (Roskoski, 1981) and *Leucaena leucocephala* (Hogberg and Kvarnstrom, 1982; Lulandala and Hall, 1986). In the agroforestry plots of the Sokoine University of Agriculture, Morogoro, Tanzania, *Leucaena leucocephala* stands of 2,500 trees ha^{-1} provided 63 kg of nodules (dry weight) ha^{-1}. In the rainy season mean nitrogenase activity was 60 nmoles C_2H_4 mg^{-1} nodule (dry weight) h^{-1} during daylight. Assuming a conversion ratio of 3:1, and mean hourly night nitrogenase activity equal to 67 percent of the mean hourly daylight activity, 197 kg N_2 were estimated to be fixed ha^{-1} yr^{-1} (Lulandala and Hall, 1986).

(iii) ^{15}N *enrichment* The direct isotope dilution method or, more properly, the ^{15}N-enrichment method, is based on the comparison of non-nitrogen-fixing and nitrogen-fixing plants grown in soil to which ^{15}N has been added (as labelled urea, nitrate or ammonium). The nitrogen-fixing plants obtain nitrogen from two sources, soil and air, and thus have a lower content in isotope ^{15}N than non-nitrogen-fixing plants which absorb only labelled soil nitrogen. The percentage of the plant nitrogen derived from nitrogen fixation is calculated from the atom per cent ^{15}N excess in non-nitrogen-fixing and nitrogen-fixing plants, respectively. The method has already been used to evaluate nitrogen fixation by trees such as *Casuarina equisetifolia* (Gauthier *et al.,* 1985) and *Leucaena leucocephala* (Sanginga, personal communication). In both examples estimates of nitrogen fixation using the ^{15}N-enrichment method were similar to the estimate obtained by the difference method.

(iv) Natural ^{15}N *abundance* This method is based on the study of small differences between the natural abundance of ^{15}N in non-nitrogen-fixing and nitrogen-fixing plants. Soil nitrogen frequently contains slightly more ^{15}N than atmospheric nitrogen. In addition, in most biological reactions, through isotope discrimination, the lighter of two isotopes is favoured slightly. Because of these two phenomena, nitrogen derived from nitrogen fixation has a very slightly lower ^{15}N content than nitrogen originating from the soil so that the natural ^{15}N abundance is lower in nitrogen-fixing plants than in non-nitrogen-fixing ones (Knowles, 1983). From the measure of the natural ^{15}N abundance in nitrogen-fixing and non-nitrogen-fixing plants it is possible to calculate the fraction of the plant nitrogen derived from fixation. This method requires access to an isotope ratio mass spectrometer and scrupulously careful manipulations, but the results are as reliable as those obtained from the ^{15}N-enrichment method (Bergersen, 1986). One of the first studies using this method was carried out on *Prosopis* in the Sonoran desert. The natural ^{15}N abundance in the tree was significantly lower than in the soil, indicating that it had fixed nitrogen though no nodules were found. *Prosopis* was presumed to develop nodules on deep roots which are not normally harvested (Virginia *et al.,* 1981).

(v) ^{15}N *depleted material* Preliminary investigations indicate that it may be possible to use ^{15}N depleted ammonium sulphate for measuring nitrogen fixation of nitrogen-fixing

trees such as *Albizia lebbeck* and *Leucaena leucocephala* (Kessel and Nakao, 1986).

(vi) Analysis of nitrogen solutes in the xylem sap   The sap ascending in the xylem of nitrogen-fixing legumes carries nitrogen compounds originating from inorganic soil nitrogen (mainly NO_3^-) absorbed by the roots and from the nodules as assimilation products from nitrogen fixation. Legumes fall into two categories: *ureide exporters* (e.g., *Vigna unguiculata* and *Glycine max*) which export fixed nitrogen as allantoin and allantoic acid, and *amide exporters* (e.g., *Lupinus albus* and *Trifolium* sp.) which export fixed nitrogen as asparagine, glutamine or substituted amides. In addition to the products resulting from nitrogen fixation, the sap contains nitrate or organic products of nitrate reduction formed in the roots.

In *ureide exporters,* much of the nitrate absorbed by the roots is passed to the shoot as free unreduced nitrate because of the low nitrate reductase activity of their roots. In non-nitrogen-fixing plants, the xylem-nitrogen is found mainly in the form of nitrate and amino acids, whereas in nitrogen-fixing plants it contains mainly ureide nitrogen and the relative abundance of ureides in sap can be used as an indication of nitrogen activity. By constrast, in *amide exporters,* only a small proportion of the nitrate absorbed by the roots escapes the reductase system of the roots, hence their sap contains mainly amides regardless of whether they are fixing nitrogen or not. This makes it impossible to use sap analysis for estimating nitrogen fixation in amide-exporting legumes (Bergersen, 1986). Preliminary studies on the composition of the sap of 35 nitrogen-fixing leguminous trees have been carried out at NifTAL by Kessel *et al.* (1987). Only two species, *Acacia mearnsii* and *Sesbania grandiflora,* showed a high relative abundance of ureides in the xylem sap (81.5 percent in *Acacia mearnsii* and 78.8 percent in *Sesbania grandiflora*). For the two species, ureides are the major nitrogen compounds in the sap, and the ureide method could probably be used for measuring their nitrogen-fixing potential.

Since citrulline is always the major nitrogenous compound in the xylem sap of *Casuarina equisetifolia,* regardless of whether it is fixing nitrogen or not, the citrulline content cannot be used as an indicator of nitrogen fixation in *Casuarina equisetifolia.* However, the abundance of citrulline compared to other nitrogenous compounds (e.g., amides or nitrate) could possibly be used as an indicator of nitrogen fixation (Walsh *et al.,* 1984).

In sum, there are a number of techniques available to measure nitrogen fixation. Under carefully controlled conditions each will give reasonable estimates (e.g., Herridge, 1982; Gauthier *et al.,* 1985; Bergersen, 1986). Whenever possible at least two methods should be used simultaneously. However, due attention should be given to the difficulties specific to perennial plants, e.g., logistic and sampling problems, variations in the nitrogen-fixing activity with the age of the trees, or interference by difference processes such as losses and redistribution of nitrogen in the different horizons or compartments of the agroforestry system.

Transfer of nitrogen fixed by the trees to associated crops

In general, the transfer of nitrogen to non-nitrogen-fixing plants intercropped with nitrogen-fixing trees and the rates of nitrogen turnover have not yet been thoroughly assessed. In one of the first studies on this problem, Sanginga *et al.* (1986) attempted to quantify the beneficial effect of application of *Leucaena leucocephala* prunings at the surface of plots grown with maize:

> Maize contained significantly more nitrogen in plots that received
> *Leucaena leucocephala* prunings. The highest maize nitrogen contents

were recorded in plots with prunings collected from *Leucaena leucocephala* previously inoculated with *Rhizobium*. Prunings from inoculated trees added 2.5 times more nitrogen than the ones from uninoculated trees, with maize yield increases of 1.4 (inoculation with *Rhizobium* strain IRc 1045) and 2.4 (inoculation with *Rhizobium* strain IRc 1050) t ha^{-1}. In plots where prunings were removed, the nitrogen contribution from *Leucaena leucocephala* roots, nodules and litter to the maize grain yield was equivalent to an average of 36 kg urea N ha^{-1}. Nodules of inoculated trees seemed to be important sources of nitrogen for maize.

Nitrogen from prunings was less well utilized by maize, probably because leaves and twigs of *Leucaena leucocephala* released 50 percent of their nitrogen within four weeks through decomposition and most of this nitrogen was rapidly mineralized and leached out.

Nitrogen transfer studies are urgently needed to improve current management practices and thus capitalize on the benefits that associated crops can get from nitrogen-fixing trees.

Mixed perennial systems

Mixed forest communities, in which non-nitrogen-fixing trees or diverse perennials are paired with nitrogen-fixing trees in tropical and subtropical countries, include the following couples: (i) *Eucalyptus* spp. with a number of nitrogen-fixing plants such as *Macrozamia* sp., which is an understorey nitrogen-fixing cycad (Halliday and Pate, 1976; Grove *et al.*, 1980), or *Daviesia mimosoides,* which is a common understorey legume in south-eastern Australia (McColl and Edmonds, 1986), or *Acacia* sp. (O'Connell *et al.*, 1979; Lawrie, 1981) or *Casuarina equisetifolia* (personal observation); (ii) *Pinus radiata* with *Coriaria arborea* (Gadgil, 1983; Silvester, 1977, 1983); (iii) *Vitas parviflora,* a species valued for its high-quality wood, with *Leucaena leucocephala* (Domingo, 1983); (iv) coffee with *Inga jinicuil* (Roskoski, 1981); (v) coffee or cacao with *Albizia falcataria* (Domingo, 1983; (vi) *Acacia nilotica,* a relatively poor nitrogen fixer with *Leucaena leucocephala,* an active nitrogen fixer (Bhatia and Kapoor, 1984).

Similar to what has been observed in the case of nitrogen-fixing trees associated with annual crops, nodules seem to constitute an important source of nitrogen for non-nitrogen-fixing species as well. The interrelation between nitrogen-fixing and non-nitrogen-fixing trees is illustrated by the fact that nodules are often found concentrated at the base of the trunk of the non-nitrogen-fixing plants, e.g., coffee-*Inga jinicuil* (Roskoski, 1981), and *Eucalyptus-Casuarina equisetifolia* (personal observation) associations.

Two types of silvicultural system which have been proposed in temperate countries (Tarrant, 1983) could be tested in the tropics: (i) a system involving nitrogen-fixing trees in mixed plantations or alternate cropping, and (ii) a system involving nitrogen-fixing plants of no economic value, especially shrubs grown as understorey. There are probably many eligible candidate plants, both leguminous and actinorhizal, in the tropics, but they have not yet been tested.

The development of research programmes on experimental mixtures comprising *Eucalyptus* spp. and *Acacia mangium, A. melanoxylon* and *Casuarina* sp. was proposed recently by Waring (1985). Some trials involving *Eucalyptus* sp. and *Acacia* are already under way in China (Wang Kwon Ming, personal communication) and in Congo (J.C. Delwaulle, personal communication). Such studies are worth pursuing, because observations increasingly indicate that mixed forests provide more stability than pure stands.

However, recommending specific tree mixtures before performing field trials can be hazardous, as was evidenced by the recent unsuccessful attempt to interplant *Eucalyptus deglupta* and *Albizia falcataria* in the Philippines; it failed because *A. falcataria* grows taller and much faster than *Eucalyptus deglupta,* the associated non-nitrogen-fixing tree (Domingo, 1983).

Conclusions

It is widely recognized that nitrogen-fixing trees and shrubs comprise a group of most promising species for agroforestry systems. However, potential direct and ancillary benefits from their introduction vary greatly depending on tree species, climate, soil, and management practices. Six concepts seem important in optimal utilization of biological nitrogen fixation in agroforestry:

1. It should be mandatory to use species, provenances and/or clones exhibiting a high nitrogen-fixing potential and adapted to the site conditions;
2. Improving nitrogen fixation is a leading research priority and, as shown, is a goal that can be achieved not only by selecting the most efficient *Rhizobium* or *Frankia* strains, but also by improving the host plant, one of the quickest methods being through clonal multiplication of elite trees.
3. Nitrogen-fixing trees improved in the laboratory, glasshouse or even in the nursery are not necessarily able to reach their maximum potential in the field. Therefore, before deciding to include a nitrogen-fixing tree or shrub in an agroforestry system, its nitrogen-fixing ability in field situations should be tested. This type of evaluation obviously presupposes the availability of accurate, simple methods to measure nitrogen fixation *in situ*. Progress has been made recently in developing such methods for annual crops, but more investigations are urgently needed to overcome the difficulties specific to perennial plants.
4. Even the least demanding nitrogen-fixing trees exhibit requirements for nutrients other than nitrogen that should be met so that their nitrogen-fixing potential can be fully realized. Fertilizer needs (mostly phosphate) are generally reduced, especially if the root system of the trees is properly infected with efficient mycorrhizal fungi.
5. Secondary effects should not be overlooked, especially those resulting in the development or the control of micro-organisms pathogenic to companion crops, or in allelopathic interferences which have been discussed in recent reviews (Koslowski and Huxley, 1983; Tarrant, 1983; Hu *et al.,* 1985).
6. Finally the dogma that research in agroforestry is a lengthy undertaking (Lundgren, 1979) should be kept in mind. Banking only on short-term effects (namely the immediate yield increase of associated crops) without recognizing long-term effects (maintenance or progressive increase of soil fertility) is like not seeing the forest for the trees.

REFERENCES

Akkermans, A.D.L. and H. Houvers. 1983. Morphology of nitrogen fixers in forest ecosystems. In J.C. Gordon and C.T. Wheeler (eds.), *Biological nitrogen fixation in forest ecosystems: foundations and applications*. The Hague: Nijhoff/Junk.

Allen, O.N. and E.K. Allen. 1981. *The Leguminosae. A source book of characteristics, uses, and nodulation*. Madison: University of Wisconsin Press.

Baggio, A. and J. Heuveldop. 1984. Initial performance of *Calliandra calothyrsus* Meissm. in live fences. *Agroforestry Systems* 2: 19–29.

Bhatia, N. and P. Kapoor. 1984. Neighbor interaction between *Leucaena leucocephala* and *Acacia nilotica* in mixed plantations in Punjab. *Leucaena Research Reports* 5: 18–9.

Bergersen, F.J. 1980. *Methods for evaluating nitrogen fixation*. New York: John Wiley.

————. 1986. Measurements of dinitrogen fixation. In *Biotechnology of nitrogen fixation in the tropics (BIOnifT)*, Proceedings UNESCO Regional Symposium and Workshop, UPM, Malaysia, 25–29 August 1986 (in press).

Bond, G. 1983. Taxonomy and distribution of non-legume nitrogen-fixing systems. In J.C. Gordon and C.T. Wheeler (eds.), *Biological nitrogen fixation in forest ecosystems: foundations and applications*. The Hague: Nijhoff/Junk.

Bourke, R.M. 1985. Food, coffee and casuarina: an agroforestry system from the Papua New Guinea highlands. *Agroforestry Systems* 2: 273–9.

Bowen, G.D. and P. Reddell. 1986. Nitrogen fixation in Casuarinaceae. In Proceedings of the 18th IUFRO World Congress, Ljubljana, Yugoslavia, September 1986 (in press).

Bowen, G.D. and P.A. Rosbrook. 1986. The management of nitrogen fixation by Casuarina. CSIRO and ACIAR (mimeo).

Budowski, G. 1983. An attempt to quantify some current agroforestry practices in Costa Rica. In P.A. Huxley (ed.), *Plant research and agroforestry*. Nairobi: ICRAF.

Carroll, B.J., D.L. McNeil and P.M. Gresshoff. 1985. A supernodulating and nitrate tolerant symbiotic (NTS) soybean mutant. *Plant Physiol.* 78: 34–40.

Charreau, C. 1985. Le rôle des arbres dans les systèmes agraires des régions semi-arides tropicales d'Afrique de l'Ouest. In *Le rôle des arbres au Sahel*. Ottawa: IDRC.

Cornet, F., H.G. Diem and Y.R. Dommergues. 1982. Effet de l'inoculation avec *Glomus mossaea* sur la croissance d'*Acacia holosericea* en pépinière et après transplantation sur le terrain. In *Les mycorhizes: biologie et utilisation* Paris: INRA.

CTFT. 1986. *Faidherbia albida*. Centre Technique Forestier Tropical, Nogent-sur-Marne.

Da Silva, G.G. and A.A. Franco. 1984. Selection of *Rhizobium* spp. strains in culture medium for acid soils. *Pesq. agropec. bras. Brasilia* 19: 169–73.

Datta, S.K. and K. Datta. 1984. Clonal multiplication of "elite" trees — *Leucaena leucocephala* through tissue culture. *Leucaena Research Reports* 5: 22–3.

Diem, H.G., D. Gauthier and Y.R. Dommergues. 1982. Isolation of *Frankia* from nodules of *Casuarina equisetifolia*. *Can. J. Microbiol.* 28: 526–30.

————.1983. An effective strain of *Frankia* from *Casuarina* sp. *Can. J. Bot.* 61: 2815–21.

Döbereiner, J. 1984. Nodulation and nitrogen fixation in legume trees. *Pesq. Agropec. Bras.* 19: 83–90.

Domingo, I. 1983. Nitrogen fixation in Southeast Asian forestry research and practice. In J.C. Gordon and C.T. Wheeler (eds.), *Biological nitrogen fixation in forest ecosystems: foundations and applications*. The Hague: Nijhoff/Junk.

Dommergues, Y.R. 1963. Evaluation du taux de fixation de l'azote dans un sol dunaire roboisé en filao (*Casuarina equisetifolia*). *Agrochimica* 105: 179–187.

Dreyfus, B.L., and Y.R. Dommergues. 1981. Nodulation of *Acacia* species by fast- and slow-growing tropical strains. *Appl. Environ. Microbiol.* 41: 97–9.

Dreyfus, B.L., D. Alazard and Y.R. Dommergues. 1984. Stem-nodulating rhizobia. In M.J. Klug and C.A. Reddy (eds.), *Current perspectives in microbial ecology*. Washington, D.C.: American Society for Microbiology.

Duhoux, E., B. Sougoufara and Y.R. Dommergues. 1986. Propagation of *Casuarina equisetifolia* through axillary buds of immature female inflorescences cultured in vitro. *Plant Cell Reports* 3: 161–4.

El Houri Ahmed, A. 1979. Effects of land use on soil characteristics in the Sudan. In H.O. Mongi and P.A. Huxley (eds.), *Soils research in agroforestry*. Nairobi: ICRAF.

Elkan, G.H. 1984. Taxonomy and metabolism of *Rhizobium* and its genetic relationships. In M. Alexander (ed.), *Biological nitrogen fixation, ecology, technology, and physiology*. New York: Plenum Press.

Escalante, G., R. Herrera and J. Aranguren. 1984. Fijacion de nitrogeno en arboles de sombra (*Erythrina poeppigiana*) en cacaotales del norte de Venezuela. *Pesq. Agropec. bras. Brasilia* 19: 223–30.

Felker, P. 1978. *State of the art: Acacia albida as a complementary permanent intercrop with annual crops*. Washington D.C.: USAID.

Fournier, L.A. 1979. Alder crops (*Alnus jorullensis*) in coffee plantations. In Proceedings of Workshop on Agroforestry Systems in Latin America. Turrialba, Costa Rica: CATIE.

Franco, A.A. 1984. Nitrogen fixation in trees and soil fertility. *Pesq. Agropec. Bras.* 19: 253–61.

Gadgil, R.L. 1983. Biological nitrogen fixation in forestry research and practice in Australia and New Zealand. In J.C. Gordon and C.T. Wheeler (eds.), *Biological nitrogen fixation in forest ecosystems: foundations and applications*. The Hague: Nijhoff/Junk.

Ganry, F., H.G. Diem, J. Wey and Y.R. Dommergues. 1985. Inoculation with *Glomus mosseae* improves N_2 fixation by field-grown soybeans. *Biol. Fert. Soils* 1: 15–23.

Gauthier, D.L., H.G. Diem, and Y.R. Dommergues. 1984. Tropical and subtropical actinorhizal plants. *Pesq. Agropec. Bras.* 19: 119–36.

Gauthier, D.L., H.G. Diem, Y.R. Dommergues and F. Ganry. 1985. Assessment of N_2 fixation by *Casuarina equisetifolia* inoculated with *Frankia* ORS021001 using ^{15}N methods. *Soil Biol. Biochem.* 17: 375–9.

Giffard, P.L. 1964. Les possibilités de reboisement en *Acacia albida* au Sénégal. *Bois et Forets des Tropiques* 95: 21–33.

———. 1971. Recherches sur *Acacia albida*. *Bois et Forets des Tropiques* 135: 3–20.

Grove, T.S., A.M. O'Connell and N. Malajczuk. 1980. Effects of fire on the growth, nutrient content and rate of nitrogen fixation of the cycad *Macrozamia riedlei*. *Austr. J. Bot.* 28: 271–81.

Halliday, J. 1984a. Principles of *Rhizobium* strain selection. In M. Alexander (ed.), *Biological nitrogen fixation, ecology, technology, and physiology*. New York: Plenum Press.

———. 1984b. Integrated approach to nitrogen-fixing tree germplasm development. *Pesq. Agropec. Bras.* 19: 91–117.

Halliday, J. and J.S. Pate. 1976. Symbiotic nitrogen fixation by coralloid roots of the cycad *Macrozamia riedlei*: Physiological characteristics and ecological significance. *Austr. J. Plant Physiol.* 3: 349–58.

Halliday, J. and P. Somasegaran. 1982. Nodulation, nitrogen fixation, and *Rhizobium* and strain affinities in the genus *Leucaena*. In *Leucaena research in the Asian-Pacific region*. Ottawa: IDRC.

Halos, S.C. 1983. Casuarinas in Philippines forest development. In S.J. Midgley, J.W. Turnbull and R.D. Johnston (eds.), *Casuarina ecology, management and utilization*. Melbourne: CSIRO.

Hayman, D.S. 1986. Mycorrhizae of nitrogen-fixing legumes. MIRCEN J. 2: 121–45.

Herridge, D.F. 1982. A whole-system approach to quantifying biological nitrogen fixation by legumes and associated gains and losses of nitrogen in agricultural systems. In P.H. Graham and S.C. Harris (eds.), *Biological nitrogen fixation technology for tropical agriculture*. Cali, Colombia: Centro International de Agricultura Tropical.

Hogberg, P. and M. Kvarnstrom. 1982. Nitrogen fixation by the woody legume *Leucaena leucocephala*. *Plant Soil* 66: 21–8.

Hu, T. W. and T. Kiang. 1983. *Leucaena* research in Taiwan. In *Leucaena research in the Asian-Pacific region*. Ottawa: IDRC.

Hu, T.W., J.C. Huang and C.C. Young. 1985. The role of nitrogen-fixing trees in soil biology. In J. Burley and J.L. Stewart (eds.), *Increasing productivity of multipurpose species*. Vienna. IUFRO.

Hutton, E.M. 1984. Breeding and selecting Leucaena for acid tropical soils. *Pesq. agropec. bras. Brasilia* 19: 263–74.

Huxley, P.A. 1983. The role of trees in agroforestry: some comments. In P. A. Huxley (ed.), *Plant Research and Agroforestry*. Nairobi: ICRAF.

IDRC. 1983. *Leucaena research in the Asian-Pacific region*. Ottawa: IDRC.

Johnson, L.A.S. 1982. Notes on Casuarinaceae. *J. Adelaide Bot. Gard.* 6: 73–87.

Kessel, C. van and P. Nakao. 1986. The use of nitrogen-15-depleted ammonium sulfate for estimating nitrogen fixation by leguminous trees. *Agronomy J.* 78: 549–51.

Kessel, C. van, P. Nakao, J.P. Roskoski and K. Kevin. 1987. Ureide production by N_2-fixing leguminous trees. *Soil Biol. Biochem.* (in press).

Khan, S.A. 1985. Nursery practices. In J. Burley and J.L. Stewart (eds.), *Increasing productivity of multipurpose species.* Vienna: IUFRO.

Knowles, R. 1983. Nitrogen fixation in natural plant communities and soils. In F.J. Bergersen (ed.), *Methods for evaluating biological nitrogen fixation.* New York: John Wiley.

Kondas, S. 1981. *Casuarina equisetifolia.* A multipurpose cash crop in India. In S.J. Midgley, J.W. Turnbull and R.D. Johnston (eds.), *Casuarina ecology, management and utilization.* Melbourne: CSIRO.

Koslowski, T.T. and P.A. Huxley. 1983. The role of controlled environments in agroforestry research. In P.A. Huxley (ed.), *Plant research and agroforestry.* Nairobi: ICRAF.

Kyle, N.E. and T.L. Righetti. 1985. *In vitro* micrografting of actinorhizal desert shrubs. In H.J. Evans, P.J. Bottomley and W.E. Newton (eds.), *Nitrogen fixation research progress.* The Hague: Nijhoff/Junk.

LaRue, T.A. and G. Patterson. 1981. How much nitrogen do legumes fix? *Adv. Agron.* 34: 15–38.

Lawrie, A.C. 1981. Nitrogen fixation by native Australian legumes. *Australian Journal of Botany* 29: 143–57.

Leaky, R.R.B. 1986. Cloned tropical hardwoods. Quicker genetic gain. *Span* 29: 35–7.

Lechevallier, M.P. 1985–1986. Catalog of *Frankia* isolates. *The Actinomycetes* 19: 131–62.

Le Houérou, H.N. 1985. Le rôle des arbres et arbustes dans les paturages sahéliens. In *Le Rôle des arbres au Sahel.* Ottawa: IDRC-158f.

Liang Zichao. 1986. Vegetative propagation and selection of *Casuarina* for resistance to bacterial wilt. *Tropical Forestry (Science and Technology) Guangazhou.* 2: 1–6 (in Chinese, summary in English).

Lindblad, P. and R. Russo. 1986. C_2H_2-reduction by *Erythrina poeppigiana* in a Costa Rican coffee plantation. *Agroforestry Systems* 4: 33–7.

Lulandala, L.L.L. and J.B. Hall. 1986. *Leucaena leucocephala's* biological nitrogen fixation: a promising substitute for inorganic nitrogen fertilization in agroforestry systems. In *Biotechnology of nitrogen fixation in the tropics (BIOnifT),* Proceedings of UNESCO Regional Symposium and Workshop, UPM, Malaysia, 25–29 August 1986 (in press).

Lundgren, B. 1979. Research strategy for soils in agroforestry. In H.O. Mongi and P.A. Huxley (eds.), *Soils research in agroforestry.* Nairobi: ICRAF.

McColl, J.G. and R.L. Edmonds. 1986. Acetylene reduction by *Diviesia mimosaides* under *Eucalyptus. Plant Soil* 96: 215–24.

Mitchell, M., D. Gwaze and H. Stewart. 1986. Termite susceptibility of Australian trees in Zimbabwe. *ACIAR Forestry Newsletter* 2 (Sept./Oct.): 2.

Nair, P.K.R., E.C.M. Fernandes and P.N. Wambugu. 1984. Multipurpose leguminous trees and shrubs for agroforestry. *Pesq. Agropec. Bras.* 19: 295–313.

National Research Council. 1977. *Leucaena: promising forage and tree crop for the tropics.* Washington, D.C.: National Academy of Sciences.

————. 1979. *Tropical legumes: resources for the future.* Washington, D.C.: National Academy of Sciences.

————. 1980. *Firewood crops: shrubs and trees for energy production.* Washington, D.C.: National Academy of Sciences.

————. 1982. *Priorities in biotechnology research for international development.* Washington, D.C.: National Academy of Sciencies.

————. 1983. *Mangium and other fast-growing acacias for the humid tropics.* Washington, D.C.: National Academy of Sciences.

Nitrogen Fixing Tree Association (NFTA) 1986. Erythrinas provide beauty and more. *NFT Highlights* 86–02.

O'Connell, A.M., T.S. Grove and N. Malajczuk. 1979. Nitrogen fixation in the litter layer of eucalypt forests. *Soil Biol. Biochem.* 11: 681–2.

Okigbo, B.N. 1984. Nitrogen-fixing trees in Africa: priorities and research agenda in multiuse exploitation of plant resources. *Pesq. Agropec. Bras.* 19: 325–30.

Orchard, E.R. and G.D Darby. 1956. Fertility changes under continued wattle culture with special reference to nitrogen fixation and base status of the soil. In *Comptes Rendus du Sixième Congrès.* Inter. Science du Sol, Paris.

Poschen, P. 1986. An evaluation of the *Acacia albida*-based agroforestry practices in the Hararghe highlands of Eastern Ethiopia. *Agroforestry Systems* 4: 129–43.

Puppo, A., L. Dimitrijevic, H.G. Diem and Y.R. Dommergues. 1985. Homogeneity of superoxide dismutase patterns in *Frankia* strains from Casuarinaceae. *FEMS Microbiol. Lett.* 30: 43–6.

Reddell, P.W. 1986. Management of nitrogen fixation by *Casuarina. ACIAR Forestry Newsletter* 2 (Sept./Oct.): 1–3.

Reddell, P.W., G.D. Bowen and A.D. Robson. 1986. Nodulation of Casuarinaceae in relation to host species and soil properties. *Aust. J. Bot* 34: 435–44.

Roskoski, J.P. 1981. Nodulation and N_2 fixation by *Inga jinicuil,* a woody legume in coffee plantations. I. Measurements of nodule biomass and field C_2H_2 reduction rates. *Plant Soil* 59: 201–6.

————. 1982. Nitrogen fixation in a Mexican coffee plantation. *Plant Soil* 67: 282–292.

————. 1986. Future directions in biological nitrogen fixation research. In *Biotechnology of nitrogen fixation in the tropics (BIOnifT),* Proceedings UNESCO Regional Symposium and Workshop, UPM, Malaysia, 25–29 August 1986 (in press).

Roskoski, J.P., J. Montano, C. Van Kessel and G. Castilleja. 1982. Nitrogen fixation by tropical woody legumes: potential source of soil enrichment. In P.H. Graham (ed.), *Biological nitrogen fixation technology for tropical agriculture.* Cali, Colombia: CIAT.

Sanchez, P.A. and J.G. Salinas. 1981. Low-input technology for managing oxisols and ultisols in tropical America. *Adv. Agron.* 34: 279–406.

Sanginga, N., K. Mulongoy and A. Ayanaba. 1985. Effect of inoculation and mineral nutrients on nodulation and growth of *Leucaena leucocephala.* In H. Ssali and S.O. Keya (eds.), *Biological nitrogen fixation in Africa.* Nairobi: MIRCEN.

————. 1986. Inoculation of *Leucaena leucocephala* (Lam.) de Wit with *Rhizobium* and its nitrogen contribution to a subsequent maize crop. *Biological Agriculture and Horticulture* 3: 347–352.

Schenck, N.C. 1982. *Methods and principles of mycorrhizal research.* Saint-Paul, Minnesota: American Phytopathological Society.

Silvester, W.B. 1977. Dinitrogen fixation by plant associations excluding legumes. In R.W.F. Hardy and A.H. Gibson (eds.), *A treatise on dinitrogen fixation.* New York: John Wiley.

————. 1983. Analysis of nitrogen fixation. In J.C. Gordon and C.T. Wheeler (eds.), *Biological nitrogen fixation in forest ecosystems: foundations and applications.* The Hague: Nijhoff/Junk.

Somasegaran, P. and H.J. Hoben. 1985. *Methods in legume-Rhizobium technology.* University of Hawaii NifTAL Project and MIRCEN, Hawaii.

Sougoufara, B., E. Duhoux and Y.R. Dommergues. 1987. Improvement of nitrogen fixation by *Casuarina equisetifolia* through clonal selection. *Arid Soil Research and Rehabilitation* 1:129–132.

Sprent, J. 1983. Agricultural and horticultural systems: implications for forestry. In J.C. Gordon, C.T. Wheeler and D.A. Perry (eds.), *Symbiotic nitrogen fixation in the management of temperate forests.* Corvallis, Oregon: Forest Research Laboratory.

Sumberg, J.E. 1985. Collection and initial evaluation of *Gliricidia sepium* from Costa Rica. *Agroforestry Systems* 3: 357–61.

Tarrant, R.F. 1983. Nitrogen fixation in North American forestry: research and application. In J.C. Gordon and C.T. Wheeler (eds.), *Biological nitrogen fixation in forest ecosystems: foundations and applications.* The Hague: Nijhoff/Junk.

Taylor, D.P. 1976. Plant nematology problems in tropical Africa. *Helminthological Abstracts* (Ser. B) 45: 269–84.

Thiagalingam, K. 1983. Role of *Casuarina* in agroforestry. In S.J. Midgley, J.W. Turnbull and R. D. Johnson (eds.), *Casuarina ecology, management and utilization.* Melbourne: CSIRO.

Torrey, J. and J. D. Tjepkema. 1979. Preface and program. *Bot. Gaz.* 140 (suppl.), i–ii.

Turnbull, J.W., D.J. Skelton, M. Subagyono and E.B. Hardiyanto. 1984. Seed collecting tropical Acacias in Indonesia, Papua New Guinea and Australia. *Forest Genetic Resources Information* (FAO, Rome) 12: 2–15.

Virginia, R.A., W.M. Jarrell, D.H. Kohl and G.B. Shearer. 1981. Symbiotic nitrogen fixation in *Prosopis* (Leguminosae) dominated desert ecosystems. In A.H. Gibson and W.E. Newton (eds.), *Current perspectives in nitrogen fixation.* Canberra: Australian Academy of Science.

von Carlowitz, P.G. 1986. *Multipurpose tree and shrub directory.* Nairobi: ICRAF.

Walsh, K.B., B.H. Ng and G.E. Chandler. 1984. Effects of nitrogen nutrition on xylem sap composition of Casuarinaceae. *Plant Soil* 81: 291–3.

Waring, H.D. 1985. Chemical fertilization and its economic aspects. In J. Burley and J.L. Stewart (eds.), *Increasing productivity of multipurpose species.* Vienna: IUFRO.

Wiersum, K.F. 1985. *Acacia mearnsii.* Multipurpose highland legume tree. *NFT Highlights* 85–02.

Williams, P.K. 1984. Current use of legume inoculant technology. In M. Alexander (ed.), *Biological nitrogen fixation: ecology, technology and physiology.* New York: Plenum Press.

Wilson, G.F., B.T. Kang and K. Mulongoy. 1986. Alley cropping: trees as sources of green manure and mulch in the tropics. *Biological Agriculture and Horticulture* 3: 251–67.

Yadav, J.S.P. 1983. Soil limitations for successful establishment and growth of Casuarina plantations. In S.J. Midgley, J.W. Turnbull and R.D. Johnston (eds.), *Casuarina ecology, management and utilization.* Melbourne: CSIRO.

Zhang, Z. and J. Torrey. 1985. Biological and cultural characteristics of effective *Frankia* strain HFPCc13 (Actinomycetale) from *Casuarina cunninghamiana* (Casuarinaceae). *Annals of Botany* 56: 367–78.

Zhang, Z., M.F. Lopez and J.G. Torrey. 1984. A comparison of cultural characteristics and infectivity of *Frankia* isolates from root nodules of *Casuarina* species. *Plant Soil* 78: 79–90.

16

Exploitation of the potential of multipurpose trees and shrubs in agroforestry

Jeffery Burley

Director
Oxford Forestry Institute
South Parks Road, Oxford
United Kingdom

Contents

Introduction

Trees and shrubs occur in a wide variety of land-use systems. For simplicity these systems can be grouped into 11 major categories.

1. Natural vegetation management

Although considerable research has been undertaken on the management of tropical rain forests, relatively little attention has been given to the natural tree and shrub associations of drier zones. Nevertheless these communities offer a significant and often the sole source of plant materials, especially for fuel and fodder; often they contain species that could have great potential as planted exotics for other sites. Generally there is more information and experience of tropical and subtropical trees than shrubs from the points of view of ecology, distribution, inventory, use and management.

2. Industrial plantations

These are large areas created and managed intensively, usually with exotic species, for the production of timber to supply sawmills, pulpmills, veneer factories, chipboard plant, etc. The plantations are usually owned and managed by state enterprises.

3. Community woodlots

These are small areas of 5–10 ha created for the benefit of village or town communities (where they are often referred to as "peri-urban plantations"), often by the state, and more recently by the communities themselves in some form of social forestry. They may be on state or community land and may yield timber, poles and fuelwood plus occasionally fruit, fodder and other products. There are often difficulties of management, protection and distribution of benefits.

4. Farm woodlots

These are small plantations of less than 10 ha, often much less, that are established by the individual farmer for the production of poles, fuel, fodder and possibly other products; multipurpose trees are thus desirable. The products supply the farmer's own needs with excess for sale and such woodlots may be established on unused or degraded land with a view to rehabilitating it.

5. Trees in crop land

Individual trees may be left or planted randomly at wide intervals in productive agricultural land to supply wood, fuel, fodder, fruit, honey and shade.

6. Alley farming

In this group of systems, one or more rows of trees are planted alternately with several rows of agricultural crop plants and the trees are hedged, coppiced or pollarded frequently. The decline in value of crop yield caused by the loss of land occupied by trees should be compensated or exceeded by the fertilizing effect of the tree leaves and other ways of soil improvement by trees, and by the value of tree products (poles and fuel). This is becoming the most widely recommended agroforestry system (see Kang and Wilson, this volume).

7. Linear planting

This includes the planting of one or more rows of trees, with or without subsequent management, along farm borders, river or stream banks, or along roads, railways or canals. They can provide the usual services and benefits.

8. Shelterbelts

These are belts/blocks consisting of several rows of trees established at right angles to the prevailing wind. They are also known as windbreaks and have significant effects on micrometeorological factors up to several times their height away from the edge. The species, age composition, canopy density, height and profile are all important determinants of their effectiveness. They also produce valuable by-products.

9. Sequential cropping

Trees and agricultural crops may follow each other on the same piece of land in planted "fallow" systems in which the trees restore the soil fertility. Taungya is a system whereby trees are planted, often at close industrial spacing, together with intercrops of agricultural species, the agricultural crops being grown for up to three or four years.

10. Silvopastoral systems

Silvopastoral systems involve the incorporation of tree and shrub management and animal husbandry. The trees may be used for fodder production, shade and pasture improvement. Intensity of the operations can vary from extensive range management in dry zones to intensive trees-over-pasture in areas of higher rainfall.

11. Protection forestry, land rehabilitation, reclamation

The use of trees in these roles encompasses many technologies. Protection forestry generally requires the management, through natural regeneration, of existing indigenous vegetation, which requires protection from grazing and damaging exploitation above all. For the reclamation and rehabilitation of degraded land the function of trees is primarily for soil conservation or improvement, coupled with production of (mainly) wood and fodder. The arrangement of the planted trees (which can be combined in agroforestry systems) should follow these major objectives.

For virtually all these systems a multipurpose plant would be considered more useful than a species fitted for only one purpose. In fact there can be very few species that, if they are used at all, are not used for several purposes, products, benefits and services. There has been considerable discussion of the definition of multipurpose trees (see the views of several specialists compiled in Burley and von Carlowitz, 1984) but the concept is now well established, largely as a result of the interests and activities of ICRAF. The common abbreviation is MPTS, which can imply "*Multi*Purpose *Tree*S" or "*Multi*Purpose *T*rees and *S*hrubs".

The attention to MPTS has developed in parallel with the growth of social forestry programmes and the research and development of agroforestry systems that can be used to meet the objectives of social forestry and integrated rural development programmes. The incorporation of MPTS into land-use systems requires significant changes in the attitude, understanding and co-operation of professional foresters, horticulturists, agronomists and various groups of extension workers.

Benefits from trees and shrubs

The products and services derived from trees and shrubs are manifold and vary between societies and environments, but they can be summarized simply as follows:

Products

Wood
Unprocessed	— fencing or building poles
Processed, solid	— saw timber
Reconstituted	— veneers, paper, chipboard
Bark	— raw and processed for various uses

Products

Energy	
Solid, raw	— firewood
Solid, processed	— charcoal
Fluid	— liquid and gaseous fuels and feedstocks
Chemical stem extractives	— resins, oils, paints, varnishes, pharmaceuticals
Leaf products	— thatch, fibre, fodder, extractives, oils, silk, smoking material, medicines
Fruit/seed products	— food, fodder, oils, drinks, medicines
Flower products	— drinks, medicines, honey, dyes, food
Root products	— fuelwood, chemical extractives, dyes

Environmental benefits

Climatic moderation (macro- and micro-)
Soil stabilization
Soil improvement
Water-flow moderation
Wildlife habitats
Boundary demarcation
Pest and weed control
Use or rehabilitation of degraded land, improving downstream environments

Socio-economic benefits

Amenity and tourism
Employment generation — especially for the landless
Income generation — including foreign exchange
Import substitutions
Public education
Rehabilitation of abandoned and degraded land, increasing production
Counter seasonality
Risk reduction
Labour saving in some situations
Improved human and animal nutrition and health.

Research needed for exploitation of MPTS potential

Exploitation is often used in a pejorative sense to indicate the utilization of a person or object for one's own selfish ends, but in fact this is the implication of man's use of multipurpose trees and shrubs. We seek species and populations that can provide the many benefits in the several land-use systems indicated above. Potential is taken here to indicate cryptic, possible value for such uses and to exploit it requires knowledge of hitherto hidden values; this necessitates research that is specific to sites, managerial systems and end-use processes.

An idealized, complete research programme for a new land-use system or site is summarized in Appendix 1.

Clearly not all of these stages are required for each site or species; many species are already at an advanced stage of genetic and managerial development either by researchers or by line managers and farmers. However, the appendix acts as a checklist for researchers and managers initiating development of new areas or systems. Not all of the stages need to be conducted sequentially; some can be undertaken in parallel or telescoped together. Nevertheless, the determination of an optimum system and combination of species requires co-operative research and implementation between a range of authorities and disciplines.

The major, specific stages and problems related to the incorporation of the MPTS themselves can be grouped into the following principal topics — genetic variation, germplasm supplies, assessment of multiple products and services, and crop management.

Genetic variation

Number of species
The large number of tree and shrub species already recorded as promising or possible (some 2,000 species were listed in Burley and von Carlowitz, 1984) is at once a potential benefit and a problem. No one species is likely to be the "wonder species" for all sites and purposes but it is difficult to cope with such large numbers in formal species trials even on efficient research stations interested in a single major product. It is infinitely more difficult to cope with large numbers on less well endowed stations or in on-farm research. Preliminary screening is thus needed and the climatic matching systems collectively known as homoclimal comparisons (and quantified by, e.g., Booth, forthcoming) permit the reduction of the number of species worth consideration for trial on a given site type. The several data bases developing within ICRAF itself will formalize and often quantify data on sources, uses and characteristics of MPTS.

However, there will always be a need for elimination and proving trials of reasonable numbers of species (20–30) on each major site type. Centrally planned co-operative trials on many sites, such as the Oxford Forestry Institute (OFI)'s international trial of dry-zone hardwoods, will allow the estimation and explanation of genotype-environment interactions and the extrapolation of results to untested sites if the environmental conditions are known. In this international network seed and herbarium material from 25 Central American species were collected (see Table 1 and Hughes and Styles, 1984); seeds are being distributed to some 60 sites in more than 25 countries. An indication of the suitability of these species for different uses in their natural range is given in Table 2. Not all of these will be found in all exotic sites and not all are equally important on any individual site, but they indicate the potential of this group of species. Similar sets of species can be found elsewhere (e.g., the many promising Australian and African *Acacia* species) and these must be compared. The collaborative (zonal co-operative) programme of ICRAF (COLLPRO) will undertake species evaluation for common site types while national programmes deal with specific locations.

Variation within species
For naturally widely distributed species, natural selection (and often man's interference) causes genetic variation between populations; for species that have been managed as exotics for several generations artificial and natural selection in the planting sites may cause the evolution of land races.

All of these different sources (termed "provenances" by foresters) should be tested for each site and management system, but clearly they compound the problems of species trials with large numbers of sources. Nevertheless, the correct choice of optimum seed source offers the major and simplest step in genetic improvement and the determination of the

Table 1 Distribution and phenological characteristics of semi-arid species in the OFI international network

Species	Family/ Sub-family	Common names	Distribution	Altitude range (m)	Tree height (m)	Thorns	Phenology		N_2 fixation
							Flowering	Seed collection	
Acacia deamii (Britten & Rose) Standl. syn. *A. picachensis* Brandeg.	Leguminosae Mimosoideae	Bisquite Orotoguaje	S. Mexico, C. America to Nicaragua	200-100	to 8	short	June	January-February	unconfirmed YES
Acacia farnesiana (L.) Wild	Leguminosae Mimosoideae	Espino blanco, Aromo, Subin	South U.S.A. Mexico, C. America W. Indies, S. America to Argentina	0-1,000	to 10	long and many	December-March	January-April	confirmed YES
Acacia pennatula (Schlecht. & Cham.) Benth.	Leguminosae Mimosoideae	Sarespino, Escambrion Huizache	Mexico, C. America S. America	300-2,000	8-10	short	February-March	January-February	confirmed YES
Albizia guachepele (Kunth) Dug. syn. *Pseudosamanea guachepele* (Kunth) Harms: *A. longepedana* Britton & Rose	Leguminosae Mimosoideae	Cadeno, Lagarto	Guatemala, C. America to Venezuela	0-800	15-20	none	December-January	January-March	unconfirmed YES
Apoplanesia paniculata Presl.	Leguminosae Papilionoideae	Baillador Made de flecha	South and West Mexico, Guatemala, Honduras, Venezuela	0-400	to 10	none	October-November	December-January	unconfirmed YES
Ateleia herbert-smithii Pittier	Leguminosae Papilionoideae	Palo de prieto	localized in Central Nicaragua, Guanacaste, Costa Rica, Colombia	0-600	to 15	none	November December	March	unconfirmed YES
Caesalpinia coriaria (Jacq.) Wild	Leguminosae Caesalpinioideae	Nacascolo	S. Mexico, C. America W. Indies to Venezuela	0-1,000	to 8	none	September-October	December January	confirmed NO

Table 1 Continued

Species	Family/ Sub-family	Common names	Distribution	Altitude range (m)	Tree height (m)	Thorns	Phenology		N₂ fixation
							Flowering	Seed collection	
Caesalpinia eriostachys Benth.	Leguminosae Caesalpinioideae	Cuayauncuavo Pintandilo	S. Mexico, Cuba	0–600	to 15	none	January-March	March-April	confirmed NO
Caesalpinia velutina (Britton & Rose) Standtl.	Leguminosae Caesalpinioideae	Aripin	S. Mexico, Guatemala to Nicaragua	0–800	to 10	none	March April	November January	unconfirmed NO
Gliricidia sepium (Jacq.)* Standtl.	Leguminosae Papilionoideae	Madre de cacao Madreado Madero negro	Mexico, C. America W. Indies to Colombia, Guianas	0–1,500	to 10	none	January-March	March April	confirmed YES
Haematoxylon brasiletto Karst.	Leguminosae Caesalpinioideae	Brasil Campeche	Mexico, C. America Colombia and Venezuela	0–800	to 8	short	February-March	March-April	unconfirmed NO
Leucaena diversifolia (Schlecht.) Benth. syn. L. brachycarpa Urb.	Leguminosae Mimosoideae	Guaje	Mexico, C. America to Honduras	300–2,000	to 12	none	December-February	February-March	confirmed YES
Leucaena leucocephala (Lam.) de Wit.	Leguminosae Mimosoideae	Guaje	Mexico, C. America to Nicaragua	0–800	to 15	none	January-March	March-April	confirmed YES
Leucaena shannoni Donn. Smith	Leguminosae Mimosoideae	Guaje	Mexico, C. America to Nicaragua	0–1,000	to 8	none	September-December	January-March	confirmed YES
Mimosa tenuiflora (Willd.) Poir	Leguminosae Mimosoideae	Carbon	Mexico to Brazil	0–1,200	to 5	short numerous	December-April	December-April	confirmed YES
Myrospermum frutescens Jacq.	Leguminosae Papilionoideae	Chiquirin	Mexico, C. America W. Indies to Venezuela	0–800	5 rarely to 20	none	February March	March	unconfirmed YES
Parkinsonia aculeata L.	Leguminosae Caesalpinioideae	Aguijote blanco Sauco	Southern U.S.A. Mexico, C. America	0–1,000	to 6	short	January-February	March	confirmed YES

Table 1 Continued

Species	Family/ Sub-family	Common names	Distribution	Altitude range (m)	Tree height (m)	Thorns	Phenology		N$_2$ fixation
							Flowering	Seed collection	
Pithecellobium dulce (Roxb.) Benth.	Leguminosae Mimosoideae	Espino de Playa Michiguiste	S. California, Mexico, C. America to Venezuela	0–1,500	to 15	short (not always present)	January-February	March	confirmed YES
Prosopis juliflora (Swartz) DC.	Leguminosae Mimosoideae	Espino real Espino ruco Algorrobo	Mexico, C. America S. America to Peru	0–1,000	to 7	long and numerous	December-February	February April	confirmed YES
Senna atomaria (L.) Irwin & Barneby syn. *Cassia emarginata* L.	Leguminosae Caesalpinioideae	Vainillo Frijolillo	Mexico, C. America W. Indies to Venezuela	100–1,000	to 10	none	January-March	February-March (1 year to ripen)	unconfirmed NO
Enterolobium cyclocarpum (Jacq.) Griseb.	Leguminosae Mimosoideae	Guanacaste Conacaste	West and South Mexico, C. America, Jamaica, Cuba, northern S. America	0–800	to 40	none	April	March-April	confirmed YES
Crescentia alata HBK	Bignoniaceae	Jicaro, Morro Calabash	Mexico, C. America to Costa Rica	0–1,000	to 6	none	September-October	March-April	NO
Alvaradoa amorphoides Liebm.	Simaroubaceae	Plumajillo Zorillo	S. America, C. America, W. Indies	0–1,400	to 15	none	January-February	April	NO
Simarouba glauca DC.	Simaroubaceae	Aceituno Negrito	S. Mexico C. America to Panama, Cuba	0–800	to 20	none	December-January	April	NO
Guazuma ulmifolia Lam.	Sterculiaceae	Guacimo Caulote	Tropical America and Caribbean	0–1,200	to 20	none	March-May	February-March	NO

* *Editors' note: Gliricidia sepium* (Jacq.) Walp. (see editors' note at the end of the References to this chapter).

Table 2 Use and managerial characteristics of semi-arid species in the OFI international network.

	Fuelwood	Fodder Leaf	Fruit	Posts	Saw-wood	Food	Shade	Live fence	Coppice ability	Pollard ability	Salt tolerance	Weediness hazard
Acacia deamii	3	-	1	3	1	1	2	2	3	2	1	-
Acacia farnesiana	2	1	3	1	1	1	1	1	3	2	2	Hazardous
Acacia pennatula	3	1	3	3	1	1	3	2	3	-	1	Slight
Albizia guachepele	2	-	1	2	3	1	2	2	-	-	1	-
Apoplanesia paniculata	2	-	1	3	1	1	1	3	2	-	1	-
Ateleia herbert-smithii	2	2	2	1	2	1	2	3	3	3	1	-
Caesalpinia coriaria	2	1	2	3	1	1	2	2	-	-	2	-
Caesalpinia eriostachys	2	2	2	2	1	1	2	3	2	-	1	-
Caesalpinia velutina	3	-	1	3	1	1	1	2	1	1	1	-
Enterolobium cyclocarpum	2	2	3	2	3	1	3	2	2	2	1	-
Gliricidia sepium	3	3	-	2	1	1	3	3	3	3	2	Slight
Haematoxylon brasiletto	3	1	1	1	1	1	-	1	-	1	1	-
Leucaena diversifolia	3	3	-	2	2	-	2	2	3	2	1	Slight
Leucaena leucocephala	3	3	2	2	1	2	2	2	3	2	1	Slight
Leucaena shannoni	3	3	1	2	1	2	2	2	3	2	1	-
Mimosa tenuiflora	3	1	1	3	1	1	1	1	3	1	1	Very hazardous
Mycospermum frutescens	3	-	2	3	2	1	1	2	3	2	2	-
Parkinsonia aculeata	2	1	2	1	2	2	2	2	3	2	3	Hazardous
Pithecellobium dulce	3	1	3	2	2	2	2	2	3	2	2	Hazardous
Prosopis juliflora	3	2	3	3	1	3	2	1	3	2	3	Hazardous
Senna atomaria	2	-	-	3	2	1	2	3	2	2	1	-
Alvadoa amorphoides	3	3	1	3	2	1	1	2	3	3	1	-
Crescentia alata	2	1	3	1	1	2	2	2	1	1	1	-
Guazuma ulmifolia	3	3	3	2	1	2	3	3	3	3	1	-
Simarouba glauca	2	-	3	1	2	2	2	3	3	2	1	-

Key: 1 = poor; 2 = acceptable; 3 = good; - = unknown.

most productive or acceptable system. Later tree-breeding efforts are wasted or reduced in value if the original germplasm source is not optimal.

Little species and provenance research has been conducted on MPTS other than the initial exploration and evaluation of arid-zone species by CSIRO, CTFT, FAO/IBPGR, NAS, NFTA and OFI.* There is great potential for national organizations and multilateral programmes to conduct systematic research on both indigenous and exotic species, including local land races that have been manipulated genetically, whether consciously or unconsciously, by local farmers. Current emphasis is on provenance variation of *Acacia, Eucalyptus, Leucaena* and *Prosopis* species, while among the most exhaustive exploration and evaluation of a single species are the studies on *Leucaena leucocephala* (Brewbaker, this volume) and the OFI international study of *Gliricidia sepium* (Hughes, 1986, 1987). The available seed sources of the latter are listed in Table 3. Preliminary assessments of field trials in several locations demonstrate considerable differences between provenances in form and growth rate up to two years.† In a trial of 10 sources in Costa Rica, Salazar (1986) observed significant differences between provenances in seed size and shape that were related to altitude of seed source; after 60 days heights ranged from 35 to 47 cm. All the earlier evidence from international and native provenance trials of industrial species such as tropical pines and eucalyptus confirms that large intraspecific variation occurs in wide-ranging species. The same situation is being demonstrated in MPTS and variations can be expected to be enhanced by the existence of local land races that developed in exotic and natural locations under man's selection.

With the exception of *Leucaena* species (covered in a separate chapter in this volume by Brewbaker) and *Prosopis* species (see several papers in Felker, 1986), for most MPTS currently in use or under research it is perhaps premature to consider selective tree breeding, but the short sexual and managerial cycles allow rapid progress to be made once individual selection begins. The principles are no different from those of industrial tree species, nor indeed from those of agricultural and horticultural species, and they were outlined for MPTS by Burley (1987).

Evidence of intra-population genetic control of one set of characteristics (biochemical) of fodder trees was found in *Prosopis* by Oduol *et al.* (1986); 14 half sib families representing several species showed intra-class correlations of pod sugar (0.3–0.4) and pod protein content (0.04–0.6) depending on planting site.

All of the above are examples of the potential of MPTS but, with the exception of *Leucaena,* they are preliminary indicators only. As highlighted by Brewbaker (this volume), the widespread attack by psyllids on *Leucaena leucocephala* and the species intolerance of acid soils stress the value of interspecific hybridization within a genus and the importance of evaluating a wide range of other species and genera.

Germplasm supplies and certification

The great interest in development of a wide range of MPTS causes a corollary problem in supplies of appropriate germplasm. Even assuming the correct genetic source has been identified, it is often difficult to obtain guaranteed supplies with the internationally recognized certificates of health, origin and physiological and physical quality. An indication of the systems of certification available was given by Jones and Burley (1973),

---

* *Editors' note:* see the full names of these institutions listed at the beginning of this book.
† Personal communication from Janet L. Stewart, Oxford Forestry Institute.

Table 3 Summary of seed collection site data for the OFI international network of *Gliricidia* species and provenance trials.

Provenance	Country	Ident. No.	Lat. (N)	Long. (W)	Alt. (m)	Rainfall (mm)	No. of trees	Seed quality	
								No. of seeds/kg ('000)	Germination %
Gliricidia sepium									
Chamela, Jalisco	Mexico	41/85	19°28'	105°05'	60–100	905	30	5.5	90
Playa Azul Michoacan	Mexico	38/85	18°04'	102°34'	0–30	900	60	6.5	97
San Jose, Guerrero	Mexico	39/85	16°48'	99°15'	30	1,400	40	4.7	83
Los Amates, Puebla	Mexico	33/85	18°28'	98°25'	1,100	650	120	5.1	96
Palmasola, Veracruz	Mexico	34/85	19°46'	96°25'	10–50	1,130	40	6.7	96
Barrosa, Veracruz	Mexico	36/85	18°20'	95°06'	100–150	2,500	20	6.0	91
San Mateo, Oaxaca	Mexico	35/85	16°13'	94°58'	10–30	950	300	6.3	96
Arriaga, Chiapas	Mexico	40/85	16°15'	93°51'	30	1,796	50	7.6	94
Tzimol, Chiapas	Mexico	37/85	16°18'	92°22'	600–700	1,030	35	7.9	94
Samala, Retalhuleu	Guatemala	14/84	14°33'	91°39'	330	3,500	75	7.8	95
Monterrico, Santa Rosa	Guatemala	17/84	13°53'	90°29'	5	1,650	200	8.1	99
Volcan Suchitan, Jutiapa	Guatemala	13/84	14°22'	89°46'	950	1,060	70	7.4	99
Vado Hondo, Chiquimula	Guatemala	16/84	14°44'	89°30'	450	830	75	7.3	93
Gualan, Zacapa	Guatemala	15/84	15°08'	89°20'	150	700	80	7.2	99
Masaguara, Intibuca	Honduras	25/84	14°16'	87°58'	825	1,100	65	7.9	94
La Garita, Choluteca	Honduras	10/86	13°26'	87°11'	450	1,200	75	7.5	93
Guayabillas, Choluteca	Honduras	24/84	13°24'	86°58'	480	1,400	180	8.0	48

Table 3 Continued

Provenance	Country	Ident. No.	Lat. (N)	Long. (W)	Alt. (m)	Rainfall (mm)	No. of trees	Seed quality No. of seeds/kg ('000)	Germination %
Piedra Larga, Esteli	Nicaragua	30/84	13°16'	86°23'	605	800	35	7.4	93
Ciudad Dario, Matagalpa	Nicaragua	31/83	12°39'	86°07'	450	900	35	8.0	92
Laguna Tecomapa, Matagalpa	Nicaragua	13/82	12°37'	86°03'	380	900	87	8.0	88
Mateare, Managua	Nicaragua	31/84	12°14'	86°27'	60	1,100	25	7.4	92
Ojo de Agua, Boaco	Nicaragua	29/84	12°23'	85°45'	220	1,200	75	7.4	91
Belen, Rivas	Nicaragua	14/86	11°37'	85°48'	75	1,650	100	7.7	94
Playa Tamarindo, Guanacaste	Costa Rica	12/86	10°19'	85°54'	0-10	1,500	150	9.0	98
El Roblar, Guanacaste	Costa Rica	11/86	10°15'	85°18'	20-100	1,000	50	9.4	97
Padasi, Los Santos	Panama	13/86	7°32'	80°04'	0-20	850	40	11.0	97
Pontezuelo, Bolicar	Colombia	24/86	10°35'	75°51'	20-50	950	150	8.5	94
Mariara, Carabobo	Venezuela	1/86	10°17'	67°43'	520	800	100	7.0	92
Gliricidia maculata									
Puerto Morelos, Quintana Roo	Mexico	42/85	20°50'	86°57'	0-10	1,500	15	13.0	57
Gliricidia guatemalensis									
Sola de Vega, Oaxaca	Mexico	1/84	16°29'	97°01'	1,800	1,500	1		

and the problems of tropical and subtropical (developing) countries in implementing such systems were described by Burley (1986). Sources of potential suppliers of a large range of MPTS were provided by von Carlowitz (1986). They show a remarkable variation in ability or inclination to provide appropriate seed-source information.

Evaluation of multiple benefits

For industrial plantation species the most important features affecting quantity or quality of products are well known and standard methods of assessment exist (for example, height, diameter, form, taper, branch dimensions, forking or wood density). For MPTS, however, many products and services need evaluation and, for some, appropriate techniques are either not yet developed or at least not known to the forestry-trained researchers who are currently conducting the bulk of the tree-related research; these include fodder-quality characteristics, soil-improving capacity, nitrogen-fixation ability, and the effect of the tree or shrub on agricultural crops or animals.

Market values may not exist for some of these traits, while others may be negatively correlated so that an improvement in one causes a decline in another. This causes particular problems for the tree breeder who then needs to develop selection indices, but rapid and simple assessment procedures are also needed for the evaluation of MPTS in species, provenance and management trials.

In addition to anatomical, chemical and morphological features, the traits of interest include phenological and physiological characters of the tree or shrub and chemical and biophysical evaluation of its effects on the soil. Many of these traits were described by Burley *et al.* (1984). The effects of MPTS on microclimates are more complex to evaluate (Huxley *et al.,* forthcoming).

When using MPTS as energy sources there is a need to estimate and partition the amount of biomass produced within and between individual plants. Classical industrial forestry methods of mensuration are not applicable to MPTS used for biomass production since they have small heights and diameters, multiple stems and branches, and a significant component in their leaves, particularly if they are to be used for liquid and gaseous fuel or chemical and dietary feedstock. There is an urgent need to develop sampling and statistical prediction methods to partition biomass among stems, branches, leaves and roots while extrapolating from small plots or individual trees (often open-grown) to large areas.

Evaluation in agroforestry systems

The initial screening and evaluation of tree and shrub species, commonly called elimination, is concerned principally with the ability of the species to survive the natural and managerial conditions of the planting site in the first few years. This stage is often carried out using standard forestry approaches with replicated designs and multi-tree plots, good protection, weeding and other site improvements.

However, MPTS must also be evaluated in the longer term to estimate their performance in agroforestry systems, both on research stations and on farms. There is therefore a need to develop designs to test for early intercropping and resource sharing, to examine the effects of trees and shrubs on the soil sustainability, and to study the response to managerial treatments such as coppicing, lopping and pruning. Such designs are currently being developed by ICRAF, particularly for its Agroforestry Research Network for Africa (AFRENA) programme. When the effectiveness of these designs is determined, and the results of systematic research become available, individual workers and institutions will be able to evaluate and capitalize on the great potential of MPTS to enhance man's survival and welfare.

Without rational research and development of acceptable species and systems the current enthusiasm for MPTS runs the risk of alienating farmers and development agencies by failing to provide socio-economically acceptable packages.

REFERENCES

Booth, T. H. Recent developments in homoclime analysis to assist species selection. In T. Darnhofer and W. Reifsnyder (eds.), *Proceedings of the International Workshop on the Application of Meteorology to Agroforestry Systems Planning And Management.* Nairobi: ICRAF (in press).
Burley, J. 1986. Problems of tree seed certification in developing countries. Invited Paper, 18th IUFRO Congress, Ljubljana, Yugoslavia. In: Proc. Div. 2, WP S2.03-14, 112-123.
————. 1987. Strategies for genetic improvement of agroforestry trees. Lead Paper, Session 8, IUFRO/ISTS Workshop "Agroforestry for Rural Needs", New Delhi, India, February 1987.
Burley, J. and P.G. von Carlowitz (eds.). 1984. *Multipurpose tree germplasm.* Nairobi: ICRAF.
Burley, J., P. A. Huxley and F. Owino. 1984. Design, management and assessment of species, provenance and breeding trials of multipurpose trees. In R.D. Barnes and G.L. Gibson (eds.), *Provenance and genetic improvement strategies in tropical forest trees.* Commonwealth Forestry Institute, Oxford and Zimbabwe Forestry Commission, Harare.
Felker, P. (ed.). 1986. *Tree plantings in semi-arid regions.* Amsterdam: Elsevier.
Hughes, C. E. 1986. Protocol for the international provenance trials of *Gliricidia.* Oxford: Oxford Forestry Institute.
————. 1987. Biological considerations in designing a seed collection strategy for *Gliricidia sepium* (Jacq.) Walp. (Leguminosae). *Commonwealth Forestry Review* (in press).
Hughes, C.E. and B.T. Styles. 1984. Exploration and seed collection of multipurpose dry zone trees in Central America. *International Tree Crops Journal* 3: 1–31.
Huxley, P.A., E. Akunda, T. Darnhofer and D. Gatama. Tree/crop interface investigations: some comments and preliminary investigations. In T. Darnhofer and W. Reifsnyder (eds.), *Proceedings of the International Workshop on the Application of Meteorology to Agroforestry Systems Planning and Management.* Nairobi: ICRAF (in press).
Jones, N. and J. Burley. 1973. Seed certification, provenance, nomenclature and genetic history in forestry. *Silvae Genetica* 22(3): 53–58.
Oduol, P.A., P. Felker, C.R. McKinley and C.E. Meier. 1986. Variation among selected *Prosopis* families for pod sugar and pod protein contents. In P. Felker (ed.), *Tree plantings in semi-arid regions.* Amsterdam: Elsevier.
Salazar, R. 1986. Genetic variation in seeds and seedlings of ten provenances of *Gliricidia sepium* (Jacq.) Steud.* In P. Felker (ed.), *Tree plantings in semi-arid regions.* Amsterdam: Elsevier.
von Carlowitz, P. G. 1984. Rapid appraisal methodology for selecting priority multipurpose tree species and criteria for determining status and research needs. In J. Burley and P. von Carlowitz (eds.), *Multipurpose tree germplasm.* Nairobi: ICRAF.

* *Editors'note:* The 1986 ICRAF "Master List" of MPTS, verified with the Royal Botanical Gardens, Kew, U.K., cites *Gliricidia sepium* (Jacq.) Walp. as the correct (current) taxonomic nomenclature for the species.

APPENDIX 1

Major steps and technology components in tree and forestry research to develop appropriate land-use systems

Background research
Social and institutional research
    Human perceptions of tree services
    Knowledge of natural environments
    Authorities and institutions
    Land use, tenure, tree ownership
    Incentives for tree planting
    Economic returns and acceptability
    Site evaluation, classification and mapping

Natural vegetation management technologies
Vegetation identification and inventory
    Ecology and climate
    Degradation rates
    Soil-loss rates
    Human/animal demands
    Regeneration methods
    Yields at different seasons and animal
        densities
    Fire incidence and control
    Products and markets

Plantation technology
Species/provenance selection
Seed technology
    Collection
    Extraction
    Storage
    Testing
    Records
Nursery technology
    Container size/type
    Soil mixture
    Season of sowing
    Fertilization
    Shading
    Irrigation
    Protection
    Records
Establishment
    Direct sowing
        Seed pretreatment
        Soil pretreatment
        Season of sowing
    Planting
        Stock type
        Ground preparation
        Planting method
        Irrigation
        Fertilization
        Spacing

Culture
    Thinning
    Pruning
    Fertilization
    Weeding
    Protection
    Irrigation
Additional agroforestry technology
    Intercropping effects
    Tree-crop interface
    Effects on soil
    Effects on micro-climate
Yield determination
    Sampling
    Major and minor yields
Tree breeding/biotechnology
    Classical breeding
        Selection
        Progeny testing
        Seed orchards
New biotechnology
    Clonal selection, testing and propagation
    Genetic engineering of trees
    Genetic manipulation of mycorrhizae
    Genetic manipulation of
        Rhizobia/Actinomycetes
Harvesting methods
    Lopping
    Pollarding
    Coppicing
    Felling
    Fodder
    Fruits/seeds
    Extractives
Conversion technologies
    Wood, fibre and forage properties
    Sawing, planing
    Reconstituted boards
    Pulp and paper
    Chemical extractives
    Stove design
    Charcoal/fluid fuels
    Ergonomics
    Work study
    Tools and machinery
    Safety
Financial and economic analysis

Basic or facilitating research
    Soil hydrology
    Soil microbiology
    Nutrient cycling
    Nitrogen fixation
    Tree physiology and biochemistry
    Taxonomy and cytogenetics
    Pathology and entomology

17

Leucaena: a multipurpose tree genus for tropical agroforestry

James L. Brewbaker

President, Nitrogen Fixing Tree Association
Professor, Department of Horticulture, University of Hawaii
3190 Maile Way, Honolulu, Hawaii 96822, USA

Contents

Introduction

The genus *Leucaena* has been popularized widely in the past decade. It has been the subject of an excellent book by Pound and Martinez Cairo (1983), of major international conferences co-sponsored by the Nitrogen Fixing Tree Association (NFTA) and the International Development Research Centre (IDRC) (IDRC, 1982) and by NFTA (*Leucaena Research Reports* 7(2)), and of several national and international conferences (Kaul *et al.,* 1981; PCARR, 1978; Indonesia, 1982).

Research on the genus is represented in over 3,000 publications, most of them included in the comprehensive three-volume bibliography of Oakes (1982–4), in the book by Pound and Martinez Cairo, and in reviews or bibliographies by Arellano (1979), Brewbaker and

Hutton (1979), Halos (1980), Olvera *et al.* (1985) and Schroder (1986). Two excellent guides to production were co-authored by leucaena experts (NFTA, 1985a, 1985b); others include those of Banco de Mexico (1980), FAO (1983, 1985) and Proverbs (1985). The present review will focus on the past decade, primarily since the review paper of Brewbaker and Hutton (1979), and will seek to identify the people and institutions closely linked to leucaena's expanded use, and ask "Why do people plant leucaena?". Many references will be made to LRR, the *Leucaena Research Reports,* an annual publication of NFTA.

Historical perspective

Leucaena has often been regarded as a kind of "miracle tree", an appellation that makes sincere scientists wince. However, such names as *subabul* in India, *lamtoro gung* in Indonesia and *giant ipil-ipil* in the Philippines reflect the genuine affection that farmers, housewives, ranchers and scientists themselves have come to hold for the arboreal leucaenas.

Leucaena is not new to scientists, and certainly not to growers. It was a significant "alley crop" hedge in Asia, far from its native Americas, more than a century ago. It clearly was a common food and soil-restoration tree of early American Indian civilizations. In the first millenium AD, the Maya of the Yucatan peninsula may have relied heavily on leucaena as a green manure and possibly as a food crop (Brewbaker, 1979). Zarate (1984) considers the genus a significant source of leguminous pods long prior to Columbus. A major species, *L. esculenta* (Moc. and Sesse) Benth., got its name in recognition of its importance as food for highland Mexican Indians. The Mexican state of Oaxaca is a centre of great diversity of the genus, and owes its name to *huaxin* or *Leucaena* spp. (LRR 7(2):6).

Leucaena crossed the Pacific Ocean in Spanish galleons in the early 1600s, and by mid-1800 was pantropical and used in a variety of ways — food, fodder, shade, soil restoration. Superb early research on its soil-restorative properties occurred in Indonesia prior to 1910, as evidenced by Dijkman's early review (1950) encouraging use of leucaena for soil-erosion control. Takahashi and Ripperton (1949) published extensive studies of fodder management and improvement of *koa haole* (leucaena) in Hawaii. E. Mark Hutton and his colleagues developed an intensive leucaena improvement programme in northern Australia in the 1950s. Review papers that cover quite thoroughly the information on this early research and development include those of Brewbaker and Hutton (1979), Oakes (1968), Pound and Martinez Cairo (1983) and the U.S. National Research Council (1984).

New prominence

Leucaena has moved ahead of other leguminous trees in the past decade in both public image and farmer adoption. Three factors that have propelled it into new prominence are:

1. Improved varieties of exceptional yielding ability;
2. Improved communication about leucaena;
3. Expanded demonstrations of leucaena's utility.

The new varieties were discovered through systematic surveys of leucaena germplasm collected from native habitats in the Americas and grown in Hawaii and Queensland, as described later in this paper. All leucaena are easily grown; they flower within two years, and are a delight to study and breed.

The improved communication began with a superb publication in 1977 on leucaena (NRC, 1984) which followed the first international conference on leucaena in the

Figure 1   Solid blocks and contour strips of leucaena serve for hill stabilization and soil restoration in the Philippines (Photo courtesy of Napoleon Vergara).

Philippines in 1976 (PCARR, 1978). This was followed in 1980 by the founding of the *Leucaena Research Reports,* an annual publication, at the University of Hawaii, with costs underwritten by the Council of Agriculture of Taiwan, Republic of China. Publication of LRR and related literature was assumed in 1982 by the Nitrogen Fixing Tree Association (NFTA), a non-profit organization presently composed of 1,200 associates. LRR has averaged 55 articles annually from more than 34 countries.

Expanded demonstrations of leucaena's growth and versatility in management and use have been the key to farmer acceptance (Figure 1). Alley-farming methods have been skilfully designed and farm-tested in Africa (Kang *et al.,* 1984; Kang and Wilson, this volume); Indonesia (FAO, 1983) and India (Kaul *et al.,* 1981).

New challenges

The role of leucaena as an agroforestry species is currently confronted with a new type of challenge — a psyllid insect pest, which has been reviewed thoroughly in LRR (7(2)). Apathy was leucaena's primary challenge prior to the 1970s, a decade that brought high oil prices and accelerated demands on tropical forests and the world's ecological health. The new challenge to leucaena is primarily sociopolitical. "Why plant leucaena — for what purpose?", will now be replaced by "Why plant leucaena — if it is not perfect?", since most laymen expect that somehow agroforestry trees must be "miracles" that have no need for research.

The test for leucaena, as the psyllid advances pantropically, will be primarily a test for policy makers asked to fund R & D on a non-food genus of major interest to the rural poor. The psyllid problem is soluble both by genetic and biological control (LRR 7(2)). It is not a significant problem in the Americas, where the psyllids are under adequate parasitization (Proverbs, 1985). Increasing commitment of funds and talents in long-range programmes on non-food crops like leucaena is essential if the constantly evolving biological challenges, such as the diseases and insects in tropical forest, are to be met.

Botany and genetics

Systematics and species collections

The genus *Leucaena* has over 50 names ascribed to it, most of them viewed as synonyms. Ten species were widely recognized as valid a decade ago (Brewbaker *et al.*, 1972), to which two have since been added (LRR 6:78). The 12 species have been described recently in some detail from our collection of about 1,000 accessions and numerous artificial hybrids we have grown in Hawaii (LRR 7(2):7–20). It is unlikely that many species will be added to this list (Table 1), although several other populations are considered incipient species (LRR 7:110). The 12 species are easily differentiated by flower colour, leaflet size (Figure 2), growth habit, ecology, distribution and other traits (LRR 7(2):7). Chromosome counts have been made on the 12 species (Table 1). None of the species or populations tested have 2n = 26 or 2n = 28, counts that characterize most mimosoids. The importance of polyploidy is thus implicit in the high base number of the genus.

Several major expeditions to collect leucaena germplasm have been made in the past decade by our teams (1978, 1985) with support from the University of Hawaii, USDA and IBPGR, by Bob Reid for CSIRO in 1980–1982, and recently by Colin Hughes for the Oxford Forestry Institute. Hawaii's collection of 905 accessions (80 percent indigenous) can be ascribed to the 12 species of Table 1 with rare exceptions. The Australian collection includes at least 700 accessions (Bray *et al.*, 1984). Unique accessions include those ascribed to the taxon *L. cuspidata*, a highland shrub of N.E. Mexico similar to *L. pallida*. Another unique accession identified by C. Hughes (LRR 7: 110) was a large tree that otherwise resembled the shrubby *L. shannoni*.

The important commercial species, *L. leucocephala* (2n = 104) has long been recognized as a polyploid, and our repeated efforts to identify segregating monogenes (e.g., isozymes) in this species have met with failure. Earlier references to simple genetic segregations in this

Table 1   Species of the genus *Leucaena*

Leucaena species	Date	Locale*	Chromo-some no.	SI†	Tree ht (m)	Leaflet length (mm)
leucocephala (Lam.) De Wit	1783	C. America (L)	104	SF	18	10
diversifolia (Schlecht.) Benth.	1842	Vera Cruz (H)	52	SI	15	5
			104	SF	18	6
pulverulenta (Schlecht.) Benth.	1842	Vera Cruz (S)	56	SI	15	5
trichodes (Jacq.) Benth.	1842	S. America (L)	52	SI	12	30
macrophylla Benth.	1844	Guerro (M)	52	SI	10	35
retusa Benth.	1852	Texas (S)	56	Sf	5	25
esculenta (Moc. & Sesse) Benth.	1875	Mexico (H)	52	SI	15	5
lanceolata Watson	1886	Chihuahua (L)	52	SI	12	30
greggii Watson	1888	Nuevo Leon (S)	56	SI	5	10
shannoni Donn. Smith	1914	Salvador (L)	52	SI	10	20
collinsii Britton & Rose	1928	Chiapas (M)	52	SI	10	10
pallida Britton & Rose	1928	Jalisco (H)	104	SI	10	10

* L = lowland, M = midland, H = highland, S = subtropical
† SI = self-incompatible, SF= self-fertile, Sf = partially self-fertile

species are now viewed with suspicion. The species is presumed to be an amphidiploid, which Sorensson (1987) suggests most logically to be built on the 2n=52 diploids *L. leucocephala* and *L. diversifolia*. Pan (LRR 7(2):6) has shown that the 2n=104 *L. pallida* (taxon conserved over synonyms *dugesiana*, *oaxacana* and *paniculata* due to priority in descriptive literature) is an amphiploid derived from the species *L. diversifolia* and *L. esculenta*. An extensive set of traits confirms this origin, including distributional, ecological, and morphological conditions. Pan (LRR 5:88) further identified 2n=52 and 2n=104 races within *L. diversifolia*, distinguished by their distribution and morphology (Figure 2).

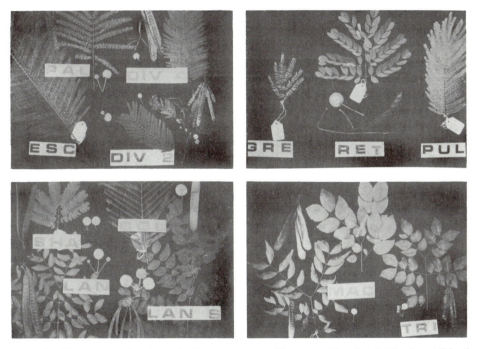

Figure 2   Morphological variation in *Leucaena* spp., with names abbreviated. Div. 2 and Div. 4 represent 2N and 4N races of *L. diversifolia*, Lan and LanS represent shrubby and arboreal variants of *L. lanceolata* (Photo courtesy of Charles Sorensson).

Breeding systems

The common leucaena is self-seeding, a fact probably underlying its early use as a food. This is both a blessing and a curse, permitting us rapid seed multiplication but also giving the species an undesirable element of weediness. Most accessions are less seedy than the common type, of which K29 and K636 are outstanding in form, but they have not been used widely due to difficulty in seed increase. The psyllid tolerance of K636 may encourage its wider use (LRR 7(2):29,84).

In contrast, nearly all other *Leucaena* species have proven self-incompatible and thus completely outcrossing, with the exception of low self-fertility in *L. retusa* (LRR 7(2):6). The tetraploid form of *L. diversifolia* is self-compatible, in contrast to the uniformly self-incompatible diploids in this species (LRR 5:88). The operation of "S" alleles of the gametophytic type was confirmed, and a theory presented for derivation of self-compatibility in the two tetraploid species (Bewbaker, 1983; LRR 7:114). Self-incompatible forms of both species might be expected following hybridization of self-compatible parents. S1 plants could be exploited in several ways, notably in production of hybrid seed and in development of broad genetically based synthetics. They might be further desirable in reducing seediness, quite probably to the advantage of wood yields.

Arboreal and shrubby varieties

Two decades ago, only *Leucaena leucocephala* was recognized to have merit in agroforestry-type land use. It was the only species distributed pantropically, represented by one variety — a seedy shrub with virtually no genetic variation. Improved strains have now replaced the common type, most of them of the "giant" or arboreal type referred to as "Salvador-type" (LRR 1:43). The first of these, K8, was released by the author in 1967 in Thailand and the Philippines, and it was described as the "Hawaiian Giant" (Brewbaker, 1975) largely because the common weedy type had been referred to as "Hawaiian" (which it is not, having come from Mexico by way of the Philippines). A later Australian release of M. Hutton, "Cunningham", resulted from crosses of the Salvador type and the more branched Peru type, and proved to be an excellent fodder variety. Batson *et al.* (1984) have collected Caribbean accessions, classifying over 90 percent as the common type.

The arboreal ("giant") and shrubby ("common") forms are now viewed as taxonomically distinct: var. *glabrata* and var. *leucocephala,* respectively (LRR 7(2):6). The "giant" or *glabrata* variety is arboreal, with glabrous branches and generally large vegetative and floral parts, and is widely distributed by man in N.E. and N.W. Mexico. The "common" or *leucocephala* variety is indigenous to a restricted area of S.E. Mexico. Hybrids are arboreal but early to flower and seedy like the commons. Similar morphological variants (Figure 2) occur in *L. lanceolata,* var. *lanceolata* (shrubby) and var. *sousae* (arboreal). Hughes (LRR 7:110) collected a variant of *L. shannoni* that makes impressive arboreal growth in Hawaii. A Panamanian tree described as *L. multicapitula* also appears to be an arboreal variant of the South American shrub, *L. trichodes.*

Genetic improvement

Genetic improvement has focused on *L. leucocephala,* a self-pollinated species whose polyploidy must be stressed as a factor impeding rates of genetic gain (Brewbaker, 1983). Improvement of biometrical traits requires large progenies and high selection indices. Most released varieties have been derived directly from accessions in Latin America. These have been marked by high uniformity of single-tree progenies. Two major breeding programmes are those of Hutton (1983) and Brewbaker (1983), the former focusing on acid tolerance and the latter on species hybridization and tolerance to cold (see later sections).

Pollination methods have been improved primarily through the work of the author's colleagues Pan and Sorensson (Brewbaker, 1983). They stress the value of early-bird activities, emasculating (as needed) at dawn and pollinating with fresh pollen taken at anthesis. Prior-day emasculation (LRR 5:29) and various detergent dips have worked less well. Dry pollen or stigmas can be moistened with glycerine to enhance results.

Many qualitative traits have been scored for their appearance in parents and hybrids of leucaena, but relatively few carried through to genetic analysis. Dominance in F1 characterizes arboreal over shrubby, early over late flowering, pink over white flowers, yellow over white flowers, pubescent branches and leaves over glabrous, flower odour over none and pendulous inflorescence over erect. A small list of gene loci has been developed for leucaena due to the work on 2n=52 *L. diversifolia* of Pan (1984), as follows:

Locus	Alleles	Description
S	Multiple	Gametophytic, personate self-incompatibility
Pub	2	Pubescent leaflets, dominant over glabrous
P x 1	2	Peroxidase, roots, co-dominant alleles
P x 2	2	Peroxidase, roots, presence *vs.* absence
P x 3	2	Peroxidase, roots, presence *vs.* absence
P x 4	2	Peroxidase, roots, co-dominant alleles

An early report of monogenic segregations for the arboreal *vs.* shrubby trait of *L. leucocephala* deserves verification. It is to be hoped that identification of genes and mapping will be accelerated as the diploids and interspecific hydrids come under increased study.

Species hybridization

One of the most exciting advances in our knowledge of leucaena has come through species hybridization in the past decade. *Leucaena leucocephala* has been crossed successfully with all other species except *L. greggii* (not tested). At least 51 species hybrids are now growing and under study (LRR 7:13), and their psyllid tolerance is shown in Table 2, following Sorensson and Brewbaker (LRR 7:13). With rare exceptions, the hybrids have been partially or highly fertile, allowing further hybridization and backcrossing for gene transfer. The entire genus is proposed to be an "effective gene pool" available for breeding improvement. Species hybrids have been made largely in our arboreta (LRR 7(2):29) and in those of Hutton and Bray (Bray *et al.*, 1984; Hutton, 1983), and many are impressive.

Table 2 Hybrids produced among species of the genus *Leucaena* as scored for resistance to *Heteropsylla cubana*

Female \ Male	COL 52	DIV 52	DIV 104	ESC 52	GRE 56?	LAN 52	LNS 52	LEU 104	MAC 56	PAL 104	PUL 56	RET 56	SHA 52	TRI 52	GCA*
L. collinsii	(1)					2	3								2.5
L. diversif. 2N	2	(5)	5			6	6	6					8		5.5
L. diversif. 4N			(6)			8	5	6		3			5		5.4
L. esculenta				(2)			3						3		3.0
L. greggii					(5)										-
L. lanceolata	3	6				(7)	5						5		4.8
L. lanc. sousae	3					5	(5)						4		4.0
L. leucocephala	3		6			7	(7)			4	7	5	6	6	5.2
L. macrophylla		3							(4)						3.0
L. pallida			3							(2)			5		4.0
L. pulverulenta	3		2			9	8	9			(8)		6		6.2
L. retusa	3		7	3				7		3	8	(4)			4.7
L. shannoni		6				6							(5)		5.3
L. trichodes						6	6							(6)	6.0
GCA	2.8	6.0	4.6	3.0	-	6.1	5.5	6.2	-	3.3	6.5	4.5	5.1	6.0	

* General combining ability = Mean of all hybrids of this parent as male or female.
Resistance scale: 1 = immune or highly resistant, 9 = highly susceptible.

The occurrence of unreduced gametes leading to polyploid progenies in leucaena has been documented by Sorensson and is a probable source of both amphiploid and autoploid populations. Large pollen grains are believed to represent such unreduced gametes, and occur with measurable frequency in many species.

Plant chemistry and physiology

Major discussions of leucaena's fodder and wood chemistry occur in other sections of this paper, notably of fodder quality, mimosine and gums. Pound and Martinez Cairo (1983) provide an excellent review of the chemistry of leucaena. Allelopathy has been attributed to leucaena leaves, based on inhibitions of seed germination with mimosine itself (LRR 3:65) and with leaf extracts (LRR 3:57). The extracts contained several common phenolics of toxic potential, e.g., ferulic and coumaric acids (LRR 3:57). Extrapolations to the field are fraught with hazards, and there has been no clear documentation of leucaena-induced allelopathy under field conditions. Leucaena suppresses nearby herbs largely through competition for light and water.

Physiological research on leucaena has been relatively limited. Photosynthetic rates increased linearly with light flux, and saturated at 40,000 lux at a rate of 25.8 mg CO_2 dm^{-2} hr^{-1} in Taiwan (LRR 2:45), and at least 40 mg in a study in India (LRR 6:42), generally similar to other C_3 plants. Leucaena has been described as a facultative shade plant that made major adjustments in stomatal density and conductance, light-saturating photo-synthetic rate, and leaf dimensions when grown under shade. Leaflet folding and stomatal responses to avoid direct sunlight and drought stress were strongly related to effective mycorrhizal associations (Huang *et al.,* 1985). Transpiration rates were calculated in the range of 0.25–0.57 ml cm^{-2} (LRR 2:45). Leaf yields as kg dry matter per hectare were linearly related to solar radiation in Hawaii by the regression $Y = -13 + 3.83X$ over the range of 8–18 MJ m^{-2} day^{-1} (LRR 6:88), roughly tripling from summer to winter in Hawaii. It is less clear how closely wood increments regress on intercepted irradiance. Early winter short-day flowering occurs in *L. esculenta* and *L. macrophylla* in Mexico and Hawaii, while other species appear day-length-neutral in flowering. *L. esculenta* has flowered irregularly when planted near the University of Hawaii's Food Science building, perhaps in response to ethylene stimulants.

Ecology and soils

Agroforestry in the tropics is often viewed as a system of band-aids or palliatives for injuries to the ecosystem. Deforested and poorly farmed tropical soils are spoken of as "degraded" and agroforestry is expected to restore them miraculously. The genus *Leucaena* is one of impressive adaptability in the tropics, but is thus being asked to grow luxuriantly on many types of generally impoverished soils. Like most plants, it grows best on the best soils, but has major problems with soils of low pH, low P, low Ca, high salinity, high aluminium saturation and waterlogging.

Acid soils

The "cerrado" soils (Oxisols and Ultisols) of South America pose a challenge to leucaena, and generally to most fodder legumes. These soils range from pH 5 down, with Ca deficiency and Al toxicity (over 50 percent saturation), and deficiencies of N, P, K, S, Mg and Zn (Hutton, LRR 3:9). Hutton (1983, 1984) has concluded that leucaena has adequate tolerance of aluminium in the soil solution, but is more sensitive to deficiencies of calcium. While other major elements are translocated both to roots and shoots, Ca is mono-directional and translocated only from roots and shoots. The Ca content is very low in the Oxisols of Brazil, and the inefficient root absorption of Ca by leucaena is considered the major deterrent to growth on these soils (LRR 3:9). Even with 1 t ha^{-1} of dolomite or 0.8 t ha^{-1} of gypsum, leucaena died within four years, evidently due to Al inhibition of Ca uptake (LRR 7:28). Gypsum was the preferred source of Ca and S to supplement rock phosphate on these acid soils (Hutton, 1984). Liming of surface soils was adequate, however, to permit good leucaena growth in low Al acid subsoils in Hawaii and in pot trials (Olvera and Blue, 1985).

Acid-tolerant leucaena include *L. diversifolia* and *L. shannoni*, marked by superior root absorption of Ca. This tolerance of acidity was transferred to their hybrids with *L. leucocephala* (Hutton, 1984). Tolerance was tested to acid, Al-toxic soils using root growth in culture by Oakes and Foy (1984), and a wide range of varietal responses noted, which did not correlate directly to Al liberation. Great reduction in growth occurred with concentrations over 4 ppm, and were little influenced by Ca concentrations (Chee and Devendra, 1983).

Acid Oxisols restrict leucaena growth severely even when Al is not a major problem (LRR 2:69; Chee and Devendra, 1983). Relative yield of leucaena increased linearly with pH on two acid soils ameliorated with lime in a continuous-function design, from about 20 percent yield at pH 5 to 100 percent at pH 7 (Munns and Fox, 1977). Manganese solubility, which is known to induce Fe deficiency, was high in the tested Oxisol, but could not be invoked as a cause of lime response. Yields of 16-month-old leucaena K 29 correlated with pH in trials near Taipei, Taiwan (Hu and Chen, LRR 2:48), ranging linearly from 3.2 t ha^{-1} (dry weight) at pH 4.7 to 15.5 t ha^{-1} at pH 8. In contrast, no effect on one-year-old trees' growth was shown for liming treatments up to 9 t ha^{-1}, that changed pH values from 4.6 to 8.4 in Indonesia, on a soil with adequate levels of Ca and P. It has been tempting to accuse acidity of causing poor leucaena growth in soils later shown to be acutely P deficient (LRR 7:117).

Saline soils

Leucaena can often be found growing on coral outcroppings very near the ocean, but never growing well on the sand dunes or on inland sodic soils, an apparent reflection of high Ca demands and low Na tolerance. Trees stagnated on saline sodic soils of pH 9.5 in India that showed salt concretions (LRR 7:66). Growth in pots was severely depressed by NaCl at 6 g l^{-1} (LRR 5:77). Irrigation with saline water (2 mmhos) in Rajasthan, India, produced only fair growth (2.5 m in one year, LRR 6:54), and 21 varieties showed small differences in growth — none convincingly related to Na tolerance.

Cold tolerance

The genus *Leucaena* includes species that vary widely in cold tolerance, as evidenced by their regions of origin (Table 1). *L. leucocephala* does not vary widely in cold-tolerance, although selected varieties have survived well, regrowing from the crown after frosts at Gainesville, Florida, USA (LRR 5:84). Outstanding tolerance of frost has been found among hybrids of *L. leucocephala* and 4n *L. diversifolia* in Louisiana, USA (Brewbaker, unpublished). *L. retusa* survives frost routinely in south Texas, when *L. leucocephala* and

L. pulverulenta are killed to the crown (LRR 5:76). Under severe frost, *L. leucocephala* was killed while *L. pulverulenta* regrew from coppice and *L. retusa* survived well (LRR 7:119).

Tolerance of low temperatures appears to differ genetically in leucaena from frost tolerance, with species like *L. leucocephala* showing poor growth under low mean temperatures. Under Hawaii's relatively low variation in temperature, *L. diversifolia* completely outgrows *L. leucocephala* at temperatures below 22°C mean annual, or where soils are acidic (Brewbaker, 1986a), and the hybrids perform as well as the *L. diversifolia* parent (Figure 3). Near Brisbane, Australia (mean temperature 19°C), R. Bray also

Figure 3   *L. diversifolia* 4N in Xalapa VC, Mexico, a fast-growing cold-tolerant taxon

observed superior cold tolerance of the diversifolias, although many accessions were low in fodder yield, while leucocephalas were nipped by frost but grew back vigorously (LRR 3:1; 5:3).

Drought and other factors

Tolerance of trees to drought and their yield response to water are subjects of much speculation but few data. Leucaena probably maximizes its yields when under no moisture stress.

Physiological studies reveal a high ability to tolerate drought, largely through avoidance responses of leaflet folding and leaf drop. Irrigation in the dry season at the Bharatiya Agro Industries Foundation (BAIF) in Maharashtra State, India, led to continuous function responses to amount of water and tree density. At this institute, Relwani and associates (LRR 4:38) calculated water use to be $320 \, l \, kg^{-1}$ of wood for trees at a 1 x 1 m spacing. They consider it a superior legume tree for reforesting the Deccan Plateau.

Leucaena tolerates fast fires and can regrow from the crown after burning to the crown by slower fires. Survival was superior to that of N-fixing trees such as *Prosopis* and *Casuarina* spp. during fires following the El Nino related droughts of 1983–1985. As agriculture increasingly overwhelms tropical forests, fire tolerance becomes a *sine qua non* for survival.

Symbionts, diseases and insects

Rhizobia

Rhizobia of leucaena have the reputation of being rather uncommon, rather specific, and predominantly fast-growing and acid-producing, the latter traits being linked inconclusively to leucaena's intolerance of acid soils. None of these features could now be considered inviolable (Bushby, 1982). Few studies seem to have contended fully with the fact that *Leucaena leucocephala* is an amphiploid (2n=104), doubtless based on amphiploid parents, with great likelihood of highly reiterated gene sequences. Intragenotypic as well as intraspecific genetic variability is to be expected for traits such as rhizobial strain specificity.

Leucaena seedlings nodulate in most tropical soils, although temperate soils and relatively sterile tropical soils often lack appropriate bacteria. Several laboratories provide inocula, and preference has been shown for strains such as NGR 8 (New Guinea), CB 81 (Australia) and TAL 1145 (Hawaii). Recent reviews of rhizobial research on leucaena include those of Halliday and Somesagaran (1983) and Schroder (1986).

The rhizobia of leucaena are certainly no longer to be considered exceptionally species specific (Trinick, 1980). Those derived from *Leucaena leucocephala* evidently nodulate all other *Leucaena* species, with some exceptional reactions involving *Leucaena retusa* (Thoma, 1983). They also nodulate a number of legumes, including *Acacia albida, A. nilotica, A. senegal, A. raddiana, Calliandra calothyrsus, Desmanthus virgatus, Gliricidia sepium, Macroptilium* and *Vigna* sp. (Halliday and Somesagaran, 1983; Pound and Martinez Cairo, 1983; Dommergues 1982; LRR 2:43). In turn, rhizobia from the following species have been able to nodulate leucaena: *Acacia farnesiana, A. mearnsii, A. senegal, Dalea* sp., *Mimosa invisa, M. pudica, M. scabrella, Neptunia* sp., *Piptadenia* sp., *Prosopis juliflora, Sabinea* sp. and *Sesbania grandiflora.*

The rhizobial reactions of *L. retusa* are most unusual (Halliday and Somesagaran 1983; Thoma, 1983). This yellow-flowered, frost-tolerant acacia-like shrub of Texas nodulates well in Texas. It did not nodulate in Hawaii until recently, and nodulated ineffectively with a strain derived from *L. pulverulenta* (Thoma, 1983), suggesting a more specific rhizobium than for other leucaenas.

Leucaena rhizobia are now distinguished from the "cowpea miscellany" group as *Rhizobium loti,* and are recognized to be quite diverse in type. Most of them have fast growth (3–5 hr generation time) and acid reactions in culture similar to temperate legume rhizobia. They are usually slightly acid or neutral in culture, but range widely in both acid formation and rate of growth. The thesis that alkali-producing, slow-growing rhizobia are necessary for acid-soil tolerance has been challenged by Halliday and Somesagaran (1983). They found only fast-growing, acid-forming rhizobia in the acid soils, and none of the isolates from alkaline soils survived in acid media (LRR 2:71). Despite inoculation with acid-tolerant strains (e.g., TAL 1145), leucaenas did not thrive in acid soils (Chao *et al.,* 1985). Nodule numbers dropped log-linearly with increasing free aluminium in acid soils (LRR 4:54). Genetic variation of leucaena's rhizobia has been reported by many authors, with wide ranges in growth rates and characteristics. Many strains have shown N-fixation as good or better than common inoculants NGR 8 or CB 81 (LRR 2:71; 6:14).

Nodule number and seedling weight are highly correlated in leucaena (LRR 2:19; 4:18, 57). Nodules appear on seedlings within five weeks if inorganic N is withheld (LRR 2:25). Transplant shock in leucaena results in loss of nodules and re-inoculation (LRR 3:91; 6:95). Strains used in inocula are often recovered with difficulty from soil after a year's growth. Rhizobial persistence is considered fairly high in soil or in compost and charcoal mixtures at room temperature (LRR 5:68; 6:73).

Mycorrhiza

Mycorrhizal inoculation may be more important than rhizobial inoculation when leucaena is transplanted into sterile soils (LRR 2:84; 6:97). Leucaena roots have poorly developed root hairs, and appear to rely heavily on mycorrhiza for nutrient uptake, notably of phosphate. Several species of *Glomus* and *Gigaspora* can infect leucaenas (LRR 7:61), and strain differences are significant but variable (LRR 4:83; 7:61, 94). Impressive responses occur for growth of shoot and root and of leaf area on sterilized soils inoculated with mycorrhiza (LRR 4:86, 6:97). Huang *et al.* (1985; LRR 4:83, 86; 5:79) have conducted extensive studies of leucaena mycorrhiza, finding *Glomus fasciculatus* best among their isolates. They used single leaflets for phosphate analysis (LRR 5:79). Mycorrhizal plants not only had improved uptake of minerals — notably P, K and Ca — but better stomatal conductance and responses of leaflet folding and orientation (Huang *et al.,* 1985).

Diseases

Although leucaenas have a reputation for high disease resistance, they are not without problems (Pound and Martinez Cairo, 1983). Three diseases have occurred with sufficient severity to attract major research interest — leafspot, gummosis, and seedling rots. Leafspot due to *Camptomeris leucaenae* (Stev. and Dalbey) Syd. has been known since 1919 and is periodically serious in Latin America. It was studied intensively by Lenne (LRR 1:8; 2:18) in Colombia, who found six *Leucaena* species to be completely resistant but not *L. leucocephala* (Pound and Martinez Cairo, 1983). *Colletotrichum* spp. also occur as secondary leaf pathogens (LRR 3:58; 7:48).

Gummosis from stems has been reported from the Indian subcontinent, and attributed to *Fusarium* spp., notably *F. semitectum* (Singh, 1981; Singh *et al.,* 1983) and *F. acuminatum* Ell. & Ev. The incidence was higher in common and Peru types, but rare in the Salvador types (LRR 3:25,33; 7:48). The fungus may be a secondary pathogen, but is probably systemic and associated with disease symptoms only under stress. Fungicide application controlled the disease (LRR 6:38). Gum exudation of the acacia-type has also been observed without evident pathogenesis, e.g., on the seedless hybrids of *L. esculenta* x *L. leucocephala* in Hawaii. These gums may have great potential value, as discussed later in this paper.

Seedling and plant damping-off occurs on leucaenas in nurseries and under poorly drained conditions, involving several common *Pythium, Rhizoctonia,* and *Fusarium* spp. (LRR 1:28; 3:58; Pound and Martinez Cairo, 1983). Fusariums were also associated with leucaena stem and root rots in Taiwan (LRR 5:64) and Colombia (LRR 3:14), while *Ganoderma lucidum* caused root rots on moist sites in India (LRR 4:35; 7:65). *Phytophthora drechsleri* infections stimulated by storm damage led to cankers and some tree death in Hawaii (LRR 1:56). A collar rot of young trees has been attributed to *Sclerotium rolfsii* in Florida.

Many different fungi and bacteria (notably *Pseudomonas fluorescens*) can occur on leucaena pods and seeds, notably following insect attack (LRR 1:8; 3:14; 4:70). None are known to be seed transmissible.

Psyllids and other insects

Insect damage to leucaena is uncommon except to pods and seeds (Pound and Martinez Cairo, 1983). Two insects have caused severe damage to leucaena trees over the past decade — ants and psyllids (jumping plant lice).

Termites or harvester ants make it virtually impossible to establish leucaenas without pesticide use in some areas such as East Africa. Ground-baiting with insecticide-treated grass is one novel solution. In contrast, leucaenas are highly resistant to root-knot nematodes (LRR 4:92; Vicente *et al.,* 1986), but may house species as yet unidentified (LRR 2:17).

Since 1980, the leucaena psyllid, *Heteropsylla cubana* Crawford, found its way out of its native Latin America, via Florida to Hawaii and thence into the Pacific and S.E. Asia (Figure 4). It has become the most severe pest of leucaena on record, despite the fact that it has been virtually ignored in Latin America due to heavy predation and parasitization (Proverbs, 1985). The psyllid was first described in 1914 and then observed attacking leucaena in the Dominican Republic in 1980 (Martinez Cairo, personal communication), in Florida in 1983 (LRR 5:86), in Hawaii in April 1984 (LRR 5:91), in the South Pacific in 1985 and in S.E. Asia and Australia in 1986 (LRR 7:6).

A section of Volume 7 of LRR was dedicated to invitational papers reviewing the status of the psyllid. W.C. Mitchell and D.F. Waterhouse (LRR 7:6) traced the spread of the psyllid. An international symposium was conducted by NFTA on "Biological and genetic control strategies for the leucaena psyllid" in Hawaii in late 1986, and a special issue of LRR was dedicated to the proceedings of this workshop (LRR 7(2): 109). Major review papers included those by J. Beardsley on the psyllids (LRR 7:2, 7(2):1), by L. Nakahara *et al.* on the predators (LRR 7(2):39; 7:9), and by Sorennson and Brewbaker (LRR 7:13; 7(2):29), Bray (LRR 7(2):32) and Pan (LRR 7(2):35) on resistance. High psyllid resistance characterized several leucaena species and their hybrids (Table 2).

The leucaena psyllid has caused a severe setback to fodder production in S.E. Asia, with

Figure 4   Psyllid-tolerant K636 (to right) and susceptible K8 (to left) under severe epibiotics in Hawaii

less obvious but significant effects on wood production. Native predators and parasites appeared to be rare or slow to find the insect. Only partial control was effected in Hawaii by two predatory beetles, both introduced early this century to control other pests. A Caribbean parasitoid, *Psyllaephagus,* however, shows specificity and excellent appetite for *H. cubana,* and may prove to be the major biological control agent of that region (LRR 7:9). It was released in Hawaii in June 1987. Genetic control is expected to provide the only viable option for some areas of the world, if predators cannot be introduced. Rapid seed increase and deployment of psyllid-resistant germplasm has been advised (LRR 7(2):88).

Seed insects are common pests of leucaena, including the seed borer *Araecerus fasciculatus* (LRR 4:70), Cathartus grain beetles and bruchids (Pound and Martinez Cairo, 1983). Other pests appear to be largely under biological control where earlier reported, such as the black twig borer, mealy bugs of pods, and the pantropical moth *Ithome lassula* whose larvae feed on the florets (LRR 2:11). Three scale insects caused damage on Taiwanese leucaenas, of which *Hemiberlesia implicata* was the most serious, involving as many as 20 percent of trees sampled (LRR 3:55). Leaf feeding by the beetle, *Apogonia rouca,* was reported in India (LRR 7:67). In Hawaii the rose beetle, *Adoretus sinicus,* may feed at night on the large-leaflet *Leucaena* spp. (Figure 2). Several types of inchworms have also been observed on leucaenas around the world, but none appear to be serious. The psyllid problem, however, should put scientists on alert for insect pests and diseases that may move pantropically and present new challenges in the future for leucaena production and improvement.

Establishment and management

Establishment and fertilization

Establishment methods for leucaena have been refined and are often shown to be site-specific. Direct seeding has been recommended where soil moisture conditions permit and economic weed control can be maintained (LRR 4:78). The mature seeds have a water-impermeable coat, carry 5–7 percent moisture, and store well at room temperature (LRR 2:59; 3:87; 4:67). Seed germination continues to be the subject of articles on methodology despite the thorough coverage in early literature (NFTA, 1985a,b). Co-planting with a crop like maize can produce excellent results (LRR 1: 19), and seeds have been successfully mixed in the planter box of machine drills (LRR 7:26). We use alachlor as a preplant herbicide for such mixed plantings. Hormonal treatments of seed or seedlings can accelerate early growth slightly (LRR 1:53; 3:83). On the farm, leucaena is often much easier to plant than to protect due to its attractiveness to grazing animals; thorn hedges and wire enclosures can be used.

The use of bare-root ("stump", "bare-stem") cuttings from seedling beds has worked variously well in Taiwan (LRR 3:45), Honduras (LRR 6:20), Thailand (LRR 4:78) and Java, Indonesia (LRR 3:45). An important trick used in Taiwan is to roll the cuttings (10 cm below ground, 25 cm above) in mud, drying them before shipping to planting site. Cuttings can be stored two weeks under wet burlap without loss (LRR 3:45).

Fertilization practices for leucaena are site-specific, but routinely involve recommendations of phosphate and lime applications to accelerate early growth. Significant responses to both Ca and P have been reported on a wide array of tropical soils (Brewbaker and Hutton 1979, Pound and Martinez Cairo, 1983). Responses to Ca and pH have been detailed in an earlier section. Responses to P are less predictable, and the effectiveness of mycorrhizal associations must be involved. No responses to K have been reported. Sulphur response can be great, and gypsum and rock phosphate are recommended for S-deficient acid soils also low in Ca or P. In low-fertility soils generally, Hutton has recommended application of 100–200 kg ha^{-1} superphosphate and 100–200 kg ha^{-1} dolomitic lime plus supplemental Mo, Zn, Cu and Bo. Mineral deficiency symptoms occurred at about 50 percent of normal leaf concentration for K, Ca, Mg and S in leucaena, and at about 80 percent of normal for N and P (LRR 1:6).

Production and tissue culture

Vegetative propagation of leucaena has been successful in relatively few locations, apparently reflecting critical environmental requirements or possibly systemic fungi. Hu and colleagues have mastered most field management problems with leucaenas in Taiwan (Hu, 1978; Hu and Kiang, 1983). Hu has directed replicated trials on every aspect of leucaena management for pulpwood, providing careful cost analysis of each step (Hu and Kiang 1983). Hu obtained excellent rooting of cuttings under mist spray in Taipei greenhouses, using 15 cm leafy twigs having terminal shoots intact from one-year-old trees (LRR 2:50). Leafless twigs from older trees failed to root; this has been the experience in Hawaii also.

Tissue cultures of leucaena have gone through the traditional phases of development, but cannot be considered field applicable. Callus growth was achieved from germinating seeds (LRR 1:54), and hormonal treatments were perfected leading to shoot differentiation and somatic embryoids (LRR 4:88). Seed-derived explants were carried without callus formation through BA (benzyladenine) and NAA (naphthalene acetic acid) media to vigorous shoots (LRR 3:81), and with added ascorbic acid to healthy plantlets (LRR 4:37). Explants from mature tissues have not been carried through tissue culture. Nodal branch cuttings will produce shoots on BA media, and rooting was obtained by subsequent transfer to media with NAA (LRR 5:22) or with auxin treatments of shoot explants prior to rooting in charcoal-supplemented media (LRR 5:37). Further refinements, including embryoid and protoplast culture and adaptation to large-scale propagation, are in progress LRR 6:32).

Grafting is a relatively simple and historic technique for leucaenas (Dijkman, 1950), but has been refined in the lab of Versace (LRR 3:3). The best results involved cleft and whip and tongue grafts, rather than bud or T-shield grafts, using four-month seedlings as stock and taping with grafting tape. Several types of interspecific grafts were successful.

Nursery methods

The most effective nursery methods employ long narrow plastic tubes with 100–500 g of medium. A potentially serious tapered open base on dibble tubes leads to aerial root-pruning and excellent taproot growth when transplanted (LRR 1:57). Dibble tubes and "root-trainers" are now extruded in several countries with ribbing to prevent coiling. A fine layer of crushed stone or sand on top of the soil or soilless mix (1 part vermiculite to 1 part peatmoss is common) prevents damping-off and weed growth. Growth is superior in larger tubes, but nursery and transplanting costs increase accordingly. Yields increased with increasing pot size (up to 40 kg per pot) in both sand and a black soil, with heights of four-month-old seedlings ranging from 20 cm in 500 g pots to 2 m in 40 kg pots (LRR 6:12). Foam rubber 3 cm cubes also worked well for leucaena seedlings (LRR 7:88).

Nursery fertilization regimes have been refined to optimize growth. Urea is minimized to levels that maintain good growth but do not fully suppress nodulation (0.5 g per seedling, LRR 7:38), and phosphate levels are critical (1 g superphosphate per seedling). Growth of 6–8 weeks under cover followed by 6–8 weeks without cover is commonplace. Full sunlight creates seedlings that are stunted but tough.

Weed control

Weed control is the major expense in tropical tree establishment, even with fast-growing leucaena at high density on good soil. Simultaneous planting for fodder with grasses such as *Brachiaria,* guinea grass (*Panicum maximum*) or *Paspalum* spp. can give effective weed control (Pound and Martinez Cairo, 1983). Establishment of solid-stand forests requires herbicides or weeding by hand or machine. Herbicide recommendations are imprecise for leucaena, in part due to differing experiences with levels of toxicity and effectiveness of weed suppression. Land preparation prior to planting is critical in reducing weed and weed-seed populations. High-input methods prior to transplanting in Hawaii involve disking, application of glyphosate (Roundup) to regrowth and use of pre-emergence herbicides.

Pre-emergence herbicides that are recommended in the literature include alachlor, bentazone, dalapon, metabenzthiazuron, monuron, nitrofen, phenoxalin, simazine and trifluralin, to which leucaena is fairly or highly resistant. Especially effective treatments have included oryzalin (Surflan) at 2.8 kg ha^{-1} (LRR 5:105), simazine at 5 kg ha^{-1}, phenoxalin at 3.5 kg ha^{-1} (Pound and Martinez Cairo, 1983), nitrofen (Tok) at 4.5 kg ha^{-1} (LRR 1:50), and trifluralin (Treflan) at 1.5 kg ha^{-1} (LRR 1:50). Post-emergence grass control with fluazifop (Fusilade) is effective at 2 kg ha^{-1} over the leucaena (LRR 6:1) and bentazone (Basagran) at 2 kg ha^{-1}. Fluazifop is currently the favoured post-plant herbicide for many types of trees. Simazine, dalapon, diuron and oxyfluorofen have also been used post-emergence for grass control (LRR 5:105).

Fodder and wood management

Fodder yields of leucaena from different varieties and different management systems are discussed in a later section. Low hedge management with 2–4 month cutting intervals is now about standard for leucaena, moisture permitting (Figure 5). Nutrient replacement regimes have not been well defined for solid fodder plantings ("protein banks") of leucaena, but must be devised to restore nutrients to the soil that are lost in harvest (Brewbaker and Hutton, 1979). Notable are the losses in fodder harvest of soil P, K, Ca, Mg and selected micronutrients. Alley farming with animal presence in the field ensures partial return of such nutrients, as do any alley-cropping systems involving complete return of leucaena foliage as green manure (although this rarely occurs).

Wood is harvested from leucaena in almost all ways known, on cycles ranging from 6 months to 10 years, and including use of roots and small branches, so that generalizations are difficult. Wood yields maximize at high population densities of 10,000 ha^{-1} on 2–4 year cycles that also minimize weed competition. Leucaenas coppice readily with rare loss of a tree, and are cut for fodder or debranched in many ways. Coppicing studies are poorly recorded in the literature. Pecson (personal communication) observed no significant differences among methods of coppicing on different varieties and densities. Leucaenas had a strong tendency to return to a fixed stand of co-dominant stems at any one site irrespective of management and varietal variables. Under Hawaii conditions of 1,500 mm rain per annum, the arboreal varieties stabilized around 15,000 co-dominant stems ha^{-1},

Figure 5 Leucaena hedge managed for fodder in Indonesia with Australian consultant Ross Gutteridge.

and numbers of coppice shoots were directly correlated with stem diameter. One-year trees coppiced at two sites in Thailand averaged about 5,000 stems ha^{-1} at one site and 19,000 ha^{-1} at a second (LRR 5:70). Branch wood is considered valuable as fuelwood in some systems, and branch numbers varied from 30–80 per tree at 11 densities in India (LRR 6:23).

Use as animal feed

Almost half the published literature on leucaena deals with its use as animal fodder. The increasing role of meat and animal products in tropical diets is clearly a cause, as is the increasing market for leucaena leaf meal (LLM) in temperate countries. Tropical poultry feeds have long used leucaena as a yolk-colouring device, and attempts are being made to improve the digestibility and energy value of LLM as feed for all non-ruminant animals. A primary interest in ruminant use has been in the toxic by-product of mimosine digestion, dihydroxypyridine (DHP).

Fodder yields

Leucaena fodder yields vary greatly for different ecosystems but less so for different management systems, and good summaries exist of the extensive early literature (Oakes 1968; Brewbaker and Hutton, 1979; Pound and Martinez Cairo, 1983). Fresh herbage yields exceed those of other shrubby tropical legumes (LRR 4:77; Brewbaker, 1985a, 1986b). Yields are comparable to the best herbaceous legumes, ranging from 40 to 80 t ha^{-1} when moisture is not limiting (LRR 2:19; 3:39; 6:40; 7:19; Brewbaker *et al.*, 1972;

Brewbaker 1976; Hedge, personal communication; Hogberg and Kvarnstrom, 1982) and from 20 to 50 t ha^{-1} in seasonally dry tropics (LRR 3:31; 4:25, 31, 69, 77, 79; 5:3) and in frost-affected subtropics (Othman *et al.,* 1985). Much lower herbage yields usually reflect serious constraints of soil fertility (Chee and Davendra, 1983).

Variety, harvest intervals, cutting heights and planting densities are major variables for the hedge management considered here. Favoured varieties are the Salvador or Peru types as they outyield common varieties by 20–100 percent (Brewbaker *et al.,* 1972; Brewbaker, 1976; LRR 2:19; 3:39; 4:3). When cut to very low stubble heights, yields are reduced for all varieties but the common types perform relatively well (Guevarra *et al.,* 1978). *L. diversifolia* produced competitive fodder yields in cooler climates with less than 20°C mean temperature (LRR 3:1), and its hybrids with *L. leucocephala* and those of *L. pulverulenta* x *L. leucocephala* (LRR 4:1) were excellent (Bray, personal communication).

Harvest intervals produce significant variations in fodder yield and quality. Some varieties branch and flower rapidly and must be harvested earlier than the favoured Salvador and Peru types. Percentage leaf in herbage dropped as cutting intervals increased (from 67 percent at 30 days to 42 percent at 150 days) in Mauritius studies by Osman (LRR 2:33,35; 3:49; 7:91). Edibility by cattle, as measured by weighing uneaten fractions, was reduced in proportion to leafiness from 100 percent at 30 days to 85 percent at 120 days (LRR 7:91). Yields maximize at 70–90 day harvest intervals (LRR 3:31; 4:25; 7:91) depending on temperature. Yields on a per-day basis were highly correlated with temperature in Australia (LRR 5:3; Bray *et al.,* 1984) and Hawaii (Brewbaker *et al.,* 1972; Hedge, personal communication).

Cutting or stubble heights have been evaluated at BAIF in India, and lead to a consistent recommendation for maintaining hedges above 60 cm (LRR 4:25,41). Cutting heights above 60 cm allow retention of some green foliage and vigorous lateral meristems. The common shrubby leucaenas can be cut to 10 cm without appreciable yield loss (Guevarra *et al.,* 1978), but the more arboreal varieties should not be cut below 25 cm (LRR 4:3, 41; 5:3; 7:91). Widely spaced high hedges can provide high-quality fodder (LRR 4:69) at the expense of yield.

High plant densities are recommended for solid fodder plantings, 50–150 x 10^3 ha^{-1}, if moisture is not limiting. Yields maximize when the trees are coppiced to produce a highly branched shrub that can grow quickly to intercept all light on the field after harvest. One metre between rows and 0.1 m between plants is a standard for comparison (LRR 3:40).

Fodder quality

Leucaenas are among the highest quality fodder trees in the tropics (Brewbaker, 1985a). Pound and Martinez Cairo (1983), Jones (1979) and Brewbaker and Hutton (1979) summarized many publications on herbage, leaf meal and seed meal. Herbage taken at peak quality has the following percentage values of dry matter: digestibility 55–70, crude protein 20–25, ash 60–10, N-free extract 30–50 (fibre 25–35, NDF 20, ADF 15, cellulose 10, lignin 5), fat 6, mimosine 4.5, tannins 1.5–2.5, Ca 0.8–1.8, P 0.23–0.27, and silica < 1.

All leucaenas are known to be highly attractive to animals as browse and feed. Introduction of cattle to the Americas probably decimated leucaena populations, although deer damage can be great (buffaloes were north of the leucaenas). Removal of feral cattle pressure can lead to explosive populations of the common seedy types. Extensive analyses have been made of fodder components of *L. diversifolia, L. lanceolata, L. pulverulenta, L. shannoni* and *L. trichodes* at the Indian Grasslands and Fodder Research Institute (LRR 7:43) and the Tamil Nadu Agricultural University, India (LRR 3:21), showing only minor variations in fodder quality and mineral contents (LRR 2:53; 7:43), often related directly to tissue maturity (LRR 4:24). Varietal differences in *Leucaena leucocephala* were also small (LRR 5:14), and appeared to be related more to stage of growth than to genotype. High digestibility and comparability of leucaena's amino-acid composition to that of alfalfa are stressed in such studies (LRR 2:53). Proximate analysis of seeds and pods revealed values comparable to foliage except for elevated fibre levels (LRR 3:21). Feed mills for the processing of leucaena into pellets or meal are becoming common, and the pellets have an international market (Manidool, 1983; PCARR 1978; FAO, 1985).

Dihydroxypyridine (DHP) and mimosine

Toxins are a *cause célèbre* for this generation, and leucaena's mild toxin mimosine still frightens many animal growers into ignoring this "alfalfa of the tropics" (Holmes, 1981). Mimosine is an amino acid found in high concentrations in leucaena tissues (2–6 percent of dry weight) whose toxicity has been covered thoroughly in the early literature. A series of definitive experiments in ruminant nutrition by Raymond J. Jones and his colleagues has clarified most doubts about mimosine and its DHP breakdown products. These studies permit accurate prediction of ruminant nutritional problems and provide a ready cure.

DHP was known to be a product of mimosine catabolism. A key discovery by Jones and colleagues was that DHP was goitrogenic (Hegarty *et al.*, 1979), its action marked by elevated levels of serum thyroxine in urine (Jones and Bray, 1983). Secondly, it was shown that further catabolism of DHP was not completed normally in ruminants of Australia, Papua New Guinea and certain other regions in the world, its accumulation being marked by loss of appetite, goitre and related toxic symptoms. It was shown that mimosine was converted to 3,4 DHP by enzymes in the leaf (LRR 2:31) and during animal chewing (Lowry, 1983), or in the rumen. A second isomer, 2,3 DHP, could also be formed in the rumen from 3,4 DHP (LRR 5:2). Megarrity (1978) then isolated a strain of bacteria from Hawaiian goats that degraded the DHP. They transferred the bacterium successfully to unadapted ruminant animals and reversed leucaena's toxicity. International service was provided by Jones to identify those regions of the world where DHP toxicity occurred, i.e., lacked the appropriate bacteria.

Fodder use is a significant feature of many, perhaps most, agroforestry species. Knowledge of toxins and their detoxification, e.g., as in leucaena, may become of great value to their deployment. The Jones' model should serve as a classic in the field of ruminant toxicity of tropical trees, many of which contain alkaloids, tannins and flavonoids at high levels as part of their defence mechanism against herbivores and tropical insects. Ruminant bacterial degradation is expected to become an option for some of these toxins, following identification and introduction of appropriate bacteria essential to control toxicity. It is probable that genetic removal of an alkaloid like mimosine would lead to unacceptable levels of insect susceptibility, but this has not been verified. Its concentration could be halved by plant breeding (Gonzalez *et al.*, 1967). Mimosine's chemical properties and activities are subjects of extensive continuing research. Mimosine is not carcinogenic or mutagenic, but inhibits many enzymes (LRR 3:67; 5:72) involving at least three mechanisms — chelation of metals, interaction with pyridoxal-phosphate enzymes, and inhibition of DNA synthesis (LRR 6:63). It does not act as a trypsin inhibitor. It can act as a fungicide on pathogens such as *Sclerotium rolfsii* (LRR 5:73), but is better known for its depilation of horses in the tropics.

Methods have been refined to quantify mimosine and DHP. The classical ferrous sulphate colorimetric method for mimosine of Matsumoto and Sherman (as modified in

Hawaii, LRR 2:66) was automated by Megarrity (1978) and modified by Jones (LRR 1:3) to collect samples directly into HCl to minimize loss of activity. Mimosine elutes with isoleucine in amino-acid analysis, and performic acid treatments destroy the mimosine, thus permitting its estimation by this method (LRR 3:78). DHP could also be estimated through ascending paper chromatography (LRR 1:16), and ion-exchange chromatography in Na citrate buffer gave an excellent estimation of mimosine and DHP (LRR 1:39). High performance liquid chromatography for both compounds was performed at very low concentration in the laboratories of Acamovic and D'Mello in Edinburgh (LRR 2:62; 6:75) and Tangendjaja and Wills (1980) in Indonesia. The extraction of crystalline mimosine and of DHP was also improved by resin column methods (LRR 5:50). Like other free amino acids, mimosine values maximize in juvenile leaves on vigorously grown plants before the dry weights stabilize (LRR 7:34).

The reduction of mimosine levels by pretreatments of feedstuffs has long been of interest, but must now be offset by the interest created in DHP. Mimosine is converted to DHP by an enzyme that is inactivated by heat or by drying in leaves (Lowry, 1983, LLR 7:77). Animal chewing thus initiates conversion that is stopped in the stomach of monogastric animals. Simple leaching with standing or preferably running water lowers mimosine dramatically. Sun drying under hot wet conditions leads to significant loss, and can be accelerated by heat under pressure (LRR 4:62). Silage processing reduced mimosine to 50 percent (LRR 1:17). The addition of sugars and organic acids accelerated this degradation down to 10 percent (LRR 7:85). Dietary supplementation to reduce mimosine toxicity is discussed in one of the later sections. Lowry (1983) stresses that conflicting leucaena toxicity data in ruminant and non-ruminant animals clearly involve mimosine and DHP.

Ruminant animals

A most thorough review of ruminant animal use of leucaena has been provided by Pound and Martinez Cairo (1983), who stress the role of leucaena as a supplement to diet rather than as its basis. A value of 30 percent of diet is adequate to fulfil its primary role as supplemental protein, roughage and mineral source. In any location it must first be determined that DHP is catabolized by ruminant animals (i.e., no elevated thyroxine level in urine, no loss of appetite on leucaena) or appropriate action taken. Even when fed 100 percent leucaena plus salt supplement, mature animals able to detoxify DHP remain thrifty and gain weight (LRR 7:26; Pound and Martinez Cairo, 1983).

A major agroforestry system with leucaena involves discrete blocks of leucaena ("protein banks") and of grass, or leucaena plants or hedges in a grass pasture. Liveweight gains often maximize on leucaena-supplemented pastures, with several studies reporting gains of 0.6–0.8 kg per day, the range influenced by dietary energy levels (Jones and Jones, 1982). Milk production in the tropics is often based on pen-fed animals receiving diets of low palatability, thus supplemental leucaena leads to increased milk yields (LRR 2:39,40; 3:21). Gains of suckling calves given supplemental leucaena and molasses/urea were comparable to controls in Mexico, Mauritius and Dominican Republic trials, but at less expense (Pound and Martinez Cairo, 1983). Leucaena hay is widely used as a supplemental feed to goats, fattening or milk cattle, and buffaloes with good response (LRR 3:99; 5:12; 7:99). Many short reports of the use of leucaena in tropical pastures or hay rations appear in the journal *Tropical Animal Production.*

Non-ruminant animals

Leucaena leaf meal (LLM) and leucaena seed meal (LSM) are prepared by grinding leaves and seeds as poultry and small-animal feeds. Both products have led to problems when fed at high levels, problems now ascribed as much to tannins and low metabolizable energy (ME) as to mimosine. Extensive early literature (D'Mello and Taplin 1978) records the widespread use and problems of leucaena leaflets or leaf meal as a poultry-feed supplement.

Broiler gains and egg laying are reduced by levels in excess of 40 g kg^{-1} of feed (LRR 2:41, 2:47). At this low intake, xanthophyll contents are so high (750 mg kg^{-1}) that egg colouration is much better than with yellow maize (LRR 6:76), even though less than 50 percent of the xanthophyll is biologically available (Taplin *et al.* 1981; LRR 6:76, 99). Use of higher levels of LLM reduced weight gains and slowed maturity rates, with few other toxic symptoms. Causes cited for these effects include mimosine, tannins, low metabolizable energy and saponins.

Daily feeding of mimosine at a level equal to that of 200 g kg^{-1} of LLM did not delay chick development, however (LRR 7:83). While other studies have indicated that ferric sulphate reduces toxic effects of LLM, including mortality and weight loss (LRR 6:70; 2:60), this may not result from mimosine precipitation *per se.* D'Mellow and colleagues in Edinburgh have conducted a series of LLM treatments to reduce nutritional toxicity. Aluminium salts given to chicks on LLM led to full excretion of mimosine, less completely so with Fe_2 and Fe_3 salts (LRR 1:38). Ferric sulphate reduced the depression in chick growth by half (150 g kg^{-1} LLM), and added polyethleneglycol (PEG) reduced this further by half (LRR 2:60).

Tannin levels of leucaena leaves are not high (1–1.5 percent), but are presumed to be neutralized by PEG. Tannins in seeds (1.5 percent) were reduced 75 percent by 30 minutes cooking in water, and over 85 percent by boiling in NaOH (LRR 7:77). The addition of cholesterol to LLM chick diets led to normal chick growth (LRR 3:72), suggesting that saponic toxicity was involved. However, LLM did not reduce plasma levels of cholesterol in the studies of Tangedjaja and Lowry (LRR 3:57) as would be expected if saponins were a problem. Trypsin inhibitors were invoked as a possible cause of the LLM toxicity, and were eliminated by heat treatments of 30 minutes at 120°C (LRR 7:97). Heat treatments of LLM under pressure are known to improve protein digestibility and weight gains of rats, also eliminating mimosine (LRR 4:62). Heated and ferric-treated LLM had 50 percent mimosine reduction (LRR 7:97). Rats showed reduced weight gains above 15 percent LLM and above 20 percent if LLM was sun dried, but ferrous sulphate treatments had no effect (LLR 4:59). Szyszka (LRR 5:5) calculated non-toxic average daily intake values (g kg^{-1}) for mimosine, based on LLM or LSM, for the non-ruminant animals — broilers 0.16, layers 0.35 and rabbits 0.30.

Metabolizable energy (ME) of LLM is very low, about 15 percent of the expected 20 MJ kg^{-1} dry matter (LRR 2:63). The causes again are elusive, and include gums, fibre, tannin, mimosine and DHP. Fat or tallow is added normally to increase ME of LLM-supplemented diets. Use of a hemicellulase to digest the gums did not affect ME (LRR 4:82).

Leucaena seed meals have been successfully fed to ducks (LRR 7:24) and chicks at modest intake levels. Lowered mortality and improved growth on the heat-treated meal (boiled one hour) occurred with chicks, but ferrous sulphate treatments were ineffective (LRR 3:66).

Swine cannot be fed high levels of LLR unless mimosine levels are very low or iron is fed, notably during gestation in gilts (Pound and Martinez Cairo, 1983). Leucaena can be fed up to 10 percent in fattening rations and provides a valuable source of vitamin A. Swine intestines have been used in permeation studies in which mimosine permeability was inhibited by ferric ions and also by low levels of lactic acid (LRR 7:86). When LLM was heat-dehydrated and ground finely, mimosine levels averaged 1.5 percent and no deleterious effects were observed in swine diets up to 16 percent LLM (LRR 2:46). Digestibility coefficients were reduced about 20 percent for dry matter, protein and gross energy on diets supplemented with 35 percent LLM (LRR 3:76). A maximum of 20 percent LLM was suggested for weaner pigs, with slight improvements in gain noted upon addition of 1.5 percent ferric sulphate or 3 percent polyethylene glycol or both (LRR 2:74).

Rabbits also tolerate low levels of LLM, which they accept readily, but feeding results are variable. Liveweight gains dropped linearly as LLM levels were raised to 100 percent of diet (Pound and Martinez Cairo, 1983). Molasses supplementation showed that low metabolizable energy was a major cause. Feeding studies with mimosine supplementation showed unusually high tolerance (>0.24 mg kg^{-1} body weight per day) (LRR 5:7). Leucaena compared favourably to alfalfa in digestibility of crude protein, fibre and cell-wall constituents, but voluntary consumption was low and urine turned red-black (LRR 2:73). It must be stressed that DHP has not been distinguished from mimosine in non-ruminant nutritional studies, and may be the basis for conflicting results from LLM and mimosine-feeding trials.

Soil improvement

Leucaena *groene bemester* (Dutch for "green manure") was the subject of many articles in the early 1900s (Dijkman, 1950), and represents perhaps the most important of leucaena's many uses in agroforestry systems. These articles referred to the use of leucaena and other legumes as companion trees ("shade trees", "nurse trees") in coffee and other plantation-tree crops. In recent times, use of leucaena in "alley farming" is also assuming great importance as the tropics are progressively denuded (Kang *et al.*, 1984; Kang and Wilson, this volume).

Alley farming

Alley farming is an agroforestry system of growing row crops between closely spaced woody hedges (Kang *et al.*, 1984). An historic practice in Indonesia and the Philippines (Figure 6) involved maize planted between contour strips of leucaena (LRR 1:13, 20). Dairies in Hawaii had an alley-farming system of pangolagrass (*Digitaria decumbens*) between leucaena hedges before 1940, allowing animals free access to both grass and legume (Takahashi and Ripperton, 1949). Agronomic research has led to management systems that optimize total yields from these systems, with demonstration of their effectiveness in a wide array of cropping systems (Kang *et al.*, 1984). Maize and leucaena are very compatible in such a system. Leucaena hedges are cut and the green manure ploughed under prior to planting, then cut again in 6 weeks to eliminate competition for light and provide added nutrient-rich green leaf manure to the maize. After maize harvest, the leucaena can be cut for fodder and fuelwood, the hedge acting to reduce erosion and build terraces. The system has been called *lamtoronisasi* in E. Indonesia, where Viator Parera has guided demonstration and improvement of the practice (LRR 1:13). Stem-girdling of the leucaena to induce litterfall and allow light penetration during the critical grain-fill period is recommended by Parera (LRR 4:45; 5:51).

Trials at the International Institute of Tropical Agriculture (IITA), Ibadan, Nigeria,

Figure 6   Alley farming by Harold Watson of grain legumes between leucaena hedgerows in the
Philippines.

have included several shrubs suitable for alley cropping; leucaena has generally been
superior. Alley widths have been varied and 4 m preferred, with intra-row spacing to create
a dense hedge. Pruning every 5–6 weeks is advised, with periodic low cutting to stimulate
branching. Tractor-pulled rotary mowers, tilted slightly, have been very effective where
available. The hedge is managed to maximize shade-control of weeds and minimize later
shading of the intercrop. Continuous alley cropping for six years showed that maize yields
could be kept at 2 t ha^{-1} with leucaena prunings alone (Kang *et al.*, 1984). Summarizing
nitrogen yields of hedgerow fodder, Torres (LRR 4:50) calculated an average of 45 g N per
one metre of row, and suggested this was adequate for yield improvement only on marginal
farms. With an average tropical maize yield of only 1.2 t ha^{-1}, however, any procedure
increasing grain yields to 2 t ha^{-1} under continuous rather than periodic cropping could
have a substantial impact on tropical maize production (Brewbaker, 1985b).

The Indian Grassland and Fodder Research Institute at Jhansi has also conducted a
series of leucaena alley-farming trials with crops such as maize, sorghum, buffel grass
(*Cenchrus ciliaris*), millet and napier grass (*Pennisetum purpureum*) with generally salutary
results (LRR 3:29, 30; 4:20; 5:24; 6:36). Competition for moisture can reduce yields of the
row crop, so the total value of crop and hedge becomes a determinant in farmer acceptance
(LRR 2:30; IITA, 1986). Similar results were obtained with leucaena and sweet potato,
where light interception by the hedge reduced crop yields but crop plus fodder made a better
total cash return (LRR 3:52). Leucaena can be used with kenaf (*Hibiscus cannabinus*) as
alley crop (LRR 6:1) and with cassava, but was not effective combined with sugarcane
(LRR 6:35).

Hedge trimmings from leucaena can be carried as excellent green manure to row crops
like maize and rice (LRR 2:41,42). Leucaena leaves mature over a period of 2–4 weeks and

the leaflets, pinnae and midribs dehisce in 3–5 months. The litter is fragile and quickly decomposed, with N half-life of 7 days if buried (Guevarra *et al.*, 1978; Kang *et al.*, 1984). Soil-incorporated leaves completed N-release in three weeks in India, with somewhat higher values under aerobic rather than anaerobic conditions (LRR 3:54). Maize grain yields to leucaena green manure regressed linearly on application rates over a wide range (C. Evensen, cited by Brewbaker, 1985b). One kg of N in leucaena mulch produced an increase of 14.8 kg of grain, while one kg of N as incorporated green manure produced an increase of 24.1 kg of grain. In other studies, excellent yield responses of maize to green manure have been reported (LRR 1:13, 22; 4:33; 5:59; Kang *et al.*, 1984).

Companion tree

Companion or "nurse" trees are planted for two major purposes with coffee, tea, cacao, teak and other tree crops — to confer shade and wind protection in early growth, and to provide nitrogen, mainly through N-fixation. Leucaena was a major coffee and cacao shade tree prior to modern use of inorganic nitrogen. Successful companion planting requires care in spacing and management. The Salvador leucaenas are generally too vigorous as nurse trees unless boldly coppiced, while *L. diversifolia* (4N) and its hybrids deserve wider appraisal. Examples are reported, not all successful, of leucaena as nurse tree to teak (LRR 1:21), mahogany (LRR 1:5, 21; 7:82, 102) and coconut (LRR 5:62, 7:72). Its use for social forestry plantings along with eucalyptus would seem well advised.

Van Den Beldt assessed both litterfall and biomass development of *L. leucocephala* in Hawaii under stand densities ranging from 10×10^3 to 40×10^3 ha^{-1} (LRR 3:95). Annual litterfall averaged 8.54 t ha^{-1} (dry matter) with no significant effects of stand density. Nutrient analysis of the litter showed that 100 kg N, 7 kg P, 16 kg K, 200 kg Ca and 12 kg S were recycled annually per hectare through such litterfall.

Tree and wood use

Leucaena production for wood use expanded most prominently in the past decade in two ways — for pulpwood in Taiwan (Hu, 1987; Tai *et al.*, 1984) and for dendrothermal energy in the Philippines (Denton, 1983). Although much less heralded, the use of leucaena in village agroforestry systems for fuelwood, for poles (e.g., propping bananas) and for housing and furniture wood has probably expanded more rapidly as a result of the spread of the arboreal varieties (NRC, 1984; NFTA 1985b).

Wood yields

Wood yields of *Leucaena leucocephala* are greater than those of species with which it has been compared in most short-duration (3–5 yr) trials. Height growth and mean annual wood increments fall in the range of 3–4 m yr^{-1} and 20–60 m^3 yr^{-1} for many trials. Only the Salvador types are now grown, with wood yields many times that of the common types (LRR 6: 49). Failures are almost entirely related to two factors — poor soil (acid, low Ca, low P) and low temperatures. Fresh-weight data from 2–4-year-old trees illustrate the limitations of *L. leucocephala* in comparison with an acid-tolerant acacia (*Acacia auriculiformis*) (Table 3). In cooler climates the use is recommended of species such as *L. diversifolia* whose hybrids with *L. leucocephala* greatly outyield both parents in the cooler tropics (Brewbaker, 1986a). *L. collinsii* is a fast-growing species that deserves evaluation in cooler tropics, while *L. lanceolata* var. *sousae* has promise for lowland tropical wood use.

Table 3   Wood yields (fresh) of *Leucaena* species and *Acacia auriculiformis* in diverse environments

Location	Age (yr)	Mean temp. (C°)	Soil pH (KCl)	Ca (meq 100g⁻¹)	Yields* Leuc.	Dive.	Auri.
					(t ha⁻¹yr⁻¹)		
Haleakala, Hawaii	2.0	19.4	4.3	2.5	2.7	18.4	14.8
Iole, Hawaii	2.5	20.1	5.1	23.0	2.8	24.9	12.1
Molokai, Hawaii	2.8	23.4	5.5	4.4	81.9	46.0	18.3
Waimanalo, Hawaii	3.0	23.9	5.2	15.5	44.4	26.6	20.1
Waipio, Hawaii	2.5	24.2	4.6	5.0	30.0	23.2	12.8
Davao, Philippines	2.0	28.1	4.9	7.5	97.6	89.9	38.1
Nakau, Indonesia	1.7	28.6	4.8	3.0	11.9	12.0	27.5

* Leuc. = *Leucaena leucocephala* Lam. de Wit

Dive. = *L. diversifolia* (Schlecht) Benth.

Auri. = *Acacia auriculiformis* A. Cunn. ex Benth.

The forte of leucaena in agroforestry systems is its ability to maintain high yields over a wide range of harvest cycles, stand densities, and systems of management and mismanagement. Van Den Beldt and associates have conducted extensive collaborative trials of arboreal varieties under diverse spacings (Hu and Kiang, 1983; Van Den Beldt, 1983; LRR 1:29, 53, 55; 3:62, 96; 4:93). Annual yields changed relatively little over a wide range of densities and age at harvest, optimizing between 10 and 20 thousand per hectare. Optimal spacing changed in relation to harvest age, with a drop in mean annual increments occurring earlier at high densities. Increased densities did not affect height appreciably; instead, the plant diameter, specific gravity and wood strengths were lower, and relative moisture and amount of bark were higher at the high plant densities (Van Den Beldt, 1983; Hu and Kiang, 1983; LRR 3:68).

Volumetric formulae for leucaena have been produced from many trials. Van Den Beldt (LRR 4:93) compared these and concluded that few deviated significantly from the formula of Kanazawa *et al.* (1983): tree volume=0.5 D² H. In this formula, tree volume is in m, D is diameter at breast height in m, and H is height in m. Leucaena trees at 1 x 2 m spacing grow with D in cm roughly equal to H in m, and thus a 10 m tree is about 0.05m³ (about 50 kg wet weight) and a 5 m tree about 0.006 m³ (6 kg).

Hu (LRR 1:29) and Van Den Beldt (LRR 4:93) calculated form factors relating stem volume to a cylinder with the same dimensions under varying conditions of tree age, density and site. These factors were very similar except for very young trees (1.5 yr), and generally fit well the Kanazawa equation.

Log transformation and corrections for intercept improved the fits only slightly. Our data as t ha⁻¹ yr⁻¹ in Table 3 use equations that recognize Y axis intercepts when relating yield in kg per tree (Y) with the formula: Y=0.5+46.5 D²H for LEU, and Y=0.72+46.5 D²H for DIV where LEU and DIV refer to *L. leucocephala* and *L diversifolia,* respectively (Brewbaker, 1986a). Hu (LRR 3:62) found the Kanazawa equation suitable for Taiwan trials in which five different Salvador varieties were statistically similar in yield (mean 32 m³ ha⁻¹ yr⁻¹). Log transformations have also been used to compute total biomass dry matter a bit more precisely: in one example: log wt. = −0.462 + 0.663 log (D^2H) (LRR 7:53).

Wood yield data are now widely available for Salvador leucaenas in different sites all over the world. They range up to 100 m^3 ha^{-1} yr^{-1} (Table 3), and would average about 15 m^3 ha^{-1} yr^{-1}. Some of the more extensive data involving varietal or spacing differences are from Thailand (LRR 4:81), BAIF in India (LRR 4:38; 40, 49, 50), Jammu in India (LRR 2:22; 3:27), the Philippines (LRR 1:27; 6:82), Taiwan (LRR 2:53; 3:59, 62), Costa Rica (LRR 3:15), Jamaica (LRR 6:60), Hawaii (LRR 1:55; 4:93) and Florida (LRR 5:84), the latter being impressive despite winter frosts that killed plants to the crown.

Fuelwood

Leucaena has become a popular fuelwood in Asia and less widely so in the rest of the tropics (Figure 7). Fuelwood properties are widely respected and well documented in the literature (NRC, 1984; NFTA 1985b; Pound and Martinez Cairo, 1983). Heating values on oven-dry basis average about 19.4 MJ kg^{-1} (4,640 kcal kg^{-1}) and the wood burns steadily with little smoke, few sparks, and less than 1 percent ash. Moisture at harvest depends on tree age, ranging from about 55 percent at 1 year to 45 percent at 4 years and 35 percent at 7 years (LRR 6:49). Specific gravities as dry matter/displacement volume average 0.5–0.6 for 4-year-old trees.

Economic conversion of wood to electrical energy remains an elusive target in the tropics, but leucaenas and eucalyptus generally remain the favoured trees due to their rapid growth and wide adaptability (Brewbaker, 1980, 1984). Only in the Philippines has an extensive dendrothermal scheme based on leucaena been developed. Denton (1983) has written in detail of this scheme, developed by the Philippine National Electrification Administration. The ambitious plan involved many plantations of about 1,000 ha in size,

Figure 7   Typical fuelwood marketing of leucaena in 50-cm-long bundles of split wood in the Philippines.

Figure 8 Overhead transport system for leucaena as fuelwood for power plant at Bolinao, Philippines.

each including a woodfuel-burning power plant of 3–5 MW peak. At least two of these became operational, but problems of plant construction and tree production have plagued the scheme. The sites were on marginal, usually deforested lands. It is certain that none were chosen because of their high site index for leucaena! Many had no leucaenas growing at all in the area, and several had been abandoned by shifting agriculturalists as worthless. Preliminary trials of species and provenances were generally bypassed, as were careful soil analysis and assessment of nutritional demands. The demands of transplanting and weeding were generally underestimated, coming as they did at the peak of annual farm activities.

The success at one site, at Balinao, provides important lessons for others (Denton, 1983). It was located on calcareous hills in an area where shrubby leucaena thrives. Local farmers were familiar with the tree, and management was excellent. Yield attained about 30 $m^3 ha^{-1} yr^{-1}$, as was predicted. Transportation to power plant has been a major cost factor, with an ineffective overhead hauling system for the small trees (Figure 8). Harvesting with machettes has led to accelerated cutting of small trees. The lack of a suitable tool for felling small trees is a limitation on such schemes. Chainsaws are too expensive and costly to maintain, heavy axes are like scythes (not designed for the steamy tropics!), and small bow saws are unfamiliar.

Charcoal production from leucaena has made little advance in recent years except in the Philippines (Guevarra in PCARR, 1978). The tree makes excellent charcoal, with heating values of 29 MJ kg^{-1} and good recovery values (25–30 percent). The Mabuhay Vinyl Corporation of the Philippines is a major producer of leucaena charcoal, with several types of improved kilns in operation. The addition of ground leucaena charcoal to fuel oil for diesel engines was found to involve no harmful agents in the ash (LRR 2:76).

Pulpwood, poles and miscellaneous

Leucaena has been evaluated highly as pulpwood, and initial commercial production in Taiwan was carefully planned and economically successful (Hu, 1987; Tai *et al.,* 1983). Pulping properties are suitable to both paper and rayon production. Compared with other tropical hardwoods, leucaena was above average in cellulose and low in lignin, low in bark (5 percent), had a medium-high specific gravity (>0.5) and a higher pulp yield, up to 75 percent under neutral sulphite semichemical processing (LRR 2:51). The wood pulp strength was greater than most hardwoods, with almost 50 percent greater ring-crush resistance. Fibre values are similar to other tropical hardwoods and produce papers with good printability but low tearing and folding strengths. Pulping data are summarized by Hu (1987) and include reports from Taiwan (LRR 2:57), the Philippines (Bawagan and Semana, 1978), India (Relwani, 1983), and the USA (unpublished data).

Poles, mine props, furniture, parquet flooring and chips for particle board are among increasingly popular uses of leucaena. Small furniture and craftwood items are strong and light in weight, and the close-grained wood finishes well and darkens to a golden brown with age (LRR 5:48.) Furniture has also been made from press-wood of glued-up 5 mm strips, and had a specific gravity of 0.58 with good strength (LRR 3:63).

"Grow your own house" could be a motto for leucaena pole and mortar houses in India, pictured in NFTA (1985b). The wood is termite-susceptible, but accepts preservatives. Poles serve a wide array of agroforestry needs, from props for bananas (PCARR, 1978) to crop-support posts for yams, pepper, and other vines.

Ornamental and shade uses of arboreal leucaenas are extensive despite seediness of early varietal releases. All taller species (Table 1) make suitable ornamentals, and seed production is lower on self-incompatible trees and on selected varieties and triploid species hybrids. Isolated self-incompatible trees are seedless, as are certain species hybrids, e.g., *L. esculenta* x *L. leucocephala*. The ability to achieve mature height in 3–5 years makes the leucaenas attractive for shade, and loppings provide useful fodder and branch wood, although juvenile growth is wind fragile (LRR 6:23). *L. diversifolia* 4N is an attractive shade tree for subtropical climates with no seed problem (LRR 5:88). The leucaenas crown out too widely to make good windbreaks, although they do not have offensive lateral root development.

Erosion control is a concomitant of alley farming, much of which is on fragile uplands. Contour strips of leucaena must be planted very densely, with small branches laid laterally along the hedge to serve effectively in erosion control (LRR 1:13; 5:51; NRC, 1984). Leucaena is not easily used on rice paddy bunds, due to problems with shading of rice and tolerance by leucaena of waterlogged soils.

Food and other uses

Food

A broadened view of leucaena that embraces all species should bring renewed attention to uses other than fodder, wood and soil amendment. The favoured species for food have traditionally been *L. esculenta* (*guaje rojo*) and *L. macrophylla* (*guaje verde*). Both are midland or highland Mexican diploids that flower heavily in short days. Pods are marketed in winter at physiological maturity of the seeds. *L. esculenta* has very large beans with low mimosine. Those of *L. macrophylla* are similar in size and mimosine content to *L.*

leucocephala, of which the "giant" varieties are also widely used for green pods, even in S.E.
Asia. Zarate (1984) considers food use of the leucaenas most important in interpreting
distributions of the species. The leucaena seeds are tasty and are often consumed fresh.
Nothing is known about human conversion of mimosine through DHP, and whether DHP
might be a human toxin of significance (Lowry 1983). Mimosine is lost on preparation
through diffusion (LRR 5:53) or following cooking or fermentation, and can be
precipitated and entirely removed in preparations involving ferrous ions, as in a rusty pot.
Other methods for reducing mimosine values were discussed earlier.

Food use of leucaena leaf tips is common in parts of S.E. Asia (Thailand, Malaysia,
Indonesia), normally served after brief stir-frying. They make a significant protein
contribution, and are often used with noodle dishes and generous portions of chilli sauce.
Leucaena seedlings ("sprouts", i.e., roots) of about three days are prepared as a food in
Java, Indonesia. They showed high protein and ascorbic- and amino-acid values with
lowered mimosine in studies of Dewi Slamet and associates (LRR 5:53). *Tempeh* is a
fermented food that is also prepared with considerable patience from leucaena seeds in
Indonesia (LRR 3:100). Phytic acid, a trypsin inhibitor, was found at a level of 7 mg g^{-1} in
leucaena seeds prepared for *tempeh,* and two-day fermentation sharply reduced these levels
by 70 percent (LRR 7:75). *Tempeh* samples prepared in our lab by nutritionist M.
Whiting were free of mimosine, but otherwise of rather controversial food acceptability.

Protein concentrates and gums

Proteins have been concentrated with difficulty from leucaena leaves, due primarily to
galactomannan gums in the leaves that require high dilution and other steps (LRR 2:81).
Precipitation of the gums caused by procyanadins or condensed tannins led to reduced
protein yields. However, protein processing removed mimosine entirely (LRR 3:93; 5:96).
The leaf protein concentrates were low in nutritional value to rats, and tannins were held
responsible (LRR 1:45).

Galactomannan gums of leucaena seeds are similar to those of gum arabic (*Acacia
senegal*), guar (*Cyamopsis tetragonoloba*) and carob (*Ceratonia siliqua*). Gum contents of
the whole seeds are in the range of 25–30 percent (LRR 5:16). Pure seed galactomannan
was isolated readily by Leshniak and Liu (LRR 2:75), and found to be of high uniformity
with 1.3 mannosyl residues to 1 galactosyl. It had high haemagglutinating activity on
human and rat erythrocytes that was further increased by reducing the galactosyl residues
(LRR 2:77). The gum was believed to be a gelling agent comparable to algal carageenan.
Mimosine contaminating the gum was easily removed by dialysis, or could be avoided
through trichloracetic acid purification with dialysis of the galactomannan (LRR 2:79).

Gums arise from the leucaena stem ("gummosis") under ill-defined conditions of injury
and disease, but notably in India with *Fusarium* infections. Anderson *et al.* (1983) have
carefully analysed these and other mimosoid gums, and found leucaena to be of potential
commercial value and very similar to gum arabic. Amino-acid fractions of the protein in
leucaena and the acacia gums were very similar (LRR 7: 108).

Miscellaneous uses of leucaena include use of seeds for ornaments, and for practical
items like hot pads. The *Leucaena* species vary widely in attractiveness to bees, and provide
pollen but not nectar for honey production. Most of the self-sterile species are actively
bee-pollinated, and those with heavy scent — *L. lanceolata, L. shannoni* — seem
particularly attractive and might make distinctive honeys.

Why do people plant leucaena?

A good example of the spread and role of leucaena in rural development is given by the work of Manibhai Desai, Director of the non-profit Bharatiya Agro Industries Foundation (BAIF) of Pune, India. He has inspired a generation of Indian scientists to dedicate their efforts to improving the livelihood of the rural poor. Leucaena is a major tool in this programme of revegetation, water management and animal improvement, a programme that earned Desai the prestigious Magsaysay Award in 1985.

Often featured in their excellent publication, The *BAIF Journal,* leucaena is the subject of extensive development and research activities at BAIF. The giant leucaenas were carefully appraised in many types of management systems prior to large-scale seed increase (to 40 tons by 1986, equivalent to 800 million seeds) and distribution among India's rural community. Desai and his scientific staff have unquestionably inspired tree planting in the past decade on a scale previously believed impossible for the small farmer.

Although fruit trees are planted by rural people of the tropics, the planting of trees for fuel or fodder is an unfamiliar practice — such trees are to be hunted and gathered. Backlash can develop, as with eucalyptus in India, when the tree planting appears to serve the rich or to bring social benefits only to a select society. The fast-grown small legume trees — leucaenas, gliricidias, calliandras, acacias — have a special attraction when they provide both fodder and wood, and can be managed so as to optimize returns from either or both (Brewbaker, 1984).

Why plant leucaena? Hedge (1985) of the BAIF staff lists the features that have made it so effective in India:

- Fast growth, rapid generation time;
- Multiple uses, easily interplanted with field crops;
- Quick returns and high profit;
- N-fixation and soil improvement;
- Strong root system, drought and salt tolerance;
- Cheap to establish;
- Few diseases and pests.

Some of the successful schemes with leucaena pictured in the US National Academy of Science publication on leucaena (NRC, 1984) illustrate Hedge's points. In the hilly islands of Cebu, Philippines, and Flores, Indonesia, leucaena is widely alley cropped to stabilize shifting agriculture and stop erosion. Markets help motivate the plantings — for pelleted leaf meal on Cebu and fodder to support *banteng* cattle production on Flores. Markets for woodfuel in pottery making or bakeries motivate leucaena production in central Thailand and the Philippines. Many plantings of leucaena, however, are small and on small farms with the immediate intention of feeding farm animals and serving as home fuel, postwood and other uses.

Leucaena is self-advertising in at least three important ways that greatly simplify its extension and demonstration:

- Animals relish the fodder;
- The rapid tree growth is astounding even to tropical farmers;
- Green manure effects are often plainly obvious.

Alley-cropping trials can show the improved colour and yield of crop plants near the leucaena hedge sufficiently well to convince the most skeptical farmer of the power of green manure. Overcoming the initial farmer reluctance to plant a non-food crop is a difficult hurdle, and almost impossible if the tree is slow-growing and the market is uncertain.

In many sites, perhaps surprisingly, the giant leucaenas have had only to be planted to tell their own story and gain wide grower acceptance. Nevertheless, the key is to inspire that first tree planting. People like Manibhai Desai in India, Harold Watson in the Philippines (also an awardee of the Magsaysay Foundation), Viator Parera in Indonesia (Parera, 1983), and B.T. Kang, G.F. Wilson and colleagues in Nigeria (Kang *et al.*, 1984) are providing that inspiration.

REFERENCES

Anderson, D.M.W., M.M.E. Bridgeman, E.I.G. Brown and J.A.M. Anderson. 1983. The gum exudate from a cultivar of *Leucaena leucocephala* Lam. de Wit. *Intl. Tree Crops J.* 2: 291–295.

Arellano R.J.A. 1979. Bibliografia sobre Leucaena leucocephala. Centro Invest. Agric. Peninsula de Yucatan, INIA, Merida, Yucatan, Mexico.

Banco de Mexico. 1980. Leucaena (huaje), leguminosa tropical Mexicana. Usos y potencial. FIRA, Banco de Mexico.

Batson, H.F., T.U. Ferguson and K.A.E. Archibald. 1984. Variability in leucaena and its potential in the Caribbean. Faculty of Agriculture, University of the West Indies, St. Augustine, Trinidad.

Bawagan, P.V. and J.A. Semana. 1978. Utilization of ipil-ipil for wood. International Consultation on Ipil-Ipil Research, PCARR, Los Banos, Philippines.

Bray, R.A., R.J. Jones and M.E. Probert. 1984. Shrub legumes for forage in tropical Australia. In E.T. Craswell and B. Tangendjaja (eds.), *Shrub legume research in Indonesia and Australia.* ACIAR, Canberra, Australia.

Brewbaker, J.L. 1975. "Hawaiian Giant" koa haole. Hawaii Agriculture Experimental Station, Misc. Publication 125.

———. 1976. Establishment and management of leucaena for livestock production. In *Produccion de Forrajes, Memoria de Seminario Internacional de Ganaderia Tropical.* Banco de Mexico, Mexico City.

———. 1979. Diseases of maize in the wet lowland tropics and the collapse of the classic Maya civilization. *Economic Botany* 33: 101–118.

———. (ed.). 1980. Giant leucaena energy tree farm: An economic feasibility analysis. Hawaii Natural Energy Institute Publication 81–04, Honolulu, Hawaii.

———. 1983. Systematics, self-incompatibility, breeding systems and genetic improvement of Leucaena species. In *Leucaena research in the Asia-Pacific region.* Ottawa: IDRC.

———. 1984. Short-rotation forestry in tropical areas. In H. Egneus and A. Ellegard (eds.), *Bioenergy* 84, Vol. I. *Bioenergy state of the art.* Elsevier Applied Science Publication 58–78.

———. 1985a. Leguminous trees and shrubs for Southeast Asia and the South Pacific. In G.J. Blair, D.A. Ivory and T.R. Evans (eds.), *Forages in Southeast Asia and South Pacific Agriculture.* Canberra, Australia: ACIAR.

———. 1985b. The tropical environment for maize cultivation. In A. Brandolini and F. Salamaini (eds.), *Breeding strategies for maize production improvement in the tropics.* Rome, Italy: FAO/UN.

———. 1986a. Performance of Australian acacias in Hawaiian nitrogen-fixing tree trials. In J.W. Turnbull (ed.), *Australian Acacias in developing countries.* ACIAR Proceedings No. 6.

———. 1986b. Nitrogen fixing trees for fodder and browse in Africa. In B.T. Kang (ed.), *Alley farming for humid and subhumid regions of tropical Africa.* Ibadan, Nigeria: IITA.

Brewbaker, J.L., D.L. Plucknett and V. Gonzalez. 1972. Varietal variation and yield trials of *Leucaena leucocephala* (koa haole) in Hawaii. Hawaii Agriculture Experimental Station Research Bulletin 166.

Brewbaker, J.L. and E.M. Hutton. 1979. Leucaena — Versatile tree legume. In G.A. Ritchie (ed.), *New agricultural crops*, AAAS Selected Symposium. Boulder, Colorado: Westview Press.

Bushby, R.V.A. 1982. Rhizosphere populations of Rhizobium strains and nodulation of *Leucaena leucocephala*. *Austr. J. Exp. Agric. An. Husb.* 22: 293–298.

Chao, C.C., C.C. Young and W.E. Cheng. 1985. Effects of rhizobia inoculant and liming on the growth and nodulation of *Leucaena leucocephala* (Lam.) de Wit in an acid slopeland. *J. Agric. Assn. China* 131: 35–41.

Chee, Wong C. and C. Devendra. 1983. Research on leucaena forage production in Malaysia. In *Leucaena research in the Asia-Pacific region*. Ottawa: IDRC.

D'Mellow, J.P.F. and D.E. Taplin. 1978. *Leucaena leucocephala* in poultry diets for the tropics. *World Rev. Animal Prod.* 14: 41–47.

Denton, F.H. 1983. *Wood for energy and rural development — The Philippine experiences.* Denton, 3181 Readsborough Ct., Fairfax, Virginia.

Dijkman, M.J. 1950. Leucaena — A promising soil-erosion-control plant. *Econ. Bot.* 4: 337–349.

Dommergues, Y. 1982. Ensuring effective symbiosis in nitrogen-fixing trees. In P.H. Graham and S.C. Harris (eds.), *Biological nitrogen fixation technology for tropical agriculture.* Cali, Colombia: CIAT.

FAO. 1983. *Leucaena leucocephala:* The Indonesian experience. FAO Regional Office for Asia and the Pacific.

————. 1985. Production of leafmeal from ipil-ipil (leucaena). FAO Regional Office, Bangkok, Field Document 4.

Gonzalez, V., J.L. Brewbaker and D.E. Hamill. 1967. Leucaena cytogenetics in relation to the breeding of low-mimosine lines. *Crop Sci.* 7: 140–143.

Guevarra, A.B., A.S. Whitney and J.R. Thompson. 1978. Influence of intra-row spacing and cutting regimes on the growth and yield of leucaena. *Agron. J.* 70: 1033–1037.

Halliday, J. and P. Somesagaran. 1983. Nodulation, nitrogen fixation and Rhizobium strain affinities in the genus *Leucaena*. In *Leucaena research in the Asia-Pacific region.* Ottawa: IDRC.

Halos, S. C. 1980. Abstract on leucaena. Forest Research Institute. (Philippines), Ref. Ser. 8.

Hegarty, M.P., C.P. Lee, G.S. Christie, R.D. Court and K.P. Haydock. 1979. The goitrogen 3 hydroxy-4(1H)-pyridone, a ruminal metabolite from *Leucaena leucocephala:* effects in mice and rats. *Aust. J. Biol. Sci.* 32: 27–40.

Hedge, N. 1985. Leucaena for energy plantation. *BAIF Journal* 5: 37–42.

Hogberg, P. and M. Kvarnstrom. 1982. Nitrogen fixation by the woody legume *Leucaena leucocephala* in Tanzania. *Plant Soil* 66: 21–28.

Holmes, J.H.F. 1981. Toxicity of *Leucaena leucocephala* for steers in the wet tropics. *Trop. Anim. Health Prod.* 13: 94–100.

Hu, Ta Wei. 1987. *Use of nitrogen fixing trees for pulpwood.* Waimanalo, Hawaii: NFTA.

Hu, Ta Wei and T. Kiang. 1983. Leucaena research in Taiwan. In *Leucaena research in the Asia-Pacific region.* Ottawa: IDRC.

Huang, Ruey-Shyang, W.K. Smith and R.S. Yost. 1985. Influence of vesicular-arbuscular mycorrhiza on growth, water relations, and leaf orientation in *Leucaena leucocephala* (Lam.) de Wit. *New Phytol.* 99: 229–243.

Hutton, E.M. 1983. Selection and breeding of leucaena for very acid soils. In *Leucaena research in the Asia-Pacific region.* Ottawa: IDRC.

————. 1984. Breeding and selecting leucaena for acid tropical soils. *Pesq. Agropec. Bras.* 19: 263–274.

IITA. 1986. *Alley farming for humid and subhumid regions of tropical Africa.* Proceedings of a Workshop. Ibadan, Nigeria: IITA/ILCA.

IDRC. 1982. *Leucaena research in the Asia-Pacific Region.* Proceedings of Workshop, IDRC and NFTA, Singapore, November 1982.

Indonesia, Government of. 1982. *National leucaena seminar.* Jakarta, Indonesia.

Jones, R.J. 1979. The value of *Leucaena leucocephala* as a feed for ruminants in the tropics. *World Animal Rev.* 31: 13–23.

Jones, R.J. and R.M. Jones. 1982. Observations on the persistence and potential for beef production of pastures based on *Trifolium semipilosum* and *Leucaena leucocephala* in sub-tropical coastal Queensland. *Tropical Grasslands* 16: 24–29.

Jones, R.J. and R.A. Bray. 1983. Agronomic research in the development of leucaena as a pasture legume in Australia. In *Leucaena research in the Asia-Pacific region*. Ottawa: IDRC.

Kanazawa, Y., A. Sato and R.S. Orsolino. 1983. Above-ground biomass and the growth of giant ipil-ipil (*Leucaena leucocephala* Lam. de Wit) plantations in northern Mindanao Island, Philippines. *JARQ* 15(3): 209–217.

Kang, B.T., G.F. Wilson and T.L. Lawson. 1984. *Alley cropping — A stable alternative to shifting cultivation.* Ibadan, Nigeria: IITA.

Kaul, R.N., M.G. Gogate and N.K. Mathur (eds.). 1981. *Leucaena leucocephala* in India. Proceedings of National Seminar at BAIF, Urulikanchan, Maharashtra, India.

Lowry, J.B. 1983. Detoxification of leucaena by enzymic or microbial processes. In *Leucaena research in the Asia-Pacific region*. Ottawa: IDRC.

Manidool, C. 1983. Leucaena leaf meal and forage in Thailand. In *Leucaena research in the Asia-Pacific region*. Ottawa: IDRC.

Megarrity, R.G. 1978. An automated colorimetric method for mimosine in leucaena leaves. *J. Sci. Food Agric.* 29: 182–186.

Munns, D.N. and R.L. Fox. 1977. Comparative lime requirements of tropical and temperate legumes. *Plant Soil* 46: 533–548.

NRC. 1984. *Leucaena: Promising forage and tree crop for the tropics.* 2nd edn. Washington, D.C.: USNAS.

NFTA (Nitrogen Fixing Tree Association). 1985a. *Leucaena forage production and use.* Waimanalo, Hawaii: NFTA.

————. 1985b. *Leucaena wood production and use.* Waimanalo, Hawaii: NFTA.

Oakes, A.J. 1968. *Leucaena leucocephala:* description, culture, utilization. *Advancing Frontiers of Plant Science* (India) 10: 1–114.

————. 1982–4. *Leucaena bibliography.* Three volumes, Germplasm Resources Lab., USDA, Beltsville, MD.

Oakes, A.J. and C.D. Foy. 1984. Acid soil tolerance of leucaena species in greenhouse trials. *J. Plant Nutr.* 7: 1759–1774.

Olvera E.M., M. Benge and S.H. West. 1985. World literature on leucaena *Leucaena leucocephala* Lam. de Wit. Monograph 11, Agricultural Experiment Station, Institute of Food and Agricultural Sciences, University of Florida, Gainesville, Florida.

Olvera, E. and G. Blue. 1985. Establishment of *Leucaena leucocephala* Lam. de Wit in acid soils. *Trop. Agric.* 62: 73–76.

Othman, A.B., M.A. Soto, G.M. Prine and W.R. Ocumpaugh. 1985. Forage productivity of leucaena in the humid subtropics. *Proc. Florida Soil Crop Sci. Soc.* 44: 118–122.

Pan, F.J. 1984. Systematics and genetics of the *Leucaena diversifolia* complex. Ph.D. Thesis, University of Hawaii, Honolulu, Hawaii.

Perera, Viator. 1983. Leucaena for erosion control and green manure in Sikka. In *Leucaena research in the Asia-Pacific region*. Ottawa: IDRC.

PCARR. 1978. International consultation on ipil-ipil research. Los Banos, Philippines: Philippines Council for Agriculture and Resources Research.

Pound, B. and L. Martinez Cairo. 1983. *Leucaena: Its cultivation and use.* London: Overseas Development Administration.

Proverbs, G. 1985. *Leucaena, a versatile plant.* Caribbean Agric. Res. Dev. Inst., Barbados, West Indies.

Relwani, L.L. 1983. Kababul (*Leucaena leucocephala*) — a renewable source of fuel, timber and pulp. In R.N. Kaul, M.G. Gogate and N.K. Mathur (eds.), Leucaena leucocephala *in India*. Dehra Dun, India: Vanguard Press.

Sanzonowicz, C. and W. Couto. 1981. Effect of calcium, sulphur and other nutrients on dry matter yield and nodulation of *Leucaena leucocephala* in a "cerrado" soil. *Pesq. Agropec. Bras.* 16: 789–794.

Schroder, E.C. 1986. Recent advances in leucaena research. Misc. Paper, Department of Agronomy, University of Puerto Rico, Mayaguez.

Singh, S. 1981. Gummosis and canker in *Leucaena leucocephala*. In R.N. Kaul, M.G. Gogate and N.K. Mathur (eds.), Leucaena leucocephala *in India*. Dehra Dun, India: Vanguard Press.

Singh, S., S.N. Khan and B.M. Misra. 1983. Gummosis, brown spot and seedling mortality in Subabul: Disease incidence and pathology of the host. *Indian Forester* 109: 185–192.

Sorensson, C.T. 1987. Interspecific hybridization and evaluation in the genus *Leucaena*. M.S. Thesis, University of Hawaii.

Tai, K.Y., Tao Kiang and K.H.K. Chen. 1983. Promotion of giant leucaena in eastern southern Taiwan. In *Leucaena research in the Asia-Pacific region*. Ottawa: IDRC.

Takahashi, M. and J.C. Ripperton. 1949. Koa haole, its establishment, culture, and utilization as a forage crop. *Hawaii Agric. Exp. Sta. Bull.* 100:56.

Tangendjaja, B. and R.B.H. Wills. 1980. Analysis of mimosine and 3-hydroxy-4(1H)-pyridine by high-performance liquid chromatography. *J. Chromatography* 202: 317–318.

Taplin, D.E., J.P.F. D'Mellow and P. Phillips. 1981. Evaluation of leucaena leaf meal from Malawi as a source of xanthophylls for the laying hen. *Trop. Science* 23: 217–226.

Thoma, P.E. 1983. Leucaena-Rhizobium compatibility and nitrogen fixation. *Desert Plants* 5: 105–111.

Trinick, M.J. 1980. Relationships amongst the fast-growing rhizobia of *Lablab purpureus, Leucaena leucocephala, Mimosa* spp., *Acacia farnesiana* and *Sesbania grandiflora* and their affinities with other rhizobial groups. *J. Appl. Bacteriol.* 49: 39–53.

Van Den Beldt, R.J. 1983. Effect of spacing on growth of leucaena. In *Leucaena research in the Asia-Pacific region*. Ottawa: IDRC.

Vicente, N.E., N. Acosta and E.C. Shroder. 1986. Reaction of *Leucaena leucocephala* to populations of *Meloidogyne incognita* and *M. javanica* from Puerto Rico. *J. Agric. U.P.R.* 70: 157–158.

Yost, R.S. and R.L. Fox. 1979. Contribution of mycorrhizae to P nutrition of crops growing on an oxisol. *Agron. J.* 71: 903–908.

Zarate, P.S. 1984. Taxonomic revision of the genus *Leucaena* Benth. from Mexico. *Bull. International Group Study Mimosoideae* 12: 24–34.

Subject index

Subject index

Acacia albida
  agroforestry system based on, 16, 58
  dry zones of Africa, 99, 100
  effect on soils, 212, 213
  fallow species, 180
  food production in Africa, 105
  N-fixation potential, 258
  N-fixing species, 246–247
  research in India, 133
  traditional African systems, 92, 105
Acacia nilotica, 134
Acacia species
  arid and semi-arid zones of Africa,
    145–146
  Indian subcontinent, 118
  mixed systems, 265
  N-fixing potential, 258
  N-fixing trees, 246–247
  species variability, 282
Actinorhizal plants
  Casuarinaceae (*see* Casuarinaceae)
  Frankia in, 251–254, 256
  Frankia inoculation, 256
Adansonia digitata, 93, 105
African agriculture
  agroforestry, 104
  food production, 104
  subsistence nature, 104
  trees and shrubs, 104
Agricultural
  by-products, 199
  extension methods, 192, 197
Agrisilviculture
  acceptance, 6
  agricultural system, 10
  agroforestry system (*see* agroforestry
    systems)
  alley cropping (*see* alley cropping)
  education, 10

  monograph, 8
  plantation crops (*see* plantation crops)
  shade trees (*see* shade trees)
  social forestry, 7
  systems (*also see* agroforestry systems)
    Acacia albida in traditional, 100
    African past, 100
    India, 127
    research results in India, 132–133
    unirrigated semi-arid, 127
    with *khejri* (*Prosopis cineraria*) (*see*
      Prosopis cineraria)
  taungya (*see* taungya)
Agroforestry (AF)
  aims, 49
  and development banks, 54–67
  biotechnology, 115
  as a catalyst for change, 48, 49
  CATIE's role, 69–87
  Central America, 69–87
  challenges, 36–39, 113
  constraints, 36–37, 43, 113
    biological, 36
    socio-economic, 36–37
  definition, 44–46, 70, 245
  development
    and ICRAF in Africa, 113
    bottlenecks, 113
    challenges, 113
    "food for AF", 19
    resource management for, 113
  development as a discipline, 48–49
  dry zones of Africa, 89–116
  ecological framework, 56
  ecological security, 25–31
  economic considerations in, 173–190
    benefits and costs of production, 174
    capital, 184–185
    development of new systems, 173